EVIDENCE-BASED PRACTICE IN EXERCISE SCIENCE

The Six-Step Approach

William E. Amonette, PhD

University of Houston—Clear Lake

Kirk L. English, PhD

University of Texas Medical Branch

William J. Kraemer, PhD

The Ohio State University

**HUMAN
KINETICS**

Library of Congress Cataloging-in-Publication Data

Amonette, William E., 1977- , author.
 Evidence-based practice in exercise science : the six-step approach / William E. Amonette, Kirk L. English, William J. Kraemer.
   p. ; cm.
 Includes bibliographical references and index.
 I. English, Kirk L., 1975- , author. II. Kraemer, William J., 1953- , author. III. Title.
 [DNLM: 1. Sports Medicine--Case Reports. 2. Athletic Injuries--prevention & control--Case Reports. 3. Evidence-Based Practice--Case Reports. 4. Exercise--physiology--Case Reports. 5. Exercise Therapy--Case Reports. QT 261]
 RC1210
 617.1'027--dc23

                              2015020547
 ISBN: 978-1-4504-3419-5 (print)

The web addresses cited in this text were current as of October 2015, unless otherwise noted.

Senior Acquisitions Editor: Amy N. Tocco; **Developmental Editor:** Katherine Maurer; **Senior Managing Editor:** Carly S. O'Connor; **Copyeditor:** Joyce Sexton; **Indexer:** Patsy Fortney; **Permissions Manager:** Dalene Reeder; **Graphic Designer:** Julie L. Denzer; **Cover Designer:** Keith Blomberg; **Photo Asset Manager:** Laura Fitch; **Photo Production Manager:** Jason Allen; **Senior Art Manager:** Kelly Hendren; **Associate Art Manager:** Alan L. Wilborn; **Illustrations:** © Human Kinetics, unless otherwise noted; **Printer:** Versa Press

Printed in the United States of America   10 9 8 7 6 5 4 3 2

The paper in this book is certified under a sustainable forestry program.

Human Kinetics
Website: www.HumanKinetics.com

United States: Human Kinetics
P.O. Box 5076
Champaign, IL 61825-5076
800-747-4457
e-mail: info@hkusa.com

Canada: Human Kinetics
475 Devonshire Road, Unit 100
Windsor, ON N8Y 2L5
800-465-7301 (in Canada only)
e-mail: info@hkcanada.com

Europe: Human Kinetics
107 Bradford Road
Stanningley
Leeds LS28 6AT, United Kingdom
+44 (0)113 255 5665
e-mail: hk@hkeurope.com

For information about Human Kinetics' coverage in other areas of the world, please visit our website: www.HumanKinetics.com

To my parents, Billy and Judy Amonette, for encouraging me, teaching the value of hard work, and instilling in me the determination to finish.

—William E. Amonette

To my parents, Martin and Karen English, who taught me, loved me, and modeled humility, determination, and joy.

—Kirk L. English

To my wife, Joan, and my children, Daniel, Anna, and Maria, for their love and support in life's journey.

To the many students and colleagues who have allowed the research process to come to life and filled my career with excitement and the joy of discovery.

—William J. Kraemer

Contents

Preface

Exercise prescription should be based on the latest, most relevant, least biased information. Although few would disagree with this statement, it is important to ask whether this is the reality in our discipline. Incorporating the latest and greatest information can be a challenging task for any practitioner. Previous generations were limited by the availability of information or difficulty in obtaining it. The current generation may be afflicted by the opposite problem—information overload! The World Wide Web, blogs, fitness magazines, and many peer-reviewed journals are available to anyone at the click of a mouse. In many cases, the information in these sources is scientifically and practically sound; in others, it is the opinion of a self-appointed "expert" or someone with access to a computer and an elementary ability to write. The problem faced by practitioners today is determining which information is useful and should be incorporated into programming and which should be eliminated as counterproductive or dangerous.

Formal study (e.g., academic degrees and professional certifications) is one methodology used to develop a sound base of information. These educational forums endeavor to diligently incorporate the latest scientific information into their curricula, but the material can be out-of-date the semester it is presented. Exercise scientists work to generate new information related to human physiology, uncovering mechanisms that underlie functional adaptations. Unfortunately, many partially understood or even misunderstood mechanisms obscure the full understanding, rendering what is currently known or taught in academia incomplete due to the complexity of the concept or physiological phenomenon. Therefore, the concepts disseminated in academia are often fundamental theories and do not include every possible scenario one may encounter in practice or the abundance of contexts in which these theories may be applied. In academia, we understand the dynamic state of knowledge, and top academics emphasize and teach the critical skills of finding and evaluating scientific information.

In the late 1980s and early 1990s, a group of physicians and health policy makers realized that many medical decisions with life or death consequences were based on outdated information learned in medical school years previously. They estimated that only 15% to 40% of decisions were based on the latest research knowledge. In an attempt to rectify the problem, they developed a new paradigm and called it evidence-based medicine (EBM). Evidence-based medicine was a process that physicians could use to ask questions relevant to their patients, find information (i.e., evidence), evaluate the quality of the evidence, and then choose to include or exclude the evidence in practice. The theoretical use of the process seemed obvious, yet it was far from routine for physicians and was in fact groundbreaking when introduced in the 1990s. Today, the literature is peppered with articles related to evidence-based medicine, and questions such as "what evidence supports this treatment" are now common in practice. In fact, most of the core curricula at top medical schools require students to complete a course in evidence-based medicine to obtain their medical degree, and there are weekend workshops to

eBook

available at
HumanKinetics.com

ix

teach practicing physicians the process. Evidence-based medicine has spread like wildfire into other allied health professions. Disciplines such as physical therapy, occupational therapy, nursing, and psychology have developed their own models. Each of these disciplines came to the same realization that medicine did—that their practices were not always based on the latest information. Instead, they were steeped in tradition, and some practitioners resisted even when scientific discovery or practice suggested a better way.

Exercise science is a broad term that is used to refer to academic programs teaching the fundamental physics and physiology of sport and fitness, exercise program design, strength and conditioning, and nutrition. Students graduating with degrees in exercise science may practice as strength and conditioning coaches, personal trainers, athletic trainers, biomechanists, exercise physiologists, or in a variety of other vocations. In many cases, the primary responsibility of exercise science practitioners is exercise program design. The essential principles of program design (e.g., progressive overload, periodization, rest and recovery) have been known for years, and students graduating with degrees in exercise science have certainly been exposed to them. Further, students obtaining upper-tier certifications from organizations such as the National Strength and Conditioning Association (NSCA) or the American College of Sports Medicine (ACSM) have demonstrated a basic understanding of these principles. However, our understanding of how to best apply these principles to advance training is ever changing.

More than at any time in the history of modern science, exercise practitioners have abundant resources to inform prescription. An excess of new exercise devices, machines, program theories, and nutritional supplements are available to exercise practitioners. When used properly, many of these resources improve practice; others, however, may be ineffective or even dangerous. Numerous journals are devoted largely to publishing scientific papers on exercise physiology, biomechanics, or strength and conditioning, for example, that document the efficacy of these resources. Additionally, educational materials developed by exercise practitioners with years of practical experience are readily available for incorporation into practice. In some cases, information from these various sources is in conflict; it is important to develop a systematic approach to interpreting and incorporating this information into practice.

The term "evidence-based practice" is beginning to appear in small doses in the exercise literature. However, as with many other concepts in exercise science, in some ways it is being deprived of its true meaning. The term "evidence-based" is often used to make something sound legitimate: "The evidence-based back rehabilitation program . . ." or "The device is evidence based." The purpose of this book is to introduce readers to the original purpose—evidence-based practice as the systematic method or process of finding and evaluating information. This evidence ultimately informs, alters, and fine-tunes practice, resulting in a better outcome for our clients, patients, and athletes.

This book is intended to serve as a primary textbook for undergraduate and graduate exercise science courses that introduce students to the evidence-based philosophy, as well as to guide them in the fundamentals of its practice. As such, it may serve as a strong adjunctive text to any course in a curriculum focusing on exercise programming, nutrition, or research methodology. It can also be used as reference material for the discipline's major certifications and as a how-to guide for practitioners who wish to gain a sound understanding of this paradigm. To improve its practical relevance, this book provides real-world examples of evidence-based practice and its usage. The book is divided into four major sections. (1) An introduction defines the theoretical rationale and philosophy of evidence-based practice; we also define "evidence" and provide an overview of different study designs to be encountered in the scientific literature. (2) Part II explains the six-part process of evidence-based practice. (3) Part III presents case studies in a variety of areas such as program design, nutritional supplementation, exercise for special populations, and exercise hardware or equipment. (4) The final section discusses the current state and the future of exercise science. These divisions will allow readers to accomplish one or more of these objectives: gain a better understanding of the evidence-based practice paradigm, learn the step-by-step process, and gain experience in the evidence-based process by working through practical examples using real-world scenarios.

This book introduces readers to a paradigm that is already used by many academics and some practitioners but that we believe will soon sweep the exercise science field. Due to our ever-changing understanding of human physiology and the increased presence of exercise science in medical and allied health science fields, the incorporation of evidence-based practice into our discipline is now a necessity. Although evidence-based practice is a concept that we hope will be introduced and taught in every exercise science classroom, its ultimate impact will be influenced by journal clubs, weight room discussions, and working group meetings where teams of individuals huddle to discuss evidence with a sincere desire to find and implement the best treatment option for their clients, patients, and athletes. We are certain that those who fail to grasp the importance of staying on the cutting edge of research and practical evidence will be relegated to the background as our field moves authoritatively forward on the firm foundation of knowledge.

Acknowledgments

I am grateful to my coauthors, Dr. Kirk English and Dr. William Kraemer, who made this project so valuable both professionally and personally. Many insights were gained through conversations, e-mails, reviews of drafts, and revisions to this text. The final product reflects a central theme of this book: The evidence-based approach is most effective when it is practiced by a team of individuals with a common goal. I am thankful to be your friend and honored to be your teammate!

Many ideas contained in this book are the result of conversations with colleagues, mentors, friends, and current or former students. I am grateful to Dr. Kenneth Ottenbacher, whose teaching molded many of the thoughts in this text and whose guidance stoked a fire for EBP and paved a new direction in my career. I am thankful to Dr. Rod Henderson, who first introduced me to the term "evidence-based." Several pages in this text arose from our coffee house chats. Dr. Barry Spiering was essential to the development of the proposal for this project. His perspectives, ideas, invaluable contributions, and encouragement drove us in the direction of this book. I am also thankful to my mentor and friend Dr. Gene Coleman. Our conversations on strength and conditioning, sport science, and performance data have taught me the real value of integrating science and practice.

Thanks to Amy Tocco, Kate Maurer, Carly O'Connor, and the Human Kinetics staff for their comments and guidance that kept us on task and helped bring this text to completion.

Most important, I would like to acknowledge the loving support of my wife, Jenny, who always seemed to understand when I needed to lock the office door and write. And my four wonderful boys, Ezekiel, William, Luke, and Isaac, remind me daily of what is truly important.

—William E. Amonette

I am thankful to Dr. Kenneth Ottenbacher for providing the spark, followed by patient caretaking of the flame, that initiated and sustained this journey. I am grateful for the friendship and professional collaboration of Dr. Barry Spiering whose contributions to this work via early papers and presentations in addition to many enlightening discussions are greatly appreciated.

To Amy Tocco, Kate Maurer, Carly O'Connor, and all at Human Kinetics who took a couple of novice book authors (and one seasoned author) from first draft to finished product—thanks for your patience and guidance!

I am grateful for the opportunity to have written this book with my best friend, Dr. Bill Amonette, whose passion and drive to make this book a reality have been impressive and inspiring. Working with Dr. Kraemer has been a special privilege; his unique perspective gained over decades of research and practice has enriched the final product.

Finally, I am immensely grateful to my wife, Charity, for her love and understanding throughout the writing process; her support never wavered. I also appreciate the patience of my daughter, Isabella, who often asked, "How many chapters do you have left, Dad?" And "Are you going to write a kid version for me?"

—Kirk L. English

I gratefully acknowledge the efforts of Dr. William Amonette and Dr. Kirk English, who worked so hard to bring this scholarly work to publication. It has been an honor and a joy to work with these two scientists, colleagues, and friends in helping to craft an approach to evidence-based practice in exercise science. This work is truly a needed part of our development in the field, especially as multidisciplinary team approaches are starting to take on more relevance in the search for answers to questions and proper directions in exercise programing and therapeutics.

I so appreciate the managerial and editorial expertise of editors Amy Tocco, Kate Maurer, Carly O'Connor, and the whole Human Kinetics staff for their patience and professionalism in helping to mediate this process.

—William J. Kraemer

PART I

Overview and Historical Background of Evidence-Based Practice

In the 1940s, a captain and physician in the Army Medical Corps, Thomas L. DeLorme, had an idea that perhaps "heavy resistance exercise" could be used to speed recovery in soldiers rehabilitating from injuries (DeLorme, 1946; Todd, Shurley, Todd, 2012). This was in stark contrast to the prevailing thought of the day, which advised patients to rest when injured. After piloting the idea in a small number of patients, DeLorme received permission to implement a clinical trial in 40 patients recovering from injuries, and the published results slowly revolutionized rehabilitation and resistance exercise. And although modern-day gymnasiums and weight rooms look dramatically different from the way they did in the 1940s, the basic principles of progressive resistance exercise that were elucidated in this study form the foundation of nearly every resistance exercise protocol; more than 70 years after it was conducted, millions of individuals who engage in strength training or rehabilitation unknowingly benefit from the results of this study. This experiment demonstrates that science can profoundly change the direction of a discipline.

Modern-day weight rooms, gymnasiums, and fitness centers are filled with equipment such as barbells, resistance machines, dumbbells, kettlebells, physioballs, medicine balls, resistance bands, chains, vibration platforms, treadmills, elliptical devices, and any number of other purported advances in the industry. In addition to prescribing training volume, intensity, duration, velocity, etc., exercise practitioners must choose which equipment they will use in their exercise programs. Most importantly, they must ask and answer the question: Will the use of [a particular piece of equipment] improve a particular outcome in my athlete, client, or patient? Physicians answer the same question on a daily basis when determining which drug to prescribe their patients; in the 1990s, a group of physicians defined a systematic methodology to answer these questions that arise in practice—they called it "evidence-based medicine." Evidence-based medicine (termed "evidence-based practice" in other disciplines) was at its core a process of asking a question, finding information from research and practical experience, determining the strength of the evidence, and then implementing the evidence into practice. In the first section of this book, we describe the history of evidence-based practice and why it is important for exercise scientists and practitioners, provide a brief overview of the philosophy of science and scientific study design, and provide a practical introduction to reading and interpreting research evidence. At the conclusion of this section, the reader should understand the importance of evidence-based practice as a philosophical basis for exercise prescription and how it can be used to find novel interventions and eliminate ineffective exercise options. After reading this section, the reader should also be familiar with basic research designs and the fundamentals of reading and interpreting a research article.

Chapter 1

THE NEED FOR EVIDENCE-BASED PRACTICE IN EXERCISE SCIENCE

Learning Objectives

1. Briefly describe the history of evidence-based medicine.
2. Describe the important characteristics of evidence-based practice.
3. Discuss the need for evidence-based practice in exercise science.
4. List five areas in which evidence-based practice is important for exercise scientists and exercise practitioners.
5. Understand the route to becoming an evidence-based exercise practitioner and define this practitioner's role on an integrated exercise and sports medicine team.

Evidence-based practice (EBP) is a term that originated from the evidence-based medicine movement of the early 1990s. A group of health policy makers and physicians, led by David Eddy, David Sackett, and Gordon Guyatt, realized that a substantial number of important clinical decisions were based on outdated information (Sackett & Rosenberg, 1995), inadvertently missing important novel clinical findings. Eddy, Sackett, Guyatt, and others championed the idea that there should be a systematic methodology employed by physicians to find, evaluate, and incorporate the newest, least biased, and most relevant information to improve patient outcomes (Oxman, Sackett, & Guyatt, 1993; Sackett, Straus, Richardson, Rosenberg, & Haynes, 2000). The movement recognized that knowledge is dynamic and that it changes with the discovery of new information. Since the 1990s, discipline-specific models of EBP have been developed for physical therapy (Maher, Sherrington, Elkins, Herbert, & Moseley, 2004), occupational therapy (Ottenbacher, Tickle-Degnen, & Hasselkus, 2002; von Zweck, 1999), nursing (Ervin, 2002), and other allied health science disciplines (Richards & Lawrence, 1995).

Exercise science is a rapidly expanding field with an evolving body of literature. Not only is the literature base changing, but also the data from the older literature are in need of reevaluation in the context of new findings, which may provide different meanings and understanding. Thus, older studies and their data are still relevant but need to be reinterpreted, not forgotten or dismissed completely. Advances in technology, novel discoveries, and accumulation of ideas may lead to stronger understandings of physiologic mechanisms which will translate to changes in practice (e.g., interpretation of growth hormone and its response to exercise; Kraemer et al., 2010).

Exercise science practitioners work in a variety of settings where their roles often include exercise prescription or testing with the goal of improving patient, client, or athlete outcomes. Similar to those in medicine and other allied health science disciplines, it is important for exercise science practitioners to adopt a discipline-specific model of EBP to ensure the best treatment of patients, clients, and athletes (Amonette, English, & Ottenbacher, 2010). In this chapter, the conceptual and philosophical rationale for EBP is introduced, as well as the need for multidisciplinary teams to search for, evaluate, and implement emerging information into practice.

EVIDENCE-BASED PRACTICE IN ACTION

On September 9, 2007, the second week of the 2007-2008 National Football League (NFL) season, the Buffalo Bills were at home for a game with the Denver Broncos. On a routine kickoff, third-year tight end Kevin Everett sprinted down the field and approached the kick returner, Domenik Hixon (Carchidi, 2008). Everett dropped his center of gravity and extended into Hixon for the tackle—a motor skill that he had presumably performed thousands of times previously. Everett led with the crown of his helmet, contacting the helmet of Hixon at a high velocity, inadvertently violating a common but too often disregarded fundamental of tackling. The consequences of the collision were devastating and nearly fatal for Everett.

Immediately upon impact, Everett fell to the ground and lay face down, motionless. The collision resulted in a fracture dislocation of his third and fourth cervical vertebrae, injuring the spinal cord and instantly paralyzing Everett (Cappuccino et al., 2010; Carchidi, 2008). The sports medicine staff reacted quickly to the injury, clearing the scene to ensure that Everett was not injured further. With an injury such as Everett's, the standard-of-care protocol provides that the helmet remain on, the player be placed on a board, the face mask removed, and the player carefully moved to a stretcher and then transported by ambulance directly to the hospital. The sports medicine staff appeared to perfectly execute the protocol, and Everett was carefully transported to Millard Fillmore Gates Hospital (Carchidi, 2008).

In the ambulance ride en route to the hospital, the neurosurgery team used an experimental therapy that they hoped would improve Everett's prognosis. Infusing a cold saline solution into Everett's veins, the medical team induced a mild hypothermic state (Cappuccino et al., 2010). The use of ice to treat an injury is not novel—ice is used acutely for virtually every orthopedic injury. The principle of inducing systemic hypothermia is similar. The cold saline solution acutely controls systemic inflammation, and the medical team apparently believed that the therapy would reduce inflammation surrounding the spinal cord injury, preventing further damage (Cappuccino et al., 2010; Mummaneni, 2010). After evaluation by the medical staff, the neurosurgeon performed an emergency surgery in an attempt to save Everett's life (Cappuccino et al., 2010). After the surgery, Everett was placed on a respirator; it was believed that if he lived, he was unlikely to ever walk again (Carchidi, 2008).

Two days after the injury, Everett was removed from the respirator; he was reported to have minimal voluntary movements in his lower extremities, suggesting a theoretical possibility that he could regain the ability to walk (Cappuccino et al., 2010). On the basis on previous cases, the medical team believed that Everett had a small chance of regaining full control of his extremities (Carchidi, 2008). In the weeks and months that followed, Everett made remarkable progress, recovering at a rate far greater than expected. He was transferred to a top rehabilitation center for neurologic injuries in Houston, Texas, that implemented a comprehensive rehabilitation program. On December 23, 2007, Kevin Everett walked onto the field of Ralph Wilson Stadium—a remarkable feat that a few short months earlier had seemed impossible.

Kevin Everett's recovery was remarkable and a testament to the professional competence and readiness of the integrated sports medicine team, who made numerous decisions improving the chances that he might walk. Everett's story is also a demonstration of personal determination and perseverance that has inspired hope. For patients suffering severe spinal cord injuries, the story validates that dramatic recovery is possible. Medical personnel are encouraged in that Everett's case shows that diligent medical and rehabilitative therapy facilitates recovery. However, there is one overriding question that surrounds Kevin Everett's case: Why did he recover and why so fast? There are a number of possible explanations:

- Everett's recovery was simply a statistical improbability—he was an outlier and his recovery a chance occurrence.
- It was the result of sheer human determination and will in an athlete who simply would not accept his prognosis and would give everything to ensure he had the best chance to walk.
- It was the result of the sports medicine team's careful execution of the standard-of-care practice to prevent further injury immediately post-accident.
- Everett's recovery was the result of the precise execution of the neurosurgery team.
- The recovery was a result of the rehabilitation protocol implemented by a leading neurorehabilitation center.
- The rapid recovery was the result of cryotherapy, an experimental method that dramatically improved his prognosis.

The case of Kevin Everett highlights an important concept, that knowledge is a dynamic phenomenon (Amonette et al., 2010). The team of doctors, physical therapists, athletic trainers, and exercise specialists involved in the acute and postacute treatment of Everett's injury made numerous decisions that positively or negatively affected his outcome. The athletic trainers and sports medicine staff identified a possible injury to the spine and ensured that the athlete was quickly transported to the hospital. A careful procedure was used to move him to avoid further injury. The physicians made a decision to use the hypothermic technique to reduce spinal swelling. The neurosurgeon decided to perform surgery, and at some point, a physician determined that it was safe to remove Everett from the ventilator. In the rehabilitative process, physical therapists along with the medical team decided which therapeutic interventions were best suited to the stage of injury and made daily decisions on when to increase, decrease, or hold parameters like the intensity, volume, and duration of the rehabilitation program.

How did the sports medicine team determine the appropriate treatment? Where did the knowledge associated with the appropriate treatment come from? Most of the individuals involved in the acute treatment and postacute rehabilitation process had advanced degrees in medicine, athletic training, physical therapy, or exercise science. In their academic training, they were provided information in lectures, books, and research articles. At certain points during their training, they completed and passed exams that demonstrated mastery and an understanding of a certain set of facts relevant to their discipline. The sports medicine staff had completed licensure and certification exams, further confirming their understanding of a core body of information learned in training. In order to practice in most professions, students participate in residency and internship programs in which they apply this information under the supervision of seasoned practitioners, confirming the practical application of their knowledge. In order to maintain licensure and certification, most disciplines, including medicine, athletic training, physical therapy, and exercise science, require continuing education credits. To fulfill these requirements, practitioners may attend workshops, conferences, and courses that teach, reinforce, or provide new information. Kevin Everett's treatment was affected by the information obtained through all of these sources.

Kevin Everett's chances of survival would have been dramatically reduced if his injury had occurred 100 years ago. In the past century, the treatment of acute injuries and life-threatening conditions has profoundly improved. One hundred years ago, there would have been no motorized transportation services to move an injured athlete to a hospital. The hospital itself would have been significantly less sophisticated, with no ability to image the spinal cord. Survival from surgery would have been reduced due to infection alone, not to mention the crude surgical techniques employed by early-1900s physicians. The medical staff would have acted based on state-of-the-art technology and the practice of the time, but the doctors and rehabilitation staff would have been ignorant of the vast discoveries, treatment options, and medical protocols of the next century.

The recovery of Kevin Everett might have been significantly different had he been injured and treated just 10 years earlier. We live in an age in which information is expanding more rapidly than at any time in human history. More specifically, the understanding of biology is changing, resulting in a greater understanding of how human systems operate. There is also a rapidly changing understanding of medicine, exercise, rehabilitation, nutrition, and so on. Of particular interest in the story of Kevin Everett is the novel cryotherapy. Cryotherapy is an experimental therapy currently under investigation to improve outcomes in a number of conditions including myocardial infarction (Dixon et al., 2002), traumatic brain injury (Aibiki et al., 1999), and spinal cord injury (Westergren, Farooque, Olsson, & Holtz, 2000). It is not a standard-of-care treatment; but if it works, shouldn't it be?

The methodology of evidence-based medicine was, at its core, a systematic approach to potentially alter stale medical treatment plans for individual patients, recognizing that advancements in medicine could change the standard of care to provide better outcomes for patients (Eddy, 1990). The changes would ultimately be grounded in evidence arising from clinical research and professional experience.

DEFINING EVIDENCE-BASED PRACTICE

An obvious question that arises is this: If medical practice was not based on evidence before the early 1990s, how exactly did physicians make medical decisions? What was the basis for their recommendations and treatments?

Due to the daily constraints of practice, physicians often make clinical decisions based on information learned many years or even decades before in medical school, residencies, or other practical experiences. In 2000, it was estimated that a substantial number of hospital decisions were not based on the most current best evidence (Ellis & Mulligan, 1995; K., Sackett et al., 2000). The decisions were instead based on experience, outdated material learned in medical school, and many years of observation in patients. Although this methodology does not always result in suboptimal treatment of patients, it may eliminate emerging treatment methodologies that are more effective, safe, or efficient. David Eddy, Sackett, and others suggested that there was a need to define a process ensuring that physicians based important clinical decisions on updated information (Sackett, Richardson, Rosenberg, & Haynes, 1997).

The methodology of evidence-based medicine was defined by a physician and epidemiologist from McMaster University, Dr. David Sackett. Drs. Sackett and Gordon Guyatt championed the idea of evidence-based medicine (EBM) as a "new paradigm in medicine." David Sackett provided the first and most often cited operational definition of EBM: "the conscientious, explicit and judicious use of current best evidence in making decisions about the care of individual patients. The practice of EBM means integrating individual clinical expertise with the best available external clinical evidence from systematic research" (Sackett, Rosenberg, Gray, Haynes, & Richardson, 1996, p. 71). Sackett's definition specified essential parameters that would create the framework for the process.

Evidence-Based Practice Is a Conscientious Process

At its core, the operational definition provided by Sackett suggests that the process of EBP is a careful and thorough process (Sackett et al., 1996). In other words, every clinical decision is important and should be thoughtfully considered so that the best decision is made for the patient at a given point in time. Sackett suggested that no clinical decision should be made indiscriminately. When deciding whether to prescribe treatment A or B for a patient, the physician should carefully consider the potential benefits and limitations of the treatment plan in light of the most current evidence so that the treatment is best suited to the individual's needs. In the case of Kevin Everett, the physician or medical team apparently made a series of considered choices. Some of the decisions were standard procedures or protocols (i.e., movement on a board, helmet removal, and so on) based on a firm foundation of evidence. Some however, were less routine and presumably based on emerging experimental research, and these thoughtful clinical decisions may have had a significant impact on his outcome.

Evidence-Based Practice Is an Explicit Process

While the process of EBP should be careful and thoughtful, the operational definition provided by Sackett suggests that the process should also be simple and clear (Sackett et al., 1996). Because practitioners are already under pressure and time constraints, the process to incorporate updated information into practice should not be cumbersome, confusing, or unnecessarily complicated. Instead, the process should be well defined and easily replicated without the need for advanced training, sophisticated tools, or long hours of researching information (Davidoff, Haynes, Sackett, & Smith, 1995).

Evidence-Based Practice Is a Judicious Process

When assessing information to inform the care of an individual patient or client, practitioners should carefully consider all possibilities or treatment plans. Evidence-based practice should be a process that fairly and objectively assesses information; the process should not be biased by the practitioner's predetermined ideas (Sackett, 1989a). In a criminal court case, a prosecutor brings evidence to a grand jury to demonstrate that a defendant has committed a crime. If the evidence is sufficient, the grand jury indicts the individual and he must stand trial for the crime. In the court proceedings, the prosecution and the defense bring evidence supporting either the guilt (prosecutor) or innocence (defense) of the defendant. After hearing the evidence presented by both parties, prosecutor and defense, a jury evaluates the evidence offered and renders a verdict of guilty or not guilty. Yet different from a court case, the jury for EBP is a team of highly trained professionals made up of physicians, scientists, and certified coaches who have the education and experience necessary to evaluate the question at hand.

Similar to the judicial process of a court, the evidence-based process requires an individual to collect all pertinent evidence before rendering a decision. After accumulating information, the evidence-based practitioner makes a judicious decision based on all the available information. Such information also includes understanding the reinterpretation of older literature in the context of new findings. (Often such literature is dismissed due to the date of the publication, and valuable insights are missed.) The decision is then incorporated into the individual's treatment plan.

Evidence-Based Practice Uses Current Evidence

When making treatment decisions for a patient, client, or athlete, the practitioner should use current and not obsolete evidence to guide, influence, and direct practice. Knowledge in any field or discipline is a cumulative phenomenon. What is known today may be better understood in the future. Typically, as more information is gathered on a particular subject, treatment plans are refined and improved; that is, in most instances, they are not completely altered. However, at times, current evidence may completely upset a treatment approach. For example, in the United States in the 1940s one could purchase cigarettes that were "physician approved," with marketing campaigns proclaiming "More doctors smoke Camels than any other cigarette" and "20,679 Physicians say Luckies are less irritating" (figure 1.1). Indeed, a large proportion of physicians, the theoretical front line of health promotion in the 1950s, were smokers (Snegireff & Lombard, 1959), and it was common practice for physicians to recommend that their patients smoke cigarettes to help relieve stress. The recommendation was likely based on observations that smoking had an acute calming effect on individuals. However, studies in the late

Figure 1.1　Actual marketing ads for cigarette companies using physician endorsement. These were common before the 1964 Surgeon General's report linking smoking to serious health risk.

From the collection of Stanford University. (tobacco.stanford.edu)

1950s and the 1960s showed a strong association between cigarette smoking and lung and mouth cancers. As the evidence mounted, the U.S. Surgeon General published a scathing report on the dangers of smoking (Bayne-Jones et al., 1964). In turn, physicians altered their recommendations and began advising their patients against smoking, which current evidence had demonstrated was deleterious to individuals' health.

Although emerging or current evidence does not always dramatically alter recommendations as in the case of cigarette smoking, it does slowly improve treatment. In the case of Kevin Everett, the medical staff used a series of standard procedures. It is likely that many of these procedures had been used for decades. However, the staff also incorporated novel treatment approaches. In conjunction with all of the other treatment decisions, the sports medicine team led by the physician used the systemic hypothermic treatment. This would not have been an option or even a thought 50 years earlier, but newer evidence suggested that it might be an effective addition to the treatment plan.

Evidence-Based Practice Uses the Best Evidence

Exercise practitioners who choose to engage in the process of EBP must strive to use the best evidence to develop the treatment plans for their patients, clients, and athletes. The term "best" denotes an important point—namely, that there are various levels of information available for the practitioner (Sackett, 1989b). In a popular television commercial for an insurance company, a young woman begins to explain her rationale for choosing an alternative insurance provider. Her friend asks where she got her information, and she responds, "I read it on the Internet—they can't put anything that isn't true on the Internet." As she finishes the statement an unshaven, overweight, and unattractive man approaches her. She introduces him as her French model boyfriend whom she met on the Internet. The point of the commercial: Don't believe everything you read on the Internet.

A plethora of information is available to practitioners to inform good decisions. However, there is also an abundance of misinformation or biased resources that may create obstacles to good decision making. Thus, practitioners must have clear and objective means of deciphering which information is best and least biased. Based on the newer, best evidence, the sports medicine team for Kevin Everett made a conscious decision to act based on what they believed was the "best" evidence. Practitioners who desire to provide the best options for their patients must remain apprised of the most current information and committed to the reinterpretation of classical research in the context of newer findings.

Evidence-Based Practice Focuses on the Patient

The ultimate product of the EBP process is the integration of clinical experience, research evidence, and patient choice resulting in the best and most appropriate treatment for the individual patient. Although this may seem obvious, it is not always the result. The health care industry, in particular, has attempted to twist the original meaning of EBP and use it for financial gain (Law & MacDermid, 2008). Some insurance companies attempt to use their interpretation of research evidence, or lack of evidence, to support or eliminate various treatments. Their motivation may be to eliminate wasteful spending by physicians, physical therapists, occupational therapists, et al., but such actions also limit the treatment options of the physician. Again, if a treatment is not supported by evidence, eliminating it is appropriate, but unfortunately research does not always exist for a given treatment. Instead, research evidence often lags behind practice. The result is that viable and effective treatment options may be excluded because of a misinterpretation of the lack of evidence. Such practice is against the spirit and true definition of EBP—to best treat the individual patient.

Evidence-Based Practice Is an Integration of Information

Finally, EBP is an integration of information from various sources. David Sackett originally defined a model in which clinical decision making was based on three sources of information: clinical expertise, research evidence, and patient preference (Lavin et al., 2005). The model was improved and adapted by Satterfield and colleagues (2009) to include the following three components: best available research evidence; available resources, including practitioner's expertise; and client characteristics, needs, values, and preferences (Satterfield et al., 2009). Further, the model suggests that all of this information must be considered in the context of the environment and organization (figure 1.2).

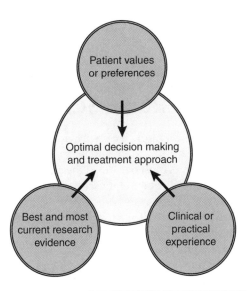

Figure 1.2 Evidence-based practice conceptual model adapted from Satterfield and colleagues (2009). All sources of evidence must be interpreted in the context of the environment or organization in which they are applied.

Adapted from Satterfield et al. 2009.

Both Sackett's and Satterfield's models indicate that information or evidence can be derived from research and clinical or practical experience. Research in most cases provides the compass for practice—it points practitioners in the correct direction for decision making. However, some information used in decision making is more appropriately derived from clinical experience and not from research evidence. For example, logistical constraints of treatment cannot be effectively defined by research. Suppose a physician finds evidence supporting a new treatment for osteoporosis. Treatment A involves daily injections of a drug that must be administered by a nurse. Treatment B involves a once per month injection of a different drug. The most current research evidence suggests that the two treatments are equally effective and similar in cost. The physician would likely make a decision based on her clinical expertise indicating that compliance will be better with drug B.

In some instances, research evidence from which to make a decision may simply not be available. In such cases, physicians will rely on their understanding of human biology, pharmacology, etc. to make a choice as to which treatment to provide the patient.

When possible, the EBP models proposed by Sackett and Satterfield suggest that practitioners should use the "best and most current research evidence" to make their clinical decisions. The research evidence should be integrated with the physician's clinical experience to determine the best treatment option. Additionally, the physician should carefully consider which evidence is best (experience or research evidence).

The physician should also carefully consider the patient's needs, preferences, state, and values, respecting individual rights in decision making. The model recognizes that treatment plans may differ based on the individual's characteristics and wishes. Although the physician may be certain, based on clinical experience and research evidence, that a particular treatment is most effective, the patient always has the right to refuse. The patient should be an equal partner in the medical decision-making process.

APPLYING EVIDENCE-BASED PRACTICE IN EXERCISE SCIENCE

Similar to medical doctors and other health care providers, exercise practitioners make decisions that directly affect the health, safety, and performance of their clients, patients, and athletes. Exercise practitioners must decide on the appropriateness or safety of testing procedures, training plans, and equipment usage and selection across a wide variety of populations. In some settings, this may involve testing clients, patients, or athletes with special needs, various disease or disability states, or functional limitations. Top exercise science programs teach the general principles associated with exercise adaptation and underlying physiological and physical mechanisms, providing the framework for sound decision making. However, it is impossible for academic programs or certifications to prepare practitioners for every possible scenario. Therefore, it is important to define a process to find information that ensures the safety or efficacy of a training or testing plan and a mechanism to find answers to the specific questions arising in practice. Figure 1.3 provides an overview of five areas in which EBP is of particular utility to exercise scientists, and in the next sections we discuss some of these areas in more detail.

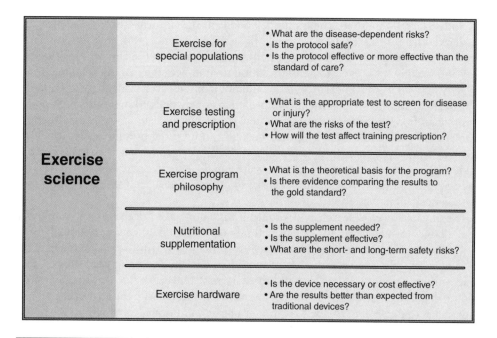

Figure 1.3   Five areas in which evidence-based practice may be particularly important for exercise scientists, with sample questions related to each domain.

Exercise Protocols and Prescription

One domain in which important decisions are made is exercise prescription. A plethora of programs are available to the consumer. Emerging concepts are passionately debated, such as the benefits of high-intensity interval training (Weston, Wisloff, and Coombes, 2014) and many other training concepts that are being examined in new studies. Many of these emerging ideas have respectable components, but one must ask whether the information is really sufficient to suggest that practitioners should fully commit to these philosophies or new approaches to exercise prescription. Evidence-based practice provides a process for evaluating whether novel protocols are really better than standard approaches and guides the exercise practitioner on where to seek information.

Exercise for Special Populations

In working with an assortment of populations, from children to athletes to people who are elderly and frail, exercise practitioners must determine the optimal exercise modes, sets, reps, velocities, durations, and so on to ensure that their clients safely and efficiently achieve their goals. In many cases, exercise prescription differs dramatically depending on the population. Certain exercises may be indicated for a mature athlete but not for a young child. Likewise, the aging process causes numerous physiological changes; as a result, certain exercise modes or methodologies should be a priority for improvement of health. In such cases, how do exercise scientists or practitioners conscientiously decide whether to employ a power clean, squat, leg press, or leg extension with a client who is elderly? How do they decide whether a client who is elderly should lift a weight at a high or low velocity or perform one, two, three, or more sets of an exercise? How do they determine which exercises are safe and which will expose their client to undue risk? Understanding and interpreting the research on special populations are essential to correctly adapt programs and to ensure the safety, proper technique, proper progression, and effectiveness of the exercise prescription.

Nutritional Supplements

Perhaps the most rapidly expanding area in which it is important to be a good consumer of information is the nutritional supplement industry. It is almost impossible to keep pace with this industry as new substances or substance combinations constantly emerge. The substances may be marketed as tools to improve performance, strength, power, endurance, hypertrophy, weight loss, or health. Indeed, many are effective and are supported by strong, credible, and well-conducted research. The particulars on how to best use these effective supplements (e.g., timing, quantity) in certain populations may emerge over time, creating a need to consistently search the literature.

As a consumer or as an exercise practitioner providing advice on nutritional supplements, it is important to understand that some are poorly studied and have little evidence to support their effectiveness or safety. In contrast to the pharmaceutical industry, there is little regulation with respect to the production of a supplement. For a prescription drug to surface on the market, the drug must be strongly supported by research. Before a drug can even be tested in humans, there must be strong support for its safety, typically evaluated in animal models. Once animal evidence supports safety and efficacy, an application may be submitted to the Food and Drug Administration (FDA) to begin human subject trials. If the FDA approves the application, data are typically collected on a small group of humans to determine potential side effects, metabolic properties, and other basic science-related questions. After evaluation by the FDA, the product can be tested in a larger human sample to determine its effectiveness in a group of patients with a specific condition. If the drug is effective in this sample, it can then move to phase III, in which it is tested in thousands of patients in different doses and in combination with other drugs to further support its efficacy and safety. If the data are strong enough, the product is approved for human clinical use. The process takes years to complete and typically millions of dollars. Despite the rigorous testing, occasionally drugs are still pulled from the shelves when it is later determined that the drug has a side effect not discovered during testing.

In contrast to this scrupulous and rigorous testing process, nutritional supplements may be developed and tested in small numbers of animals and humans and then made available for purchase and human consumption. In many cases, effectiveness and safety are actually tested in the consumers who buy the supplement. The process has both benefits and drawbacks. In most cases, supplements are not dangerous, and reputable companies fund laboratories to independently and objectively test their products. If the results on a nutritional supplement are published in a respected peer-reviewed journal, this provides a level of evidence for understanding its effects. Yet some supplements on the market have not been subjected to any independent experimental research in humans. This allows supplements to reach the shelves at a more rapid pace and helps to reduce the end cost for the consumer. However, the lack of regulation can be problematic. In some cases, quality control is poor, and supplements not tested by a credible third-party testing entity may contain traces of products that are illegal or banned in certain sports. Supplement companies may misrepresent research findings, as marketing claims are not regulated. Some may also move a product to market too quickly, before its safety is well evaluated.

Another concern with the nutritional supplement industry is that individuals tend to respond differently to a given stimulus. Because the sample size for most nutritional supplement studies is far smaller than for most drug studies, results of studies may not generalize to all individuals and populations. Thus, the number of responders and non-responders in a particular niche study on a nutritional supplement is an important piece of evidence. The probability of a study not detecting dissimilar subject responses could be magnified with small sample sizes. Poor study design can exacerbate this effect (e.g., not matching subjects before randomization into groups or use of within-study designs to control genetic influences). For these reasons, exercise practitioners must be careful in generalizing to populations not represented in a particular nutritional supplement study (e.g., if the sample consists of men, one should be cautious in generalizing to women).

Exercise Equipment and Devices

Exercise practitioners must learn to evaluate the effectiveness and safety of exercise equipment. An abundance of exercise devices, equipment, and gadgets are marketed as important or necessary to improve fitness, health, and human performance. Television is peppered with late-night infomercials on devices that seem effective and marketing materials are designed to lure individuals desperate for an easy fix for a complicated health or fitness problem. Often these devices are backed by Hollywood testimonials, fitness industry gurus, or remarkable individual success stories about use of the product. Sometimes the devices are even backed by "research evidence" and marketed as supported by "university research." The consumer, either from outsized hope, a lack of desire to investigate the marketing claims, or general indifference, buys in. The products are often ineffective at best; at worst, they are dangerous and lead to injury. This wasteful and injurious process can be avoided if individuals are good "consumers of research" (Amonette, English, Spiering, & Kraemer, 2013).

Although late-night infomercial marketing is an easy target for a book on evidence-based practice, the industry also preys on educated and well-informed individuals. While in many homes the treadmill is used as a clothes hanger and "core training" devices are stored on the top shelf of the closet, the fitness centers of corporations, hospitals, and physical therapy clinics, as well as high school, college, and professional weight rooms, may also house devices that are not effective or are potentially dangerous. The devices may have been purchased, with good intentions, on the advice of a colleague or a respected industry salesman or with year-end budget surplus. Because many fitness industry jobs are time-consuming, people may not have the time to thoughtfully and carefully consider these situations, or they may not be trained to consume research evidence and make the best decisions. In some cases, elite programs, teams, and corporate facilities are looking for a competitive edge; they may purchase products with limited research support in hopes that they will provide such an advantage. This practice is understandable considering the competitiveness of the industry, but when products are selected based on flimsy or misused evidence, problems may result. Moreover, adopting products before their safety is fully evaluated can expose clients to injury and the exercise practitioner to litigation. There must be a systematic, reproducible, easily accessible way to evaluate products so that exercise practitioners can ensure the tools used in training are effective and safe. Evidence-based practice provides a reliable process for evaluating equipment and making sound and reasonable decisions to implement equipment into or eliminate it from prescription.

TEAM APPROACH TO EVIDENCE-BASED PRACTICE

The rapid propagation of research information and growth of the knowledge base within many disciplines and subareas have created a unique opportunity for collaboration and teamwork. The various vectors of knowledge influencing client, patient, or athlete health, fitness, and performance are broad and deep with respect to information (figure 1.4). It is impossible for any one individual to be trained in each area; moreover, a single individual cannot remain current with emerging evidence in each area. Therefore, there is a need for discipline-specific experts to work in unison, bringing unique and valuable ideas for discussion and implementation from developing research in their specific subarea. The exercise practitioner is a valuable constituent on the integrated team, searching for evidence supporting treatment options related to strength, conditioning, fitness, testing, supplementation, and exercise prescription.

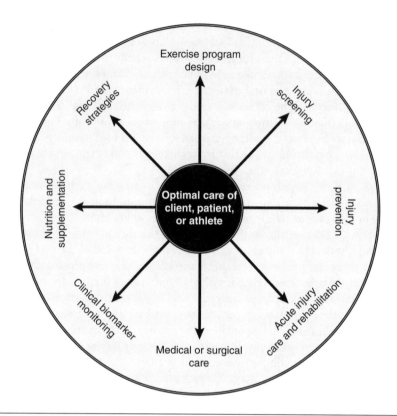

Figure 1.4   Nonexhaustive list of knowledge vectors associated with care of a client, patient, or athlete. This highlights the need for an integrative team approach to care.

BECOMING AN EVIDENCE-BASED PRACTITIONER

How does a practitioner decide which exercise to prescribe or whether a testing protocol is safe for an individual? What if the individual has a functional limitation or disability? Is there a way to decide if an exercise treatment is more or less appropriate for that individual? The answers to these questions are found in the methodology of EBP.

Are your programs evidence based, and are you an evidence-based practitioner? Both of these questions are commonly asked in the medical and health fields. Unfortunately, the questions are confusing, presenting an either/or scenario that violates the true spirit of the process defined by David Sackett and others. Unless your programs are completely random or the decisions you make are based on nothing—then yes, your program is evidence based. The real and fundamental question is: How current and sound is the evidence from which you are developing your decisions? Simply speaking, an evidence-based practitioner is an individual who is committed to the process of finding, evaluating, and incorporating the "best" evidence to steer the treatment plans for patients, clients, and athletes.

A common misconception about EBP is that anyone who has an education (i.e., academic degrees) should be able to intuitively engage in the process. This is not necessarily the case. Most academic degree programs, especially at the undergraduate level, focus on teaching basic sciences and learning a set of "facts." These facts are memorized and restated on an exam, and if students memorize 70% of the information provided by the instructors they are eventually awarded a degree. To further their education, graduating exercise practitioners may choose to complete certification exams, accreditation, or continuing education from reputable organizations such as the National Strength and Conditioning Association (NSCA), American College of Sports Medicine (ACSM), Australian Strength and Conditioning Association, and UK Strength and Conditioning Association. For such certifications, individuals are required to memorize more information in a specialized area. A student who memorizes enough of the information to pass the test is awarded a certification credential. The problem with degrees and certifications is that they demonstrate mastery of information at one point in time and represent only "minimal competency." They are important for developing the foundation for learning, but they are just the beginning of the learning process. Because scientific information constantly accumulates, the evidence-based practitioner must be committed to an ongoing process of learning: consistently finding new research and in some cases reinterpreting old study findings within the context of the new information.

In the spirit of Sackett's definition, the process should be simple. However, as with any other skill, practice and training are required. In order to become a competent, effective evidence-based practitioner, you must commit to learn and practice the process. The purpose of this book is to provide a systematic methodology to access and incorporate knowledge into practice. Ideally, this information will improve the outcomes of patients, clients, and athletes.

CONCLUSION

Evidence-based practice is a "new" process that was introduced in the early 1990s by David Eddy, David Sackett, and others. Although using evidence to improve outcomes was not new per se, the novelty of Sackett and Eddy's proposal was the definition of a process combined with a commitment to use the most current and best evidence. Exercise practitioners make daily decisions that affect the health, fitness, or performance of their patients, clients, and athletes. These decisions are based on information, but one must ask whether the information is the most current and best available. The purpose of this book is to help define the process of EBP as it relates to exercise science, providing a new model that is specific to the exercise discipline.

Chapter 2

THE PROCESS OF EVIDENCE-BASED PRACTICE

Learning Objectives

1. List the original five steps in the evidence-based practice process.
2. Briefly describe each of the five original steps in the process.
3. Discuss the benefits of an evidence-based approach to exercise science, testing, and prescription.
4. List some of the classical criticisms of evidence-based practice and provide a rebuttal.

Similar to the situation in medicine and other allied health science fields, use of the term "evidence-based" is becoming common in exercise science. The term is frequently used as an adjective to describe a product, philosophy, or device that is supposedly based on a strong set of research studies (English, Amonette, Graham, & Spiering, 2012). But are the products, philosophies, and devices we implement on a daily basis really evidence based? More importantly, what is the strength of the evidence to support their use? The term "evidence-based" was coined to denote an active, dynamic process—not to serve as a window-dressing adjective. It describes a methodology by which a clinician systematically evaluates information to provide patients with effective and safe treatment plans.

The methodology of evidence-based practice (EBP) was originally described by David Sackett in the early 1990s. After providing his seminal definition, Sackett, along with Guyatt and others, authored a series of articles that appeared in *Journal of the American Medical Association* (Guyatt et al., 2000; Guyatt, Sackett, & Cook, 1993, 1994; Guyatt et al., 1995). The articles were the result of an evidence-based medicine working group that described the process and recommendations for incorporating evidence into the treatment of individual patients. From these articles and other writings, a five-step approach to EBP emerged (Sackett, Straus, Richardson, Rosenberg, & Haynes, 2000):

1. Ask a question.

2. Find evidence.

3. Evaluate the evidence.

4. Implement the evidence.

5. Periodically reevaluate the question or evidence.

In this text, we have added a sixth step to the original model: confirming the efficacy of the evidence in the individual case. This novel step and its importance for exercise practitioners are discussed in detail in chapter 10. In this chapter we provide an overview discussion of the traditional five-step approach.

THE IMPORTANCE OF EVIDENCE IN EXERCISE SCIENCE

Problems arise in exercise science as a result of unique or novel situations in practice. Through academic training, exercise practitioners acquire a background of information from which they develop their testing and training prescriptions. The role of academic preparation is to disseminate stable bases of knowledge such as laws and principles; this information is supported by a rich body of replicated research and is unlikely to change (Kraemer, Fleck, & Deschenes, 2016). For example, basic physical principles associated with lever systems, gravity, force, and power are based on nearly 300 years of scientific thought and evidential support. Likewise, biological theories such as the size principle and sliding filament theory are stable sources of information. They are unlikely to change, and the understanding of these basic constructs provides a firm foundation of general knowledge that allows a student to begin investigation of the details related to exercise physiology, biomechanics, and training.

Other sources of information, such as the details of exercise prescription, are less stable than laws or principles, and academic training is limited in that it can convey only a limited amount of information at a single point in time (Amonette, English, & Ottenbacher, 2010). It is impossible to complete a course or series of courses in an academic training program that prepare a student for every possible situation that may occur in practice. Practitioners are likely to encounter unique patient populations, training devices and philosophies, nutritional supplements, and novel case situations in which they must think critically and apply previously learned information. In many cases, the answer of how to best approach these situations cannot be found in lecture notes or textbooks used in previously completed courses. Thus, there is a need for practitioners to develop a systematic approach to learning once their academic career is finished.

Even if the answers to questions arising in practice are found in a textbook or lecture notes, they may be outdated and potentially wrong (Fletcher & Fletcher, 1997). Because the body of scientific knowledge is constantly evolving, recommendations for exercise practice are also constantly changing. Although the evidence does not often suggest extensive changes to practice, it does stimulate subtle modifications over time. A classic example of these subtle modifications can be seen through position statements.

The American College of Sports Medicine publishes position stands on topics such as resistance exercise, exercise in adults who are older, exercise and type 2 diabetes, and exercise and fluid replacement. When these position stands are written, they represent authoritative statements by the College based on the best available evidence at the time. However, these position stands are updated on a regular basis because it is understood that new evidence accumulates over time, necessitating modifications to the position (American College of Sports Medicine, 2009; Kraemer et al., 2002). The same can be said of many academic courses teaching the details of exercise prescription. For instructors who teach based on the most current evidence, academic courses change by semester. Instructors incorporate newer textbooks with more up-to-date information. Newer, more relevant research papers are incorporated into lectures and provided for student reading. However, this information may be outdated by the time it is printed, particularly in the case of textbooks, which take considerable time to publish. Moreover, the material may not cover every specific situation that is encountered in practice.

APPLYING THE FIVE STEPS OF EVIDENCE-BASED PRACTICE

An evidence-based approach to practice provides practitioners with a method to incorporate newer information into practice as a complement to academic training. Thus, when encountering a novel situation or circumstance, the practitioner is equipped with the tools to find the information so the best decision can be made for the client, patient, or athlete. The approach also allows practitioners to confirm information learned through academic training, conference presentations, and certification material and determine whether newer information overturns what was previously taught.

Suppose you are the director of an exercise physiology program at a hospital. You have a program in which patients discharged from physical therapy are offered the opportunity to enroll in your transitional exercise program. The program primarily uses free weight training, and the facility is well equipped with barbells, bumper plates, platforms, and dumbbells and has experienced success, developing a strong referral base from a variety of physical therapy clinics in the community. As the director, you have authority to purchase equipment for your facility but have a limited budget. Thus, you are careful to purchase only equipment that is cost-effective and impactful for your patients. One of your employees attends a conference where he hears a presentation on the use of a novel exercise device ("product X"), titled "An Evidence-Based Approach to the Use of Product X to Promote Improvements in Bone Mineral Density." At your facility, 75% of your client population is over the age of 50, thus at increased risk for low bone mass. Moreover, a large percentage of the older population at the facility are women, most of whom are postmenopausal. If the product works, it would indeed be an important addition to practice. However, the product is expensive, and purchasing the device would represent a significant portion of the budget.

Step 1: Ask a Question

The first step in the process of EBP is to develop a fundamental or background understanding of the problem in order to formulate a specific question that can be answered by the research. In this hypothetical example, product X is a new device, and it is likely that you know little or nothing about it. For the sake of illustration, suppose that product X is a resistance training device that uses elastic cords to create resistance. The elastic cords attach to a bar that can be placed across the shoulders, and an individual can perform exercises such as squats, heel raises, and deadlifts. Of course, there is much evidence supporting the use of heavy resistance and high-impact exercise to promote bone mass in a variety of populations (Cussler et al., 2003; Dickerman, Pertusi, & Smith, 2000; Humphries et al., 2000; Witzke & Snow, 2000). In particular, these exercises promote increases in bone mass in the spine and hip—two common skeletal regions for osteoporosis-related fractures (Bruyere et al., 2009; Hedlund & Gallagher, 1989; Ross, Genant, Davis, Miller, & Wasnich, 1993; Wilk et al., 2014). Product X uses an elastic cord to impart resistance to the user. Unlike the predictable force curves associated with the acceleration of mass using free weights, elastic resistance provides force proportional to the stretch of the cord. Moreover, elastic resistance exhibits a reduction in force when one relaxes and returns the cord to its resting length (hysteresis). In short, the force curves between free weights and product X are different. Thus, the question at hand is, Does product X promote an increase in bone mass, and is it more effective than traditional resistance such as free weights, particularly for women who are older? From the hypothetical example, the following direct question can be constructed:

In postmenopausal women, are squats, deadlifts, and/or heel raises performed with product X more effective than traditional free weight exercises to improve or mitigate the loss of bone in the hip and spine?

Step 2: Find Evidence

After constructing a question, the next step in the process of EBP is to search for and find evidence (i.e., information) that helps answer the question. Before starting, you already have one piece of evidence—the conference presentation suggesting that product X is effective in improving bone mineral density. Many practitioners would stop at this step and simply assume that the presentation is sufficient to support the product. However, there are many issues with a single piece of evidence, especially when it is simply someone's opinion. First and most importantly, the information may be intentionally or unintentionally biased by the individual's own preconceptions. In any case, it is prudent to look for less biased information to support or refute the use of the product.

The best evidence would be a randomized controlled trial in which product X is compared to free weights (figure 2.1). Ideally in such a study, postmenopausal women would be randomly assigned to a no-exercise, free weight, or product X training group after matching for activity background, prior resistance training experience, nutritional profile, and bone density. The free weight and product X groups would perform an equal number of sets and reps at identical intensities, velocities, and so on over the same period of time. These two groups would also consume a similar diet and be equal in as many ways as possible except for the independent variable (i.e., product X vs. free weights).

Unfortunately, this study does not exist. In fact, it is rare that an individual searches the literature and finds the perfect study to answer the question. After searching for information, it may be that you find no research whatsoever to support product X. However, this does not mean that product X is not effective; it simply means that there is no scientific support for this particular product at this point in time. Some might consider the lack of direct evidence a reason to not use or purchase the device. However, a more thorough approach might be to rephrase the question in such a way that you

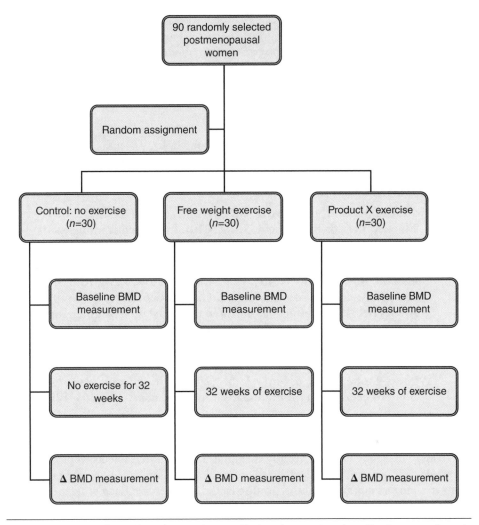

Figure 2.1 Hypothetical randomized controlled trial assessing the effectiveness of product X as a tool to improve bone mineral density compared to the "gold standard," free weight exercise.

find evidence supporting or refuting the effectiveness of the two types of resistance and their ability to improve bone mineral density or mitigate its loss. The question could be rewritten this way:

> **Are spinal loading exercises performed with elastic resistance more effective than traditional free weight exercises to improve bone mineral density or mitigate its loss in the hip and spine?**

In the rephrased question, the population is broadened to include any individual. Also, the specific exercises (squats, deadlifts, and heel raises) are broadened to include any spinal loading exercise. Finally, product X is replaced with any elastic polymer-based resistance device. There are potential problems associated with broadening a question; we discuss these in chapter 6.

Using the new question, the following search phrase is entered into an online search engine: "elastic resistance training and bone mass." The search returns four research articles. Of the four articles, one employed a rodent model (Swift et al., 2010), one used blood flow occlusion in combination with elastic resistance (Thiebaud et al., 2013), and two implemented a combined elastic resistance with free weights or some other form of dynamic constant external resistance (DCER) (Judge et al., 2005; Winters-Stone & Snow, 2006).

> Judge, J.O., Kleppinger, A., Kenny, A., Smith, J.A., Biskup, B., & Marcella, G. (2005). Home-based resistance training improves femoral bone mineral density in women on hormone therapy. *Osteoporos Int*, 16, 1096-1108.
>
> Swift, J.M., Gasier, H.G., Swift, S.N., Wiggs, M.P., Hogan, H.A., Fluckey, J.D., & Bloomfield, S.A. (2010). Increased training loads do not magnify cancellous bone gains with rodent jump resistance exercise. *J Appl Physiol*, 109, 1600-1607.
>
> Thiebaud, R.S., Loenneke, J.P., Fahs, C.A., Rossow, L.M., Kim, D., Abe, T., Anderson, M.A., Young, K.C., Bemben, D.A., & Bemben, M.G. (2013). The effects of elastic band resistance training combined with blood flow restriction on strength, total bone-free lean body mass and muscle thickness in postmenopausal women. *Clin Physiol Funct Imaging*, 33, 344-352.
>
> Winters-Stone, K.M., & Snow, C.M. (2006). Site-specific response of bone to exercise in premenopausal women. *Bone*, 39, 1203-1209.

After examining the reference lists of these articles, we find another paper that compares an elastic polymer system to traditional free weight training (Schneider et al., 2003). This study used 28 healthy men and women. The subjects were randomly assigned to a free weight group, a lower-volume elastic training group, a higher-volume elastic training group, and a control group that did not exercise. Each group performed squats, heel raises, and deadlifts with progressively increasing resistance. The protocols for the free weight group and the low-volume elastic resistance group were identical. The higher-volume elastic training group performed identical exercises but with twice the volume. Before the investigation, bone mineral density (BMD) of the hip, spine, and calcaneus was assessed. The results of the 16-week training study indicate that there were no changes in BMD in either elastic training group; however, the free weight group experienced a significant increase in BMD of the lumbar spine (Schneider et al., 2003).

Step 3: Evaluate the Evidence

As noted earlier, in this hypothetical example there were no studies directly supporting the use of product X to improve BMD. Most of the studies using elastic resistance to improve BMD used another training modality in combination with the resistance training intervention and thus do not provide a good comparison. The one study using an elastic polymer-based system with a direct comparison to free weights demonstrated that free weights improved BMD in the lumbar spine and that the elastic-based system did not. Thus, the evidence supporting the use of product X is the opinion of one expert in a presentation. The evidence against the use of an elastic-based system to improve

BMD is a single randomized, controlled trial demonstrating that free weights are a more effective tool. Neither the argument for or the one against the use of product X or elastic resistance is strong, but the argument against has less potential for bias. Also, there is a rich body of literature supporting the use of free weights to improve BMD in a variety of populations (Dobek, Winters-Stone, Bennett, & Nail, 2014; English, Loehr, Lee, & Smith, 2014; Gray, Di Brezzo, & Fort, 2013; Suominen, 2006; Winters-Stone et al., 2011, 2014).

Step 4: Incorporate the Evidence Into Practice

After a careful examination of the evidence, there is no compelling evidence supporting product X at this point in time. The evidence supporting free weight training alone to improve BMD is robust. Since purchasing the device would require a significant financial investment and would not likely make a commensurate improvement in clinical outcomes, purchasing product X would not be a prudent decision. The use of free weight resistance training is still a viable and well-supported option for your facility.

Step 5: Reevaluate the Evidence

It is important to note that the verdict is specific to evidence at the time the question was asked. In the future, studies may arise that strongly support the use of product X. It is possible that product X is engineered to exhibit properties different from the devices used in prior studies. However, at present, no evidence exists to demonstrate that it is different from previously tested devices. Thus, product X should be excluded for now, but the question should be reassessed in the future in case new information materializes. Figure 2.2 summarizes the steps we have described to apply EBP to answer the question whether product X is effective in increasing BMD.

THE CASE FOR EVIDENCE-BASED PRACTICE

Our hypothetical example highlights the importance of the process of EBP. It standardizes the way we acquire, review, and evaluate information, removing, when possible, bias and emotionally charged arguments. Additionally, the EBP methodology creates a standardized terminology for discussing information with other allied health science professionals. It is not purely science-based practice but instead is the integration of scientific information with practical knowledge. The integration of EBP into practice will undoubtedly improve the credibility and prestige of exercise science as a discipline.

Provides a Standardized Method to Answer Practical Questions

Evidence-based practice provides a unique and standardized methodology for assessing professional questions. As discussed in chapter 1, questions in exercise science arise primarily from five different domains: evaluation of novel exercise devices, nutritional supplements, exercise program theories, testing and measurement, and exercise for special populations. Each of these areas routinely presents new and often unpredictable scenarios. As questions arise, it is important for exercise practitioners to have a simple method for assessing the information that results in a predictable solution. In other

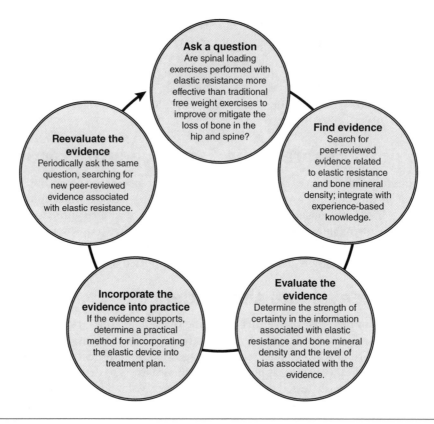

Figure 2.2 The five-step approach of evidence-based practice as proposed in the medical literature.

words, if two exercise practitioners from two different countries were to have the same question, if they use the same method they should arrive at a similar conclusion. In the example just presented, the exercise practitioner asked a very specific question, searched for evidence in a publically available database using defined search terms, evaluated the evidence based on its applicability to the question, and arrived at a solution based on the available information.

Refutes Misconceptions and Misinformation

Exercise is prone to misconceptions and easily swayed by industry trends. Because exercise is a common activity and behavioral pattern, everyone has an opinion on the "correct practice"—even those with no formal exercise training! Many individuals begin exercise at a young age and are often taught exercise theory by individuals whom they respect but who may not have an adequate academic background to assess exercise practice. Although these people earnestly seek to help, they may provide misinformation that is reinforced by many years of practice. The theories taught by these individuals are taken as truth, and it is difficult to accept that these ideas may not be correct.

Society also perpetuates misinformation by promoting exercise philosophies that appear correct but do not stand up under the scrutiny of science. A classic example can be seen in public perceptions of the "correct practice" in designing programs for adults who are older. Aging is associated with numerous deleterious physiological adaptations including losses in skeletal muscle mass, muscle strength, muscle endurance, aerobic capacity, balance and coordination, and walking velocity. All of these negative effects of aging result in a reduced capability to perform functional tasks and a concomitant decline in independence. Due to the decline in physical capabilities associated with aging and the perceived increased chance of injury, exercise programs for people who are elderly often involve light resistance training and careful, slow movements. However, a thorough review of the research literature strongly suggests that heavier resistance exercise is beneficial for improving strength and power (Candow, Chilibeck, Abeysekara, & Zello, 2011; Hakkinen & Hakkinen, 1995; Macaluso & De Vito, 2004) and muscle mass (Candow et al., 2011). Moreover, there is strong evidence to suggest that high-velocity resistance exercise improves physical function in those who are elderly. Research supports the use of lighter-weight, high-velocity resistance training as a method to improve physical performance in tasks such as stair climbing, rising from a chair, and rising to a stand from the ground (Pereira et al., 2012). Additionally, the research suggests that the light, slow protocols promoted for people who are elderly are minimally beneficial. Despite the evidence, it is difficult for many individuals to accept that heavy and fast resistance exercise is an appropriate prescription for persons who are elderly (Kraemer & Ratamess, 2004).

Misconceptions are also perpetuated by perceived "experts" within the field of exercise science. These individuals often appear on late-night infomercials or radio shows and write in non–peer-reviewed sources. These media are ideal for disseminating information rapidly to large numbers of individuals. However, there is no filter to ensure that the information is accurate. So individuals who are portrayed as experts appear in these marketing campaigns and express opinions that are based on poor, flimsy information. As they hide behind a long list of lower-level certifications, the public has little or no ability to determine whether these individuals are truly experts.

Misinformation is a very real and current problem within exercise science. Many times, the information is harmless; using the example discussed earlier, elastic resistance is not ideal to improve BMD, but it is unlikely to cause any negative effects. Although the device would not be the most efficient use of time and money, it is probably safe— albeit ineffective. Likewise, light resistance exercise is suboptimal for ideal functional outcomes in those who are elderly, but it too is not dangerous. On the other hand, at times we find that product or programming recommendations can be dangerous. Consider, for example, the use of certain weight loss supplements. In the 1990s, popular weight loss supplements included the combination of ephedrine, caffeine, and aspirin, popularly known as the "ECA stack." The idea behind the ECA stack was that ephedrine and caffeine were both powerful stimulants that theoretically would increase systemic metabolism and increase exercise capacity. Moreover, adding aspirin to the combination would thin the blood and potentially increase the potency of the supplements. Use of the combination of these supplements was widespread among bodybuilders and those in the gym who desired to decrease body fat. Non–peer-reviewed sources lauded the use of these supplements and reported their potential positive benefits (Phillips, 1997). Supplement manufacturers began to capitalize on these trends by developing fat loss supplements that included an ECA stack along with other potentially effective weight loss agents such as appetite suppressants. Some of the most popular supplements in the 1990s contained these combination drugs.

Along with non–peer-reviewed sources and "expert recommendations," several studies supported the use of ECA as an effective weight loss cocktail. In particular, a double-blind, randomized controlled trial of adults who were obese demonstrated that a low-dose ECA stack was effective in producing modest short-term weight loss (Daly, Krieger, Dulloo, Young, & Landsberg, 1993). Moreover, the data suggested that the weight loss resulting from the ECA stack was sustainable when the supplement was used over a long-term period. Interestingly, the weight loss occurred without caloric restrictions. The weight loss resulting from the low-dose ECA stack was modest but statistically significant. Over an 8-week period, the experimental group experienced approximately 1.5 kg greater weight loss than control; at 5 months, the between-group difference in weight loss was 5 kg. It appeared that there was evidence to support the use of the ECA stack, without caloric restriction, to promote sustained weight loss in humans who were obese.

The obvious implications of these findings and anecdotal evidence of positive weight loss from the ECA stack would be that individuals who are obese would benefit from use and may in fact realize positive health outcomes. However, in the early 1990s the health risks had not been fully evaluated. In retrospect, it seems that the health risks of the ECA stack should have been apparent even without research evidence. It is intuitive that combining two powerful stimulant drugs with a blood-thinning agent might pose risks to cardiovascular function—more is not always better (Gurley, Steelman, & Thomas, 2014).

Throughout the 1990s there were news stories reporting cardiac emergencies arising from ECA use. Between 1995 and 1997, the Food and Drug Administration (FDA) was reported to have received more than 900 calls on side effects due to ephedrine use or misuse; 37 of these reported serious adverse events, including deaths (U.S. Food and Drug Administration, 2004). In 1997, the FDA began requiring supplements containing ephedra to include a warning label outlining the potential adverse effects. Although it is impossible to determine if the adverse effects were due to legitimate use, misuse, or combining ephedra with other powerful stimulants, in 2003 the FDA issued a press release announcing that over-the-counter ephedra use would be banned in 2004 due to the overwhelming information concerning negative health risks (U.S. Food and Drug Administration, 2004).

Unlike the hypothetically harmless but ineffective product X, the ECA stack was minimally effective but dangerous. In short, the small beneficial responses reported from ECA use did not outweigh the potential deleterious health effects. In this case, the misinformation perpetuated by the exercise community had a significant negative effect.

Creates a Common Language

Another potential benefit of EBP is that it creates a common language for exercise professionals and health providers. As alluded to in the introduction to this chapter, it is now common for exercise professionals to use the term "evidence-based" or to say that there is "evidence supporting" this protocol or that device. Ironically, it is true that there is always some form of evidence to support programming theories—if there were not, our recommendations would be purely random. The strength of the evidence-based philosophy is that it provides a standardized methodology for describing the strength of the evidence. Instead of using vague statements like "The lower back therapy program is evidence based," true evidence-based practitioners say, "There is level A evidence to support the use of this lower back training program." The level of evidence provides a clear definition of the strength of certainty associated with the recommendations and provides the consumer or fellow practitioner with a clear understanding of the supporting information.

Integrates Scientific and Practical Knowledge

Evidence-based practice provides a standardized methodology for integrating scientific and practical knowledge. Three different terms have been presented in the literature to describe the knowledge source on which practice is based: experience-based practice, science-based practice, and evidence-based practice (Cook, 2004; Koukoura & Hajiioannou, 2014; Krieger, Newman, Parse, & Phillips, 1994). Implicit in each term is the recognition that information is gained from multiple experiences, sources, and observational opportunities (figure 2.3).

Experience-Based Practice

Valuable lessons are learned through experience—in some cases, experience may be the only way to acquire certain information. For example, the logistical and practice-based knowledge needed to work one-on-one with a client or athlete is not typically obtained through the scientific process. If you observe a practitioner who has been working in a field for 30 years, you will see skills that just are not developed in a newly graduated student. Through practice and years of repetition, practitioners are able to develop and fine-tune procedures and techniques. The perfection that comes with experience is the result of trial and error—mistakes and successes that lead to the development of

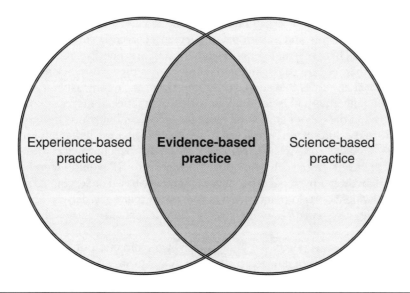

Figure 2.3 Evidence-based practice integrates information obtained through science and practice to formulate the best treatment plan for the client, athlete, or patient.

knowledge. Many times, if you ask experienced weightlifting coaches why they use a certain exercise or drill to correct a mechanical flaw in a lifter, they cannot point to a peer-reviewed article or a textbook. They just know the drill works because they have used it hundreds of times and observed success. Experience is important and necessary for success in any vocation; however, it can lead to bias and subjectivity in practice unless it is balanced with other, more objective information.

Science-Based Practice

Science-based practitioners center most of their decisions on research and published sources of data. There is strength to this type of practice as it removes, to the extent possible, subjectivity from decision making. Science-based practitioners find published information to support their training philosophies, protocols, prescriptive recommendations, and so on. The result is programming that is based on rigorous scientific recommendations. When you ask a science-based practitioner why he is using a certain technique or protocol, he often refers to a paper demonstrating that the technique was more effective compared to a contrasting technique or control. The weakness of the science-based philosophy is that it does not allow for flexibility or the integration of any information learned outside of published science.

Evidence-Based Practice

In contrast to experience-based or science-based practice alone, EBP integrates information obtained from both practice and science (figure 2.3). Philosophically, EBP recognizes that knowledge is gained from many sources of information, observation, and experience. However, it provides a systematic way of dealing with conflicting forms of information and a way to standardize and objectively answer questions. For example, sometimes professional opinions are the only information available to answer a question; this is often the case in the practice of any discipline. As discussed in chapter 1, the exercise industry is rapidly expanding, frequently at a much faster rate than science. There are tools, protocols, and nutritional supplements currently employed in practice that are effective. There may not yet be a study to support their use, but practitioners have observed their effectiveness and are confident that they work. Scientists simply have not performed studies that support this effectiveness. As long as the tool or philosophy is not dangerous and is not the foundation of practice, it makes good sense to incorporate it. On the other hand, some tools, protocols, and nutritional supplements are currently in use and sincerely believed to be effective because judgment is clouded by bias. A plethora of strong scientific evidence may exist that contradicts the personal belief of the practitioner. In such cases, the philosophy of EBP would suggest that the practitioner disavow her personal experience and practice based on less biased scientific information. Evidence-based practice leaves room for experience-derived knowledge, but also provides a systematic way of checking knowledge when more reliable information is available.

Science in many cases points the practitioner in the direction of ideal practice; it may not define the details of how to arrive at the model protocol. Practical experience leads to many of the specific components of training, with science always serving as the navigation tool to ensure that the practitioner is traveling in the correct direction. Thus, both science and practical experience play essential roles in the journey to optimal outcomes in a client, patient, or athlete (figure 2.4).

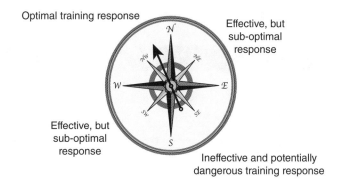

Optimal training response

Effective, but
sub-optimal
response

Effective, but
sub-optimal
response

Ineffective and potentially
dangerous training response

Figure 2.4   Evidence-based practice serves as a compass and provides direction in practice. Research evidence points practitioners in the direction of optimal interventions, but many decisions along the journey are influenced by experienced-based knowledge.

Increases the Credibility of Exercise Science

Evidence-based practice is important to the future of exercise science as a professional discipline and a key contributor to the allied health sciences (Amonette et al., 2010). A glance at the scientific literature suggests that EBP is a trend in nearly all health professions. Much is written about the integration of science with experience-based knowledge in many disciplines, including medicine, nursing, occupational and physical therapy, and athletic training. The purpose of bringing EBP into these disciplines is the recognition that many important decisions in the care of patients are based on tradition or flimsy theories. In each of these fields, practitioners have recognized that in order for the discipline to flourish, there is a need for an introspective look at practice and a fresh approach to clinical decision making. When strong scientific rationale exists to subtly (or dramatically) alter practice, incorporate new techniques, or eliminate techniques, clinicians need to have the courage to change for the sake of their patients.

Such an introspective look is important for exercise science. Many of the protocols, techniques, devices, and so on that have formed the foundation of practice for exercise scientists are effective and have strong scientific evidence to support their use. The support for some components of exercise practice, however, is not so strong. If we want exercise science to be a leading discipline in the allied health sciences, it is essential to eliminate elements contraindicated by science. This will result in a stronger and leaner discipline, one that is well-respected among our allied health peers.

CRITICISMS OF EVIDENCE-BASED PRACTICE

While the philosophy of EBP has many strengths, some have proposed criticisms. The criticisms highlight important limitations and suggest the need for continued evaluation and development of the process. Nonetheless, we contend that the strengths outweigh the limitations. Table 2.1 provides a list of common criticisms of EBP. The next section presents a discussion of the criticisms and rebuttals to these objections.

Table 2.1   Criticisms of Evidence-Based Practice and Rebuttal Arguments

Criticism	Rebuttal
Shortage of relevant, consistent scientific evidence	There is a strong body of literature to support many questions—where research is limited, the practitioner should rely on experience.
Difficulties in applying evidence	The evidence-based practitioner should implement the principles derived from research, not the protocols.
Limited time and resources	If the process is important, the practitioner should make time. Research is significantly easier to access with the widespread availability of online resources, and most academic programs teach the basic skills of finding research. Exercise practitioners should use these skills consistently to develop proficiency.
Paucity of evidence that evidence-based practice "works"	This is an absurd, impractical argument.
Evidence-based practice ignores clinical expertise	On the contrary, evidence-based practice integrates experience with science.
It disregards patients' values and preferences	The original models proposed integrating patients' preferences with science and experience.
It promotes a "cookbook" approach to practice and undermines the art of prescription	Science provides the "ingredients"; the practitioner can creatively integrate these ingredients into practice.
It leads to negativism or inaction in the absence of evidence	It results in a leaner, more effective program that minimizes ineffective techniques, devices, and theories.

Based on Straus and McAlister 2000.

Lack of Consistent Scientific Evidence

One of the most common criticisms leveled against the process of EBP is the lack of strong scientific evidence supporting relevant practical questions (Straus & McAlister, 2000). In other words, there is often no evidence to support or refute a proposed question, so why even engage in the process? If evidence exists, it is often from poorly designed studies using protocols dissimilar to real-life prescriptions implemented on the field, in the weight room, or in the clinic. Although this argument has merit, it is irrational to avoid the process of searching for evidence simply because strong evidence may not exist. Instead, this argument supports the very premise of the evidence-based philosophy—the integration of scientific literature and practical knowledge. Evidence-based practitioners search for evidence in the scientific literature and sometimes find strong, well-designed research that either supports or repudiates current or proposed practices leading to modifications. Other times, the search returns weak or no evidence. In such cases, the evidence-based practitioner is to use the only evidence available—practical experience—to make a decision.

This criticism also highlights an important opportunity for the scientific and practicing exercise community. Strong communication is needed between scientists and practitioners to ensure that scientists are investigating cutting-edge protocols, techniques, and devices used in practice. There is also a need for the scientific community to clearly communicate results to consumers and practitioners in a timely manner. Often there is a disconnect between researchers and practitioners or consumers; scientists have historically neglected to communicate their findings in an understandable fashion to the lay public. The importance of this has been recognized by many professional scientific and medical organizations; many (e.g., American Association for the Advancement of Science) now offer workshops to teach scientists how to better communicate with the general public and other more unique groups, such as legislators.

Difficulty Applying Evidence in Real-Life Situations

Another criticism of EBP is that it is difficult to apply evidence derived from scientific literature to the individual patient (Straus & Sackett, 1999; Straus & McAlister, 2000). Because the goal of many scientific protocols is to isolate a single independent variable, the protocols are often dry and not practical to implement in a patient. The goal of searching the literature should not be to find a protocol to implement but instead to find a principle. For example, the strength training literature provides evidence supporting the use of heavy resistance exercise to potentiate an increased power response in an explosive movement. Some of the literature suggests that if you perform a near-maximal load on a back squat followed by a vertical jump, the heavy back squat may increase the power output of the jumping exercise (McCann & Flanagan, 2010; Crewther et al., 2011). The response is greatest if the vertical jump is performed 7 to 12 min after the heavy exercise—but there is a small but significant effect for jumps performed immediately after the heavy squat (Jensen & Ebben, 2003). In most training situations, it is not practical to rest 7 to 12 min between sets. The protocol may not apply, but the principle certainly does: Heavy exercise performed before light explosive exercise may increase power production. Therefore, it may be beneficial for a strength practitioner to implement contrasting heavy and light exercises during the power phase of a periodized scheme, as the evidence suggests that this technique may result in greater power production during an individual workout.

Limited Time and Resources

Some have argued that the implementation of EBP requires the practitioner to develop a new skill (Straus & McAlister, 2000). Most exercise science programs teach students how to search for research literature; therefore this is not really a new skill but a novel approach to implementing the skill. Many allied health science programs include a course in EBP. In the next 10 years, it is likely that this trend will continue to emerge in exercise science; to facilitate this, the next generation of exercise scientists must be trained in the specific language and techniques of EBP. Practitioners who have been working for many years in the field have opportunities to attend workshops, conferences, and courses focused on EBP. Several lectures discussing the process of EBP have been given at recent National Strength and Conditioning Association and American College of Sports Medicine conferences. In the future, these lectures and workshops are likely to be more common, providing ample opportunities for more seasoned practitioners to learn the skill.

When David Sackett proposed the concept of evidence-based medicine in the early 1990s, it was difficult to find and access research information (Sackett et al., 2000). Acquiring a single article usually required a trip to a medical library and an arduous process of finding a reference, digging through a card catalogue, finding a journal on the shelf, photocopying the article, and then repeating the process for the next article. For a practitioner with a busy schedule, this may not have been practical or even possible, depending on available library resources. However, the development of the Internet and widespread dissemination of electronic information have simplified the process. An exercise practitioner who wants to find an article today can get on the Internet, go to a publicly available peer-reviewed search engine such as PubMed or Google Scholar, enter search terms, and then sift through an abundance of articles. Abstracts are freely available, and in many cases the entire article is publicly available free of charge. Today evidence-based practitioners face a different problem than in the early 1990s—we have access to so much information that it can be difficult to find the best evidence. In chapter 7, we discuss in detail how to find the best articles and to ensure that searches are limited to the most relevant information.

Lack of Evidence Supporting Evidence-Based Practice

Ironically, some have proposed that there is no evidence that EBP works (Cohen, Stavri, & Hersh, 2004; Davidoff, Case, & Fried, 1995; Sackett et al., 2000; Straus & McAlister, 2000). In other words, where are the systematic reviews of randomized controlled trials demonstrating that EBP is more effective than experience-based practice? It is unlikely that such studies will ever be conducted, nor would they be feasible or even ethical. In rebuttal to this argument, we propose the following question: What is the antithesis of EBP? The opposite of EBP is practice based on nothing or pure randomness. Experience-based and science-based practice are both based on evidence. So, people who argue that there is no evidence that EBP works are likely evidence-based practitioners themselves! They simply are not practicing based on the highest levels or the least biased forms of evidence.

Evidence-Based Practice Denigrates Clinical Expertise

One of the most common complaints against EBP is that it devalues clinical expertise (Cohen et al., 2004; Sackett, Rosenberg, Gray, Haynes, & Richardson, 1996). As discussed in detail previously, EBP in no way minimizes the importance of clinical expertise—in fact, it emphasizes its importance in the overall scheme of knowledge acquisition.

However, in the process of evaluating evidence, the EBP philosophy recognizes that clinical expertise may be biased and often lacks a valid comparison that would determine its effectiveness. Clinical experience is still extremely important in the treatment of a patient or client.

Evidence-Based Practice Overstandardizes Prescription

Some have proposed that EBP minimizes or devalues a patient's values or preferences (Sackett et al., 1996). In David Sackett's original model, scientific evidence, clinical experience, and patients' preferences were all equal contributors to the evidence-based knowledge process. This is an important concept, especially in the prescription of exercise. Patient preference and values are essential to maximizing adherence. An exercise scientist can develop the perfect protocol for health and fitness, but if the patient does not buy into and use the program, it is ultimately ineffective. Therefore, the values and preference of the patient are just as important as the scientific and experience-derived evidence. Perhaps a talent needed by a good practitioner is the ability to "sell" the evidence to the patient. If the evidence—scientific and experience-based—clearly supports a program or technique, the exercise practitioner needs to be skilled at convincing the patient to implement and ultimately adhere to the program.

Another interesting argument against the use of EBP is that the search for evidence results in a standardized "recipe" for prescription (Sackett et al., 1996). For example, if an evidence-based practitioner is developing a program for an individual with diabetes, he goes to the shelf, pulls a journal, finds a relevant paper, and copies the protocol with the patient. The result is that practitioners in China, India, Germany, or Texas are implementing the exact same protocol because they are following the same "recipe." As we discussed earlier, the goal of the process is not to copy a protocol but to implement principles. To borrow from the cookbook analogy, there are thousands of recipes for cakes, but they all have common ingredients. Most cakes contain eggs, flower, and sugar, along with other ingredients. You mix them in a bowl, put the batter in the oven, and out comes a cake. As long as you follow the general principles of baking, a cake is the result. Even though bakers follow the same general principles and use similar ingredients, there are hundreds of types of cakes. The same is true of EBP; while practitioners in China, India, Germany, and Texas follow the same general principles outlined by scientific and experienced-based evidence, the exercise prescription will vary but nevertheless elicit similar results as long as the principles are followed. Ultimately, the prescription of exercise is both an art and a science; the process of EBP still leaves room for unique programs. Science forms boundaries for prescription, but what is done within the boundaries is up to practitioners and their creativity.

Evidence-Based Practice Leads to Pessimism

A final argument against EBP is that it leads to negativism or skepticism regarding prescription (Smith & Pell, 2003). There may be little data supporting practice or techniques routinely used in practice (figure 2.5). Exercise practitioners may be left to wonder, "If I limit my techniques to only those supported by scientific evidence, what can I actually do with my client?" Indeed, developing prescriptions based on the best-supported forms of exercise may limit or eliminate certain techniques. If a technique is shown to be ineffective or if evidence to support it is weak, do practitioners

BMJ 2003, 327(7429), 1459-1461.

Parachute Use to Prevent Death and Major Trauma Related to Gravitational Challenge: Systematic Review of Randomised Controlled Trials.

Smith, G.C., Pell, J.P.

Abstract

Objectives: To determine whether parachutes are effective in preventing major trauma related to gravitational challenge.

Design: Systematic review of randomised controlled trials.

Data sources: Medline, Web of Science, Embase, and the Cochrane Library databases; appropriate internet sites and citation lists.

Study selection: Studies showing the effects of using a parachute during free fall.

Main outcome measure: Death or major trauma, defined as an injury severity score > 15.

Results: We were unable to identify any randomised controlled trials of parachute intervention.

Conclusions: As with many interventions intended to prevent ill health, the effectiveness of parachutes has not been subjected to rigorous evaluation by using randomised controlled trials. Advocates of evidence based medicine have criticised the adoption of interventions evaluated by using only observational data. We think that everyone might benefit if the most radical protagonists of evidence based medicine organised and participated in a double blind, randomised, placebo controlled, crossover trial of the parachute.

Figure 2.5 Sarcastic paper published in *British Medical Journal* using an absurd example to highlight the fact that there may not always be evidence to support recommendations; sometimes good clinical judgment is sufficient.

Reprinted from Smith and Pell 2003.

really want such techniques to be the foundation of their programs? Of course not! Leaner, stronger, more effective programs should be the goal of exercise scientists and practitioners; if certain techniques are minimized or eliminated due to lack of evidence, practice will be all the better for it.

As a side note, skepticism is a double-edged sword—it can be a good and a bad thing. When determining what is best for patients, clients, and athletes, practitioners should not quickly change course and implement every new trend, device, or philosophy. Instead, the foundation of prescription should be those devices and philosophies solidly and firmly rooted in the strongest forms of evidence. While novel techniques and devices with minimal support may be used to a limited extent, they should never be the foundation of prescription. Practitioners and scientists alike must remain skeptical but flexible, with an open mind that is willing to change course when the preponderance of evidence suggests that such a change will benefit our clients, patients, and athletes.

CONCLUSION

Much has been written about the process of EBP, but the basic steps are key to understanding the methodology: Ask a question, find evidence, evaluate the evidence, implement the evidence, and reevaluate the evidence. In the discipline-specific model described in this textbook, we add an additional step after implementing the evidence: confirm the evidence in the individual. Some have been critical of the process, suggesting that it is too time-consuming, results in a cookbook approach to prescription, devalues clinical experience, and requires the development of new skills. Most of the criticisms are based on misconceptions and a misinterpretation of the philosophical backbone of EBP: the integration of science, experience, and patient preference. In contrast to what the critics say, EBP provides a standardized, reproducible technique to support or refute anecdotal evidence and unsupported theory, which are all too common in exercise science. It also provides a common language to communicate with professionals from other disciplines. The widespread incorporation of EBP into exercise science is important to the field's credibility and future as a growing allied health science discipline.

Chapter 3

PHILOSOPHY, SCIENCE, AND EVIDENCE-BASED PRACTICE

Learning Objectives

1. Understand why epistemology is an important foundation of science.
2. Gain familiarity with selected ancient and Western philosophers.
3. Understand deductive and inductive reasoning.
4. Understand rationalism and empiricism.
5. Know the founders of evidence-based medicine.
6. Understand different applications of evidence-based practice in exercise science.

Knowledge seems like a fairly simple, straightforward concept—the word denotes the fact or condition of knowing or being aware of something. But can anything really be known, and how can we know it? Although it is not a part of our daily experience to ask these questions, humans have done so for millennia, with well-preserved records of early theories dating from the time of the ancient Greeks. By the modern era (around the 16th century), there was an explosion of theories and philosophies regarding the nature of knowledge and scientific knowledge in particular. These theories and philosophic principles form the underpinnings of modern science.

Exercise prescription is based on information that has been codified as knowledge. This knowledge may be related to concepts such as the intensity or volume needed for optimal strength gains or the safest and most effective protocols to promote improvements in functional exercise capacity in clinical populations. In both of these examples, there was a time when no answer to the question existed—in fact, there was not even a question. Through some theoretical pattern, a question was formed and an answer obtained. However, the answer may have changed significantly over time, and the answer often leads to more questions, which is typical of scientific study. How do exercise practitioners develop questions and find answers that lead to knowledge? How do we know that an answer is correct—can we ever know that we are correct? Although it is easy to take for granted the answers to these foundational questions, exercise practitioners and scientists benefit from examining the pillars of knowledge and philosophy upon which modern science and thought are built.

EPISTEMOLOGY

The idea that anything can be known is not foremost in the minds of individuals living in the 21st century. However, the nature of knowledge is a controversial philosophical concept that has been debated for millennia. The study of knowledge, termed *epistemology*, seeks to define the nature of knowledge, how knowledge is acquired, and the true extent to which a particular phenomenon can be known.

Epistemological thought has changed throughout time, but there are two general themes: that knowledge is either *a priori* or *a posteriori*. A priori knowledge proceeds from reason and is independent of experience or study (McGinn, 1975): "All senior citizens are old" can be elucidated based on logic and definitions without observation of a single senior citizen. In contrast, knowledge that is a posteriori is gained from experience or observation. A statement such as "Many senior citizens live active lives" necessitates observation or scientific study for validation.

Ancient Epistemology

Answers to crucial philosophical questions about knowledge have changed throughout human history. In this chapter, we discuss the acquisition of knowledge through the eyes of some key philosophers of science, beginning with ancient Greek thinkers.

Plato

The ancient Greek philosopher Plato (424-348 BC) believed that all knowledge was innate (a priori), ingrained within the minds of humans at birth (Gulley, 2013). As humans mature and learn, they are simply recalling or elucidating facts that were already known; this is seen clearly in his Theory of Forms. This theory argues that the highest and most fundamental truths and reality are ethereal, abstract "Forms" or ideas that, existing independently, are recognized and elucidated by humans through philosophy and study; this recognition is possible because as humans we have been imprinted at birth with these forms (Gulley, 2013). Plato's writings create a clear distinction between knowledge and opinions. Knowledge, Plato contends, is pure and right. Opinions are knowledge that has been clouded by the experiences and observations of humans. In this difference between knowledge and opinions, Plato equates knowledge with a recognition and elucidation of the forms, and opinion with a belief that is distorted by human observation. Thus Plato disparaged human observation as impure and inferior to the acquisition and understanding of the forms through reason.

Aristotle

Aristotle (384-322 BC) took a somewhat different view on these matters, a difference that is well depicted in Raphael's "The School of Athens" (figure 3.1). In the fresco, Plato and Aristotle are shown walking together side by side; Plato gestures toward the heavens to represent his ascription to the Theory of Forms, while Aristotle's hand

Figure 3.1 "The School of Athens" by Raphael. The painting depicts the ancient philosophers Plato and Aristotle walking side by side in discussion.

Raphael's, The School of Athens, ca. 1511. Photo from Wikipedia. Available: https://upload.wikimedia.org/wikipedia/commons/9/94/Sanzio_01.jpg

is stretched out, palm down toward the earth, to convey his devotion to knowledge through empirical observation. As Plato's most famous student, Aristotle subscribed to the concept of Forms; but unlike Plato, who believed that the Forms (the universal) could be elucidated by reason and used to inform and understand our world (a deductive, top-down approach), Aristotle advocated for empirical observation of the world around us (Irwin, 1995). This would lead to an understanding of the universal (an inductive, bottom-up approach), which he believed did not exist as a separate, ethereal Form but only in an "instantiated," concrete way.

Western Epistemology

For well over a thousand years, classical Greek philosophy formed the basis of educated Western thought. By the 16th century AD this changed, with the explosion of the Renaissance in Europe affecting art, literature, religion, and philosophy. Renaissance philosophers expanded on the ideas of Plato and particularly Aristotle, developing new streams of thought and an increased emphasis on a posteriori knowledge.

Descartes: Rationalism and Deduction

Rene Descartes (1596-1650), a French philosopher, believed that knowledge was obtained through deductive reasoning, a view known as *rationalism* (Gibson, 1967). Deductive reasoning, in short, is a logical process whereby an individual begins with broad concepts (premises) and applies them to reach a specific conclusion (i.e., top-down logic). It involves the logical construction of true arguments to arrive at a true statement for a specific case. An example of deductive reasoning can be seen in figure 3.2.

Two important concepts associated with deductive reasoning are validity and soundness. A valid argument is one whose conclusion cannot be false if its premises are true. On the other hand, soundness refers to the "trueness" or accuracy of the premises; if a premise is false, then the conclusion is not sound. An example of a false premise can be seen in figure 3.3.

In this case, the argument is valid (the conclusion cannot be false if its premises are true), but not sound. The lack of soundness is due to the falseness of the first premise: All humans do not have two hands. Due to birth defects, some are born missing one or both hands; also, humans can lose a hand in an accident or a medical amputation. They certainly do not cease to be humans because of the loss of a hand! Thus, this line of reasoning is not sound due to the faultiness of its first premise.

The method of deductive reasoning is limited in that its conclusions are definitive with no provision for uncertainty; while this may seem like a positive, in many cases it is not. For instance, deductive reasoning requires absolute premises for which there are no exceptions. Often, we are unable to make an absolute statement, either due to lack of knowledge or simply because the state of a matter cannot be encapsulated in a single, absolute statement. Either way, the deductive approach is potentially problematic either because of its inability to account for uncertainty or our inability to provide sound premises from which to deduce.

Bacon: Empiricism and Induction

Francis Bacon (1561-1626) was the father of empiricism, an epistemological view that emphasizes the importance of sensory inputs in the acquisition of knowledge (Anderson, 1971); empiricism is the direct opponent of Plato's epistemological views, which considered human observations and experiences inherently subjective and flawed. Similarly, John Locke (1632-1704), another empiricist and the "Father of Classical Liberalism," asserted that all humans are born with a blank slate *(tabula rasa)*, knowing and understanding nothing (Sproul, 2000). All information gained through sensation and observation is imprinted on the human mind and leads to knowledge.

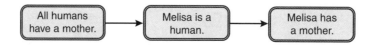

Figure 3.2 An example of deductive reasoning as a tool for acquiring knowledge.

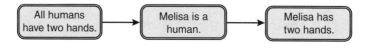

Figure 3.3 An example of invalid deductive reasoning. Because the initial premise is false, the entire argument is untrue.

Bacon advocated inductive reasoning—inferring or elucidating broad principles via observation of specific phenomena (Bacon, 1999). This bottom-up approach is essentially the reverse of deductive reasoning. Inductive reasoning is also based on premises that lead to a conclusion (a rule or generality). A critical difference between induction and deduction is that inductive conclusions are uncertain. They are based on probability, and although many inductive conclusions have thousands of accurate and reliable observations to support them, inherent to inductive reasoning is an inability to ever state with absolute certainty that something is so. Nevertheless, inductions based on empirical data can have a very high probability of being correct, and the chance that they might be incorrect can be mathematically quantified. As you will see throughout this book, practitioners should operate with the best knowledge (conclusion) available based on the probability of it being true and consideration of the consequences of it being untrue. The following is an example of inductive reasoning:

1. All humans that we have observed breathe air (i.e., air moves in and out of their chest).

2. Therefore, no human can live without breathing.

This statement could be viewed as a certainty, but technically it is not. That is to say, although we have never observed a human living without breathing, we cannot completely rule it out. This goes to the idea that you cannot prove or demonstrate a negative. We can demonstrate that something is not true (e.g., "All bears are brown") by observing the contrary (i.e., we see a black bear and thus disprove this contention); but even if all the bears that we have ever seen are brown, we cannot rule out that a variant we have never seen may not be brown. To continue with a more unlikely example, we cannot induce (know) for certain that pigs cannot fly. It is possible that a pig could have a genetic mutation or eventually evolve to fly, although this has never been observed. In the meantime, we induce that because we have never seen a pig fly (or even gotten close!), it is highly improbable that pigs will fly. From a practical standpoint, due to the infinitesimally small probability that a pig will fly, we live and operate as if we are certain that pigs cannot fly.

Induction is a key component of the scientific method (Medawar, 2013; Reichenbach, 1968). But as discussed previously, inductive reasoning does not provide certainty; it provides a level of probability based on these elements:

1. The number of observations made to support the induction

2. The accuracy of the observations

3. The reliability of the observations

4. The face validity of the observations or their harmony with other well-demonstrated, highly probable phenomena

In the "pig flight" example, this fourth point provides particularly strong support for our level of certainty (i.e., the probability) regarding the ability of pigs to fly. Based on our understanding of the laws of physics and the nature of genetic mutations, we can estimate the probability of a pig flying, at least in our lifetime, as being fantastically low.

This is the ground level of empirical human knowledge: We make numerous observations with accurate, highly reliable measurement techniques over a vast array of disciplines (e.g., physics, astronomy, anatomy and physiology, pharmacology, pathology), and with the findings construct models or theories that are consistent with our observations. Future observations confirm, refine, or disprove the model. Importantly, we can never "prove" that the model or theory is valid because tomorrow a new observation may disprove it; however, we can say that the model or theory is consistent with most (or all) observed data and thus is our best (most probable) explanation for a given phenomenon.

Popper: Falsifiability and Science

In the 20th century, Karl Popper (1902-1994) provided yet another distinct voice offering a unique perspective on the nature of knowledge, particularly knowledge derived via scientific means. Popper was keen to separate science and the unscientific and asserted that falsifiability is the "criterion of demarcation" between the two. A statement or theory that is falsifiable is one that can be tested and (potentially) shown to be false (Corvi, 1997). As a strong proponent of falsifiability in science, Popper noted the paradoxical effect of falsifiability on the "growth of knowledge": Science is viewed as the principal means to increase knowledge, yet the falsification that he advocated serves, in effect, to reduce knowledge by eliminating inaccuracies from the canon of understanding (Popper, 2014).

Popper employed his notion of falsifiability to solve the problem of induction. Induction has been historically criticized for two weaknesses:

1. Observing characteristics of a sample or portion of a population does not necessarily tell us about the entire population. The classic example is swan color; although we may have observed only white swans (leading to an inductive conclusion that "all swans are white"), that in no way precludes the existence of non-white swans.

2. Past observations do not necessarily predict future performance. The universal example invoked is that of the sun rising each day; that it has risen every day for as long as we have observed does not mean it will rise tomorrow morning.

Popper argued that the best theory is one that is simple and easily falsifiable; because a theory cannot be proved with finality through induction but can easily be disproved, Popper thought that the best theories would be those that could be easily disproved. A theory framed so that it could easily be disproved would stand a better chance of being correct if it held up over time (Corvi, 1997). Thus, Popper sidestepped the criticisms of induction and, in doing so, tacitly endorsed it. He advocated a new scientific standard of falsifiability in place of positivism, which argued that the best theory is the one that is most likely to be true. In effect, Popper considered it more important to recognize and discredit a false theory than to enshrine a theory that is very likely to be true (but, like all theories, can never be indisputably proven) but is very difficult to disprove. To put this in statistical terms, Popper preferred highly specific theories to highly sensitive ones.

EPISTEMOLOGY OF MODERN SCIENCE

From this discussion, it should be clear that epistemology and the thought processes of deduction and induction are quite complicated. Philosophers and great thinkers disagree on many of these abstract concepts, but despite these conflicts with regard to deep theoretic principles, modern science has been formulated as a basic structure (whose characterization could also be endlessly debated) comprising an assortment of the previously discussed tenets.

Modern scientific thought is based on a posteriori knowledge gained through observation and experimental evidence. Systematic investigation yields a collection of related observations (e.g., a dataset), which is employed to infer a broad, governing principle that represents new or refined knowledge. Modern science is a highly inductive, bottom-up affair. It is important to note that science relies on the concept of probabilities. For example, inferential statistics are mathematical techniques used to generalize the findings in a study sample to an entire population and predict the degree of certainty with which the generalization can be made.

THE PARADIGM OF EVIDENCE-BASED PRACTICE

The epistemology of evidence-based practice (EBP) is based on modern scientific thought, but also borrows concepts from Karl Popper (Shahar, 1997). The goal of the evidence-based philosophy is not to find "absolute truth." There may be no such thing, or we may be limited by the technology of the time, which simply does not allow us to observe a phenomenon in sufficient detail to describe its exact mechanism. Instead, EBP searches for the best solution for the patient, athlete, or client based on the available observations. However, the philosophy of EBP recognizes the possibility that a theory or particular technique may be falsified in the future or that an individual's reaction may be different from the population mean. Thus, EBP philosophically intertwines empirical and Popperian thought.

The Rise of Evidence-Based Medicine

Evidence-based medicine (EBM) and its philosophy began in the mid-1970s, although the term did not appear in the medical literature until 1990 (Eddy, 1990). The term was coined and the concept originally published by David Eddy, a physician and mathematician. Evidence-based medicine was an outgrowth of his realization in 1974 that many medical decisions were based on outdated information and, in many cases, purely historical precedent (Eddy, 2011). Initially, Eddy devoted his EBM efforts to updating (or in many instances, developing) national guidelines, particularly in the realms of screening and treatment. For screening, this involved determining via published research which tests were appropriately sensitive and specific to a given pathology and who should be screened and how often. Besides the obvious problem with a screening test that fails to detect a large number of diseased persons (i.e., low sensitivity), Eddy highlighted the problems associated with low-specificity tests (i.e., tests with a high false-positive rate). These include patient apprehension and unnecessary financial costs and potential physical harm caused by either the initial screening or the follow-up testing (incorrectly) indicated (Eddy, 2011). Similar problems were apparent with low- or no-efficacy treatments; patients were at risk for additional physical harm (e.g., painful radiation or chemotherapy treatments that did not remit the cancer) as well as incurring the often considerable financial costs associated with medical treatment.

Throughout the 1990s, Eddy expanded his evidence-based work into the areas of medical insurance coverage policy and performance outcome measurement (Eddy, 2011).

These natural extensions of the EBM premise endeavored to instill the highest-quality evidence-based information into national guidelines that all physicians consult, medical insurance policies that ensure tests and treatments that work are covered and those that do not work are not covered, and performance measures that ensure decisions in the aforementioned areas are based on empirical data. While Eddy continued to work and publish in these areas for the next several decades, a new application for EBM emerged, led by David Sackett, a physician and epidemiologist.

Sackett saw the potential to extend the EBM paradigm to a more individual level; while Eddy had worked to develop national guidelines, shape insurance coverage policy, and evaluate these high-level implementations via performance outcome measures, Sackett labored to bring the power of evidence to bear on the treatment and care of individual patients. To this end, he founded the first-ever "clinical epidemiology" department, which, as the name implies, focused on bringing population-based research (epidemiology) down to the individual (clinical) level. He and his collaborators published extensively on this topic throughout the next decade or so (Sackett, 1995, 1997; Sackett, Richardson, Rosenberg, & Haynes, 1997; Sackett & Rosenberg, 1995; Sackett, Rosenberg, Gray, Haynes, & Richardson, 1996; Sackett & Straus, 1998). Sackett acquired the moniker, "Father of Evidence-Based Medicine" (Schoenfeld, 2008), and his focus on decision making and care of the individual patient came to dominate the EBM paradigm. Sackett and others rightfully asserted that potentially lifesaving or life-altering decisions should not be based on antiquated information. Because knowledge is a dynamic phenomenon, physicians should develop and implement methods to stay current on information related to their profession. Sackett characterized EBM as a systematic process of asking questions, finding information, evaluating the validity of the information, implementing the information, and then consistently reevaluating the information to stay current. Explicitly, he defined EBM as "the conscientious and judicious use of current best evidence from clinical care research in the management of individual patients" (Sackett et al., 1996, p. 71). As a historical footnote, Eddy published a later paper that presented his views on a unification of the population-based (national guidelines, insurance policies, etc.) and individual-based facets of EBM: "EBM is a set of principles and methods intended to ensure that to the greatest extent possible, medical decisions, guidelines, and other types of policies are based on and consistent with good evidence of effectiveness and benefit" (Eddy, 2005, p. 16).

Evidence-Based Practice in Exercise Science

In the ~25 years since its introduction in the field of medicine, EBP has made its way into numerous other allied health fields, such as nursing, dentistry, physical therapy, and orthopedics (Compton & Robinson, 1995; Fox, Kay, & Anderson, 2014; Hanson, Bhandari, Audige, & Helfet, 2004; Madhok & Stothard, 2002; Mordecai, Al-Hadithy, Ware, & Gupte, 2014; Partridge, 1996; Partridge & Edwards, 1996; Poolman et al., 2007; Shorten & Wallace, 1996; Simpson, 1996). Following Eddy's lead, the most prominent professional societies in the exercise science field (e.g., the American College of Sports Medicine [ACSM] and the National Strength and Conditioning Association [NSCA]) have adopted EBP in a variety of ways. The NSCA has incorporated EBP into its motto, "Bridging the gap between science and application," signifying the importance of using research evidence to inform practice. The ACSM has published nearly two dozen position stands that are available to the public on its website on a wide range of practical topics such as exercise and diabetes, exercise and adults who are older, exertional heat illness, and the female athlete triad. Each position stand is a comprehensive review paper that cites hundreds of peer-reviewed articles and provides definitive guidelines on the given topic. Similar to position stands, ACSM also publishes team physician consensus statements, which are evidence-based recommendations on issues ranging from nutrition for athletes to concussion (mild traumatic brain injury) and the team physician. These general sources of guidance can be thought of as the exercise science field's equivalents to the national guidelines that Eddy promoted in evidence-based medical practice.

However, on the individual athlete level that Sackett pioneered in the medical field, relatively little has been seen in exercise science beyond vague notions that practitioners should provide "evidence-based" programs to their athletes and clients and any slow, trickle-down effects of the publication of the aforementioned evidence-based guidelines by the large national professional societies. The present authors have published several papers (Amonette, English, & Ottenbacher, 2010; English, Amonette, Graham, & Spiering, 2012) and a book chapter (Amonette, English, Spiering, & Kraemer, 2012) advocating the adoption of EBP in the exercise science field at the individual level and outlining the basic steps for implementing EBP at the individual athlete and client level. Another idea we have advanced is the inclusion of EBP as a fundamental component of undergraduate exercise science coursework either in the form of a single class or as an entire undergraduate program built around learning via EBP methods. Either approach would serve to prepare future practitioners for a successful career founded on a dynamic knowledge process rather than a static set of memorized facts. This book is a continuation of the work of equipping practitioners for the important task of implementing the best available knowledge and research evidence into exercise programs for their patients, athletes, and clients.

CONCLUSION

As exercise practitioners and scientists, how do we know what interventions are most safe and effective? Where does that knowledge ultimately come from? The answers to these crucial questions lie in the realm of epistemology. Early philosophers believed that knowledge was innate within the human makeup. However, more recent philosophical thought has hinged on observation—we know because we have seen, touched, tasted, smelt, or heard. The problem with this philosophy is that we have not seen all that there is to observe, so the explanations we propose are never completely certain even though there may be a very high probability that they are true. Science and knowledge are dynamic and are shaped by new discoveries. Often, new work simply adds to the knowledge base, prompting only minor adjustments to the underlying theory. However, some new discoveries are so disruptive and unexpected that they necessitate discarding the current theory and developing a new one—a paradigm shift. The philosophy of EBP draws heavily on empiricism and recognizes the dynamic and continuous flux in knowledge resulting from new clinical and scientific discovery.

Chapter 4

SOURCES OF EVIDENCE

Learning Objectives

1. Know the three sources of knowledge.
2. Discuss the strengths and limitations of learning from professional experience, academic preparation, and scientific research.
3. Know the two broad types of evidence.
4. Gain basic familiarity with various research designs.
5. Provide examples of research design that exercise practitioners may encounter when reviewing evidence.

The previous chapter discussed epistemology and its relationship to evidence-based practice. Knowledge, and specifically its acquisition, is an important concept as it relates to human biology. We unquestionably have more information and understand more today about human biology than we did 100, 50, 20, 10 years, or even 1 year ago. If an individual has a myocardial infarction today, his survival chances are vastly improved on the basis of numerous medical advances. First, the general public has been well educated regarding the symptoms of a heart attack; second, immediately upon recognition of symptoms, we can call for an ambulance, and within minutes medical personnel will arrive. Even before professional assistance arrives, many public places now have automated external defibrillators (AEDs) that can rapidly restore normal cardiac rhythm and peripheral perfusion for individuals with ventricular tachycardia and ventricular fibrillation. When an ambulance arrives, emergency medical technicians insert an intravenous line and immediately begin infusing fluids or drugs that improve survival chances. When the patient arrives at the hospital, a medical team is prepared to begin treatment; after stabilization, the team can use imaging techniques to precisely identify the cause of the problem. If they determine that an artery is occluded, they can perform an angioplasty or cardiac bypass surgery to resolve the problem. After release from the hospital, a patient will be prescribed cardiac rehabilitation and given dietary recommendations. Together, these advancements improve both short- and long-term survival. If someone were to have had the same cardiac emergency in 1850, what would he have done? There were no AEDs, motorized transportation, or telephones, and hospitals were few and far between. One's best hope for survival was to remain in a comfortable position and hope for the best.

There have been tremendous technological advancements in the previous century in many areas related to human biology, including exercise science. These advancements are the result of new knowledge facilitating practice improvements that have enhanced outcomes for clients, patients, and athletes. In this chapter, we first discuss the primary forums in which knowledge is acquired. Second, we discuss the two types of evidence—non-peer-reviewed information (e.g., expert opinion) and peer-reviewed scientific research (i.e., observational and experimental research)—as well as the various research designs that fall under these categorizations.

SOURCES OF KNOWLEDGE

Knowledge is acquired from a variety of sources. Each of these sources makes a unique contribution to what can be known at a given point in time. It is essential to understand, as discussed in chapter 3, that evidence-based practice is a continual, dynamic process and thus that "what can be known" on a particular subject is highly time dependent (Amonette, English, & Ottenbacher, 2010). Practitioners never reach a point of completion, because we often don't know what we don't know! With this in mind, practitioners can and do attain competent levels of understanding that permit them to practice with a high level of fidelity; a good example is our understanding of the resistance exercise program elements (e.g., progressive overload) necessary for hypertrophy and increased muscle strength. In combination with these already established models, scientists must continually gather new data and integrate work from other disciplines; this will result in an evolving model that progressively approaches a more comprehensive scientific understanding. In most areas of science, complete or comprehensive understanding is still far in the future; nevertheless, practitioners should strive to practice with the most contemporary level of knowledge. This is possible only with an integration of all three sources of knowledge.

Academic Training

Academic training may be the most obvious source of knowledge. From primary school to graduate school, knowledge is passed on from teacher to student; most readers are quite familiar with the organized, pedagogical model for gaining knowledge. During undergraduate college work, basic skills are built upon with introductory-level general courses (e.g., college algebra, anatomy, and physiology) and several years of focused instruction in a given field such as exercise science or kinesiology. Graduate work comprises increasingly advanced courses with a narrower focus and greater depth. Academic training may be thought of as the foundational source of learning. The information and understanding gained during formal academic training serve as the base upon which later learning is erected and into which it is integrated. It would be very difficult to attempt to learn all the "basics" of a field via on-the-job training (professional experience) or personal study of the scientific literature. An undergraduate or graduate degree in a specific discipline provides an efficient process by which to accrue broad, fundamental knowledge in a particular discipline or field.

It might be argued that academic training is a subset of professional experience. While this is a defensible assertion, the two sources can be delineated by defining academic training as the finite set of knowledge that one accumulates during formal academic study and defining personal experience as the cumulative knowledge of a career in practice (e.g., as an exercise physiologist, strength and conditioning coach, or athletic trainer).

Professional Experience

Professional experience is the most universal source of learning. While many exercise practitioners have little or no academic training and even fewer regularly draw from scientific research, all possess some measure of professional experience. Professional experience is critical; if all practice was governed by a canonized dogma derived from research evidence, practice would be little more than a robotic replication of techniques applied by every other exercise professional. Professional experience is the product of both the volume and quality of a person's cumulative experience. Obviously, cumulative time—decades of working as an exercise professional—can provide invaluable knowledge, a wealth of experience to inform future practice. But total time is only one factor in the professional experience equation; a diligent and ongoing commitment to the quality of one's professional experience is essential.

Professional experience is also the most subjective source of learning. This feature is a double-edged sword, as professional experience is both very important for the unique perspective it provides and potentially misleading due to its strong potential for bias. The incorporation of professional experience into exercise programming decisions is very much the "art" of our science.

As discussed in the previous section, ideally the lines between the sources of knowledge will blur somewhat. For instance, an exercise scientist should regularly gather data on her clients, patients, and athletes; these data can be used not only to improve her training prescription, but also to inform her overall training philosophy (i.e., her quality or depth of professional experience). In this way, the collection of scientific data also becomes an invaluable component of the practitioner's professional experience.

Scientific Research

Scientific research is the root source of learning—truly the "primary" or first source that feeds the other two. Yet scientific research is made up of many building blocks that comprise the body of knowledge in a particular area, and many parts remain under construction with large gaps of understanding to be filled in. Such is the body of knowledge in the ever-expanding universe of human understanding. Academic training is but a manageable distillation of vast amounts of scientific research. There could be no academic training without scientific research to inform and develop the curriculum. And although it may be an unnecessarily slow process, scientific research continually trickles down from laboratories to inform the domains of practice (professional experience) and academic training. So, although academic training is the foundation of a young practitioner's knowledge and training, scientific research is the foundation of that foundation—the fundamental understandings on which academic training, future personal experience, and indeed further scientific research are based. Ironically, scientific research is often the knowledge domain that practitioners are the least cognizant of, and herein lies a central theme of this text: Although academic training is essential to provide a foundational, working body of knowledge, and personal experience is the natural (and necessary) product of a career in practice, both the classic and the most recent scientific research supply the other two sources of knowledge and, more importantly, enable the practice of an objective, self-correcting, and progressive field of science.

Integration of the Sources of Knowledge

Perhaps the clearest example of the integration of personal experience, that is, the "art" of science, into exercise programming is in the training of an elite athlete. Top-level competitors have undoubtedly implemented all the well-understood and researched training techniques—those born of academic training (fundamental knowledge) and scientific research (newer, emergent knowledge). What they need are innovative, novel ideas that can provide small, perhaps even immeasurable 1% improvements. These variations are the difference between 1st place and 5th . . . or 25th place. Attention to the finest of details and innovative training developments are what fuel winning world-class performances. The key ingredient in these small but critical performance advancements is the personal experience of the coach. Coaches of elite athletes should have many years of experience, because elite athletes already know what they are doing and will not benefit much (if any) from a coach who simply has a firm grasp of training fundamentals. Years of experience with a particular athlete or previous experience with other athletes—an immersion in the sport—fosters an encyclopedic knowledge and high level of competence that allows the coach to make subtle adjustments to an athlete's program and introduce novel programming elements that result in meaningful improvements.

Figure 4.1 depicts three exercise practitioners with different levels of competence in the three sources of knowledge; two of these are subpar exercise practitioners, while the third represents a well-rounded professional. Figure 4.1*a* represents an exercise practitioner who has extensive professional experience, minimal academic training, and no regular interaction with or implementation of scientific research into practice; 4.1*b* depicts a practitioner who has significant academic training and who reads and implements scientific research but has no real-world experience in the judicious application of this knowledge; 4.1*c* represents the ideal exercise professional with deep, balanced competence in each of the three sources of knowledge.

TYPES OF EVIDENCE

Whereas sources of knowledge is a somewhat theoretical construct, the division of evidence into various types is a much more practical, concrete categorization. In this text, we partition evidence into two broad categories: (1) non–peer-reviewed information (e.g., expert opinion) and (2) peer-reviewed scientific research (i.e., observational and experimental research). Although there is a degree of correspondence between the sources of knowledge and the types of evidence, the concepts are based on different perspectives. Sources of knowledge are categorized based on the vehicle, medium, or setting whereby one acquires knowledge (i.e., formal schooling, professional vocational experience, targeted search of the scientific literature); types of evidence, as the term implies, represent a systematic characterization of the evidence itself without particular regard for where the evidence was acquired.

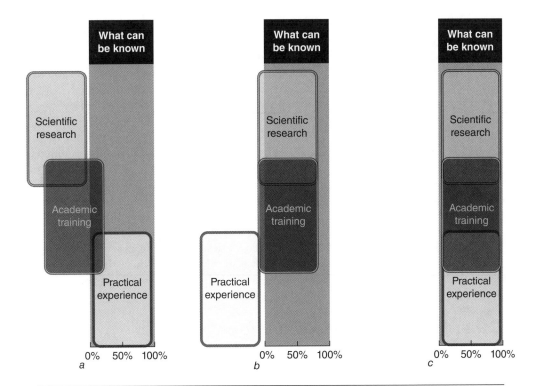

Figure 4.1 Theoretical example of exercise practitioners with different levels of competency in the three sources of knowledge.

Based on Amonette, English, and Ottenbacher 2010.

Non–Peer-Reviewed Information

Non–peer-reviewed information is the most readily available but most biased type of evidence. Thus, practitioners must carefully evaluate non–peer-reviewed information to ensure that it is sound.

Expert Opinion

Expert opinion is the evidence type that corresponds to professional experience in the sources of knowledge. Because it represents one individual's synthesis of a career's worth of observations and experience, expert opinion is prone to bias; that is, the individual may introduce inaccuracies in multiple ways, as follows. (1) The mentally gathered "data" may be inaccurate. (2) The information may be inaccurately recalled. (3) Even if the information is remembered correctly, the synthesis may be faulty, overemphasizing some outcomes while inappropriately minimizing others; this could be done purposefully to advance an agenda or inadvertently. These potential biases are serious and can lead to significant errors in exercise programming for an athlete or client. Despite this potential for error, as touched on in the previous section with regard to personal experience sources of knowledge, expert opinion forms "the tip of the spear" when it comes to developing well-rounded, evidence-based exercise programs for athletes and clients. Expert opinion should never be the foundation of an exercise prescription; however, it can and should elevate a solid, well-developed exercise program from effective to exemplary.

Unfortunately, some exercise science practitioners use expert opinion (their own or others') as the primary (or only!) evidence for their programming decisions. Depending on the expert, this can lead to programs ranging from subpar to disastrous. If one uses expert opinion appropriately, that is, as "icing on the cake" of an already well-grounded program, one carefully considers the qualifications of the expert, recognizes the expert's areas of proficiency, and avoids applying information or opinion beyond the expert's sphere of expertise or oversubscribing to a particular expert just because the individual is very knowledgeable in a single area. Last, practitioners must be humble in assessing their own level of expertise. While trial and error is an indisputable component of strength and conditioning and thus of professional development, as practitioners gain this experience over the course of a career they must be aware of where they lie on the continuum of expertise at any given time; this will help prevent poor decisions based on an overestimation of their own level of expertise.

Other Non–Peer-Reviewed Information

"Other non–peer-reviewed information" is a catch-all category that consists of materials created by individuals who do not rise to the level of an expert in their field. Such material is everywhere: Local bookstores, pharmacies, and grocery stores have an abundance of bodybuilding, fitness, and other health-related magazines. The authors have been selected by the magazine to write a story for the purpose of sales. Articles may be written by experts in the field using legitimate sources of knowledge; they may also be written by non-experts expressing opinions based on poor or inaccurate information. The reader must decide which is which, as there is no peer review process to safeguard the accuracy of the information.

The Internet also contains an abundance of non–peer-reviewed evidence on a variety of subjects related to exercise science. A Google search using the search phrase "resistance exercise protocols" returns over 1 million links to articles ranging from sources such as PubMed, WebMD, and Wikipedia to personal websites and blogs. Some of the information is legitimate and useful; much is poor and inaccurate or without any context for use. There is no way to tell the difference without searching for the evidence upon which these non–peer-reviewed contributors based their opinions. Non–peer-reviewed information can be entertaining to read, but it is often unsubstantiated, and exercise practitioners should never form opinions based on these sources alone.

Peer-Reviewed Scientific Research

Scientific research is typically published in peer-reviewed journals. Before acceptance for publication, the paper is examined by discipline experts for soundness in study methodology, data collection techniques, and interpretation of the results. Of paramount importance to solid peer-reviewed science is the process of blind reviews, which have been threatened in recent times. When implemented appropriately utilizing blind reviews by legitimate discipline experts, this process helps to safeguard the information appearing in journals and results in stronger sources of information. The peer review process is not infallible; papers rejected by scientific journals have subsequently gone on to win the Nobel Prize. Likewise, poorly designed scientific studies sometimes appear in journals, espousing misinformation. While no study is perfect, one must realize that with the abundance of scientific journals available today and the limited number of field-specific scientists to carefully review every submitted paper, articles can be published that have small or even fatal flaws in design. Therefore, the evidence-based practitioner must carefully evaluate all information, including peer-reviewed papers.

Scientific research can be broadly categorized into two distinct classes: observational research and experimental research. The fundamental difference between the two is control of the independent variable. In research, dependent variables are the outcome measures being tested to determine the effectiveness of an intervention. They are outcome measures because changes are dependent on the independent variable. In an exercise intervention study, one might be interested in outcomes such as strength, power, aerobic endurance, or flexibility. The independent variable is manipulated by the researcher and differs between groups; it is the variable of ultimate interest and in exercise science may be a novel exercise intervention, device, protocol, or supplement. In observational research, the investigator does not control the independent variable; exposure to it occurs naturally. In experimental research, the independent variable is closely controlled by the investigative team, and the dose or type of exposure is precisely prescribed and monitored. In the following sections we discuss both subtypes of research.

Observational Research

In contrast to expert opinion and other non–peer-reviewed information, which as already discussed are prone to subjectivity and bias, observational research is an evidence type that is firmly based on the rigorous acquisition of objective scientific data. Nevertheless, observational research is not immune to bias. Bias in observational research is subtle in that it is not subjective but is an undesirable feature of the research design itself.

As the name suggests, observational research is characterized by an investigator's observing—not intervening in—one or more groups of subjects. In some multiple-group designs such as case–control studies, the observations made are compared between the groups. These studies are often cross-sectional; that is, the comparison between groups is made at a single point in time. In contrast, prospective cohort designs typically employ a series of longitudinal measurements to characterize change over time. In single-group designs, individuals (case studies) or a number of individuals (case series) are compared against themselves over time but without comparison to another group.

The field of study underlying much observational research, particularly multiple-group study designs, is epidemiology. Epidemiology, meaning "the study of what is upon the people" (Greek *epi*, upon; *demos*, people; *logos*, study or word), is a field that examines the causes and effects of health and disease in specific populations (Gordis, 2009). Epidemiologists often serve as a "first line of defense" against diseases or unhealthy practices. In addition to the formal research designs described later, epidemiologists also use what are called "natural experiments." A classic example of a natural experiment is a study by John Snow that determined the source of a cholera outbreak in London in 1854 (Centers for Disease Control and Prevention, 2004; Paneth, 2004; Stanwell-Smith, 2002). Snow noticed that cholera cases were highly concentrated in a particular geographic area and hypothesized that the cause was contaminated water (vs. the popular "miasma" theory of the day, which blamed illness on toxic air). After consulting maps and interviewing many cholera sufferers, Snow was able to link a large majority of the cholera cases to usage of the Broad Street water pump, which was sourced from a part of the Thames River where sewage was dumped. Using his data (e.g., maps and case interviews), Snow was able to convince the authorities to remove the handle of the pump to disable it. Cholera cases quickly declined and the outbreak was stopped. This is called a natural experiment because Snow did nothing to set up or organize the study. He simply investigated a natural phenomenon using the working hypothesis that cholera was a fecal-borne illness transmitted via contaminated water—a hypothesis that was well confirmed due to the association Snow showed between use of a single contaminated water source and a high incidence of the disease.

The major limitation of observational research is its inability to demonstrate definitive cause and effect (Flanders & Longini, 1990). Because the investigator is neither controlling the participants nor actively intervening, outcomes can only be associated with proposed causes. Nevertheless, observational research is a powerful tool for studying, in a relatively economical manner, associations between factors such as lifestyle choices and occupational exposures to important health outcomes (Gordis, 2009). Here we discuss four basic types of observational study design.

Cohort Study

A cohort study can be either prospective or retrospective in design (Gordis, 2009). A prospective cohort study follows a group of unaffected or non-diseased individuals over time, waiting for some number of them to develop the outcome (e.g., disease) of interest (figure 4.2). The investigators have a hypothesis for what causes the outcome, and after enough individuals develop the outcome, they look to see if a disproportionate number of those who developed the disease also were exposed to the hypothesized cause. If so, a link between the exposure and the disease is established. For example, if a cohort study was conducted to evaluate the association between vigorous exercise and colon cancer, investigators would recruit a group of individuals that was relatively homogenous for factors that might affect the outcome of interest (e.g., subjects would be similar in age, overall health status, exposure to coal dust) but included both exercisers and non-exercisers (figure 4.2). All subjects would be screened to confirm that they did not have colon cancer or any precancerous growths. Then the investigators would follow up with the subjects annually for the next 10 or even 20 years. When the follow-up period ended, investigators would calculate the incidence proportion of new cases using the following formula and compare between exercisers and sedentary subjects (Riffenburgh, 2012):

$$\text{Incidence proportion} = \text{New cases} \div \text{Population at risk}$$

Let's say that 10,000 people were in each group. After 10 years, 50 new cases were diagnosed in the sedentary group but only 2 new cases in the exercise group. Although this is a simplified calculation (and thus slightly inaccurate), the annual incidence in the sedentary group would be

$$\text{Annual incidence} = (50 \text{ cases}) / (10 \text{ years}) \div 10,000 \text{ people}$$
$$= 5 \text{ cases per year per } 10,000 \text{ sedentary}.$$

The annual incidence for exercisers would be

$$\text{Annual incidence} = (2 \text{ cases}) / (10 \text{ years}) \div 10{,}000 \text{ people}$$
$$= 0.2 \text{ cases per year per } 10{,}000 \text{ exercisers}.$$

The incidence of colon cancer in the sedentary group was 25 times greater than in the exercisers. Statistical analysis would reveal that the probability of this difference occurring due to random chance is very low, and thus the investigators would conclude that exercise is strongly associated with a decreased risk of colon cancer. Due to the research design, causation cannot be determined; it would be technically incorrect to state, based on the results of this study, that a sedentary lifestyle causes colon cancer. What can be said is that a sedentary lifestyle is associated with a 25 times greater likelihood of developing colon cancer. So although we cannot definitively establish causation, we can clearly demonstrate that sedentary behavior is strongly linked to colon cancer and that exercise may dramatically decrease one's risk of developing colon cancer. Other potential weaknesses of prospective cohort studies (and indeed all observational research designs) include the use of self-reported data and bias resulting from subjects lost to follow-up.

A retrospective cohort design is similar in that the investigator does not control the level of exposure to the independent variable, exercise. Suppose an exercise scientist wanted to examine the same study question, the association between colon cancer and exercise, using a retrospective cohort. If the investigative team had access to a large historical database of medical records including the documentation of colon cancer cases and noncases with a secondary question to each patient concerning physical activity and exercise, the same incidence calculations made in the prospective design could be performed (figure 4.3) to determine whether there was a correlation between exercise habits and colon cancer. Of course, there are problems associated with this design,

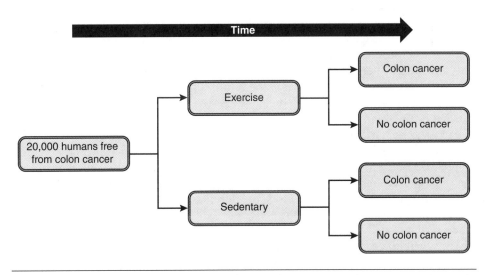

Figure 4.2 Prospective cohort design evaluating the association between exercise and colon cancer.

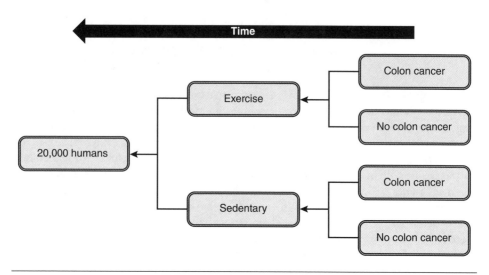

Figure 4.3 Retrospective cohort design evaluating the association between exercise and colon cancer.

namely that exercise is self-reported and that there may be no information concerning its intensity, volume, duration, and so on. However, the strength of the design is that the question can be answered with existing data and the investigative team can obtain a relatively quick answer to their research question.

The cohort study design is very useful in epidemiologic research but not particularly applicable to exercise intervention research. Cohort studies typically evaluate categorical outcomes (often binary; e.g., having a disease or not) whereas exercise science uses interval/ratio scale outcomes, for example, maximum strength. Everyone possesses some amount of strength—the investigator is interested in how much strength one group has compared to another. Cohort studies are not conducive to making these determinations. There is, however, an emerging movement that may facilitate greater use of and access to cohort research in the future. The Exercise is Medicine project of the American College of Sports Medicine (ACSM) is a global public health movement initiated by ACSM in partnership with the American Medical Association (AMA). One goal of Exercise is Medicine is to bring the importance of exercise and physical activity to the forefront of health care. Along with encouragement to provide general exercise counseling to patients, the ACSM and AMA are pushing for the inclusion of standardized questions concerning exercise during all physician visits. In other words, when you go to the doctor, your height, weight, blood pressure, and temperature would be measured and you would be asked about your exercise habits. Combined with the increasingly widespread use of electronic medical records, the intent is that the Exercise is Medicine initiative will lead to a large database of health outcomes and exercise habits that can be used to examine associations between exercise and disease using cohort designs.

Case–Control Study

A case–control study is a retrospective observational research design that divides a subject sample into two groups, cases (those with the outcome of interest) and controls (those free of the outcome of interest), and then looks back to determine if an individual was exposed to the hypothetical cause of the outcome. Case–control studies are similar to retrospective cohort studies in that they begin with individuals who have the disease (cases) and those who are healthy (controls) (Levin, 2006). One difference is that the cases are matched to the controls using important characteristics that may also influence the outcome (e.g., age, sex, body mass index [BMI], race, ethnicity); this reduces bias (Cerhan et al., 2011). The case–control research design applied to our previous exercise and colon cancer example would yield a study like this: (1) Investigators would obtain access to a large medical record database and identify groups of both colon cancer patients and those without the disease; (2) individuals with the disease would be matched closely by sex, BMI, race, and ethnicity to those without the disease; (3) the investigators would determine whether the individual was an exerciser or nonexerciser from medical records or an interview; and (4) exercise prevalence in both the cases and controls would be statistically compared to determine whether lack of exercise was associated with developing colon cancer (figure 4.4).

Great care must be taken with the selection of the control group for a case–control study (Levin, 2006). It can be argued that unbiased, appropriate selection of the control group is the critical step in the proper execution of a case–control study. Biased selection can distort the findings of the study by making an exposure and outcome that are not truly linked appear to be associated or by obscuring a true exposure–outcome link. For instance, what if a case–control study for exercise and colon cancer was conducted without controlling for other confounding factors? Suppose the investigative team determined that sedentary lifestyle was associated with colon cancer. However, the group of sedentary individuals, with the greater incidence of colon cancer, also included a higher proportion of smokers, were more obese, drank more alcohol, ate diets higher in saturated fats, and had stronger family histories of colon cancer compared to the control group. Each of these is a risk factor for colon cancer (Kono, Toyomura, Yin, Nagano, & Mizoue, 2004). Since the study was poorly controlled, it is impossible to truly determine whether the increased colon cancer prevalence was caused by lack of exercise, smoking, obesity, alcoholism, or diet. Thus, it is critical that cases and controls be as similar as possible (except for the exposure of interest) and that confounding variables be tightly controlled.

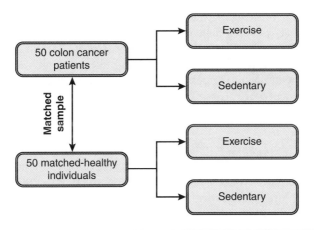

Figure 4.4 Case-control study design to determine the association between exercise and colon cancer.

Case Study

A case study is a research design that describes a single person (the case) (Kennedy, 1979). In contrast to cohort and case–control research designs, case studies are actually very useful and applicable to the exercise science field. In the medical field, a case study is usually reported on a somewhat remarkable case, either a patient with a rare disease or an uncommon presentation of a disease or a patient who responded unusually (either positively or negatively) to treatment (Vandenbroucke, 2001). In the medical research literature, case studies enlighten other physicians to an off-nominal or peculiar condition, treatment plan, or patient response and inform their diagnosis or action plan when faced with a similar patient. As such, case studies are very careful to systematically convey all details relevant to the case, often comparing and contrasting the reported case with what is typically seen in the traditional patient.

This research design is very applicable to the exercise science field, particularly in the study of high-level athletes. As previously discussed, athletes' training programs are based on physiological understandings that have vast research evidence. However, to stop here would limit all athletes of a particular sport and level of development to a single generic program; in fact, perhaps there would be no need for exercise practitioners and a computer could generate all athlete training programs. Why is this scenario not a reality? Because each athlete is an individual, and although all are members of one species and governed by the same general physiology, small differences in individual physiology (not to mention differences in history and environment) play a significant role in training adaptations, particularly at the upper levels of sport where seemingly miniscule differences separate winners and also-rans. In many ways, case studies are a documentation of the art of our science, the creative application of sound physiological principles to the training programs of individual athletes.

The writing and publication of useful case studies in the exercise science field requires much of the exercise professional. While coaching and training may be done in an ad hoc fashion, to properly and accurately convey the training course of an individual patient, client, or athlete, the practitioner must relate the program in great detail and include all relevant information to ensure that other practitioners have a well-developed picture not only of the program but also of the particular athlete (e.g., training history) and the training environment (e.g., off-season vs. preparatory vs. in-season). A good example of a published case study of an elite athlete is a paper in *Journal of Applied Physiology* (2005) that described changes in Lance Armstrong's physiological performance over the first seven years of his professional career (Coyle, 2005). Later revelations of performance-enhancing drug use notwithstanding, the study provides unique insight into the physiology of one of the best cyclists of all time. A significant omission in this paper is any reference to Armstrong's training program, although this is understandable given that he was still competing at the time of the above noted publication.

Case Series

Case series are very similar to case studies but measure multiple patients, clients, or athletes over time (Kooistra, Dijkman, Einhorn, & Bhandari, 2009; Martyn, 2002) (figure 4.5). In contrast to case studies, which are retrospective (i.e., an individual performs well and then the nature of his program becomes a matter of interest), case series studies are prospective and deliberate in their execution. Practically speaking, all exercise practitioners are perpetually conducting a case series study with their patients,

Figure 4.5   Case series research design testing the effectiveness of exercise to improve outcomes in colon cancer patients.

clients, and athletes even if it is not rigorously documented and instead exists mostly in the practitioner's head. Ideally, every exercise practitioner would formally track each individual and thoroughly document what is observed. Understandably, very few case series are submitted or published in the scientific literature. Nevertheless, over time these records provide invaluable information for the practitioner himself, both later in the same client's career or to guide the program of another client.

Experimental Research

Experimental research is the second branch of peer-reviewed scientific research. Whereas the first branch, observational research, has advantages such as relatively low costs on a per subject basis and even the capability to conduct a study via review of existing data, it also has disadvantages such as a moderately high likelihood for research bias and the inability to demonstrate causal relationships between exposures and outcomes. Experimental research in the exercise science field is the polar opposite of this as it is generally rather costly to perform, requires human subjects (or animals or cells), when well designed and conducted minimizes the risk of bias, and most importantly, facilitates the drawing of causal inferences between exposures and interventions and outcomes of interest. Generally, experimental research is hypothesis driven; that is, the investigators develop an experimental hypothesis, design a research study to test the hypothesis, and based on the data either accept or reject the experimental hypothesis.

Randomized controlled trials (RCTs) are the gold standard of experimental research (Meldrum, 2000). They are designed to empirically evaluate a novel hypothesis (called the alternative or experimental hypothesis) in a rigorously controlled environment. Although many variations exist, the basic RCT is a remarkably simple research design. A random sample, or a matched random sample in small sample size studies, is drawn from a defined population, and the individuals in the sample are randomly assigned to a treatment group. The simplest study will consist of a control group and an experimental group. All things should be equivalent between the groups with the exception of the independent variable, which the investigators manipulate to test its effects (a series of outcomes, or dependent variables) on the two groups. For instance, previously sedentary individuals could be assigned to an experimental group that performs resistance exercise for 8 weeks; they would be compared to a control group that remained sedentary for the duration of the study. This sort of comparison (one in which the control group receives no intervention or treatment) is appropriate for an intervention with little evidence for its efficacy; for more well-studied treatments such as resistance exercise training, it is more appropriate to compare the novel training to the current standard of training to determine if the novel training method is superior (or at least equivalent). In the field of exercise science, an experimental hypothesis could be, "Training program B will increase muscle strength to a greater degree than traditional training program A."

An effective RCT should consider several factors in its design. (1) The groups should be similar in all factors that could affect the outcome of interest and matched for these factors before randomization (e.g., physiological, anthropometric, and performance variables such as age, sex, height, weight, training status, aerobic fitness, strength, and sport). (2) These same factors should be chosen to maximize the likelihood of observing an effect due to the intervention (this is particularly true of initial investigations of novel interventions), or, alternatively, to ensure that if an experimental intervention fails to equal or supersede the traditional one, it is not because the conditions were unfavorable. (3) Investigators should carefully select the population to be studied, as physiological responses to a particular intervention often vary widely between populations. Randomized controlled trials reduce the likelihood for bias by randomly assigning subjects to a treatment group; this contrasts with a prospective cohort design in which subjects determine their own group assignment based on their choices (e.g., exercise or sedentary), which may lead to bias. Random assignment is a unique characteristic of experimental research.

The controlled environment of RCTs allows scientists to infer causation if in fact an effect of the intervention is observed, because the groups are similar with respect to all relevant factors and the only difference in how they are treated is the independent variable—the experimental intervention. It can be inferred that any poststudy differences between the groups are caused by the intervention. There is much more to the design of a good research study than this, but these are the primary factors. The degree to which changes in the dependent variable can be attributed to the intervention is called *internal validity*. Well-designed studies with the characteristics already detailed (in addition to several others) have a high degree of internal validity. An example of a component of good experimental control in a study comparing two resistance training protocols would be the provision of a specific research diet or, at the least, dietary records, as it is well understood that total energy and protein intake have an effect on resistance training adaptations. If a study documents enhanced adaptations in one group versus the other and can also show that their diets were not different from each other, we have increased certainty that the superior adaptations were a result of the training intervention and not of a confounding factor such as diet. Randomized controlled trials are thus the highest form of original research evidence.

The simplest randomized design compares a single intervention to a single control group (Gordis, 2009). For example, one might test the effectiveness of a particular exercise protocol in reducing the loss of skeletal muscle mass in patients undergoing treatment for colon cancer (figure 4.6a). In this case, after baseline muscle mass testing, 50 patients would be randomly assigned to an exercise group and 50 patients would be randomly assigned to a control nonexercising group (for the purposes of this and subsequent examples, we will set aside any potential ethical concerns about experimentally withholding a treatment—exercise—that is likely to be effective). At the conclusion of a preselected time period, muscle mass would again be tested, and the difference in means and variations between groups would provide evidence for or against the exercise protocol. The same study design could be used to assess one or many interventions (figure 4.6b).

Although experimentally it is simple to compare single exercise or nutritional interventions, in practice it is rare that a single intervention is being implemented in a client, patient, or athlete. Instead, it is typical that some combination of exercise and nutritional intervention is being utilized. One strength of a randomized controlled study is that it

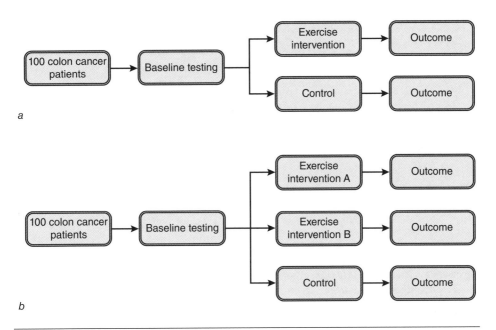

Figure 4.6 Randomized controlled trial designs testing *(a)* the effectiveness of a single exercise intervention to improve outcomes in patients with colon cancer, *(b)* two different protocols to improve outcomes in patients with colon cancer.

allows for the simultaneous comparison of more than one intervention (Gordis, 2009). What if a scientist hypothesized that a certain dietary intervention in combination with exercise would synergistically reduce muscle mass loss in colon cancer patients? This research question would be evaluated with a factorial study design: After baseline muscle mass testing, patients would be assigned to either a control group (normal diet and no exercise), an exercise intervention group, a nutrition intervention group, or a combined exercise and nutrition group. At the conclusion of a predetermined time period, muscle mass would be tested, and the between-group difference would again be ascertained to determine the efficacy of the two interventions alone and in combination (figure 4.7).

Systematic Review

Systematic reviews are a synthesis of peer-reviewed research evidence; as such, they are a unique branch of research evidence because, unlike observational and experimental research, they are not original investigations. A systematic review is intended to be a thorough, objective, and authoritative answer to a research question; however, instead of gathering new research data to address the research question, a systematic review uses the published work of others as its data. Thus, it attempts to synthesize all the relevant information on a particular topic into a definitive answer (or as definitive as possible given the current knowledge in the area) to the relevant question.

As the name implies, systematic reviews are conducted in an objective, reproducible manner; this differs from typical narrative review papers, which, although based on published research evidence, are somewhat subjective and editorial in their content, with the authors under no compulsion to include or substantively discuss any particular

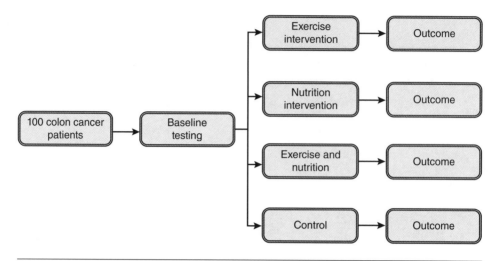

Figure 4.7   Factorial study design, testing the effectiveness of exercise or nutrition alone versus a combination treatment to improve outcomes.

study. Briefly, systematic reviews begin with one or more focused questions followed by an exhaustive literature search using specific, relevant search terms. Using defined inclusion and exclusion criteria, systematic review authors must then sift through the results of the literature search to compile a final list of studies that merit inclusion in the review. Next, a summary of the results of all of the included studies is prepared; this is usually one or two primary outcome measures common to all of the studies and part of the systematic review's research question. These data are often presented in a large table with one column devoted to simple categorical outcomes (e.g., "increased," "decreased," or "unchanged"). Some systematic reviews include a meta-analysis, which is an aggregate statistical analysis of one or more outcome measures from all of the studies included in the review. Systematic reviews must also account for different levels of evidence, as conflicting findings from two different study designs should not be considered equally. Thus, because of the rigor with which systematic reviews are conducted, they can constitute very high-quality evidence and represent the combined strength of all the studies that they review (Sackett, Straus, Richardson, Rosenberg, & Haynes, 2000). However, they can also be weak or unsubstantial forms of evidence if the scientific papers reviewed are poorly constructed or fatally flawed in their design.

CONCLUSION

Knowledge is the essential foundation on which exercise scientists make programming decisions. It is obtained through academic lectures and seminars, professional experience, and scientific research. Information on exercise is widely available for public consumption. It can be high quality, based on peer-reviewed sources, or it can be faulty, based on poor, biased information. Even peer-reviewed sources have flaws, and depending on the research design, different studies merit more or less credence. In order to become an effective evidence-based practitioner, it is essential that practitioners and scientists alike gain the skills needed to identify and understand research designs.

Chapter 5

READING AND INTERPRETING RESEARCH EVIDENCE

Learning Objectives

1. Discuss the concept of a "consumer of research" and describe who should be a consumer.
2. Discuss the basic components of a research manuscript.
3. Understand the types of statistical error and understand how to interpret statistical results.
4. Learn to interpret figures and graphs.
5. Learn how to read a scientific manuscript to minimize personal bias in interpreting the results.

Traveling to a foreign country can be an intimidating task. However, the difficulty of the situation is compounded if you cannot speak or understand the language. Language is a communication medium that is used to convey ideas and thoughts. Languages have alphabets that are made up of letters; certain sounds are associated with each letter. When the letters are arranged in a certain order they create a word. When words are sequenced in a particular order they make up a sentence. Letters, words, and sentences are used to communicate ideas and can be expressed in written or oral form. The intricacies of each language's alphabet, sounds, and grammatical structures can be incredibly complex, requiring many years to master; but with only a small vocabulary, you can communicate in a foreign language. It may be difficult, but with minimal learning you can perform simple tasks like ordering food, asking questions, and reading road signs. Such minimal language ability affords you the capacity to effectively operate in a foreign country. With dedicated study of the language and years of immersion in the culture, competencies increase, and the scope and breadth of accomplishable tasks increase. However, increased competency requires patience and practice.

Novices first reading research evidence may feel as if they are learning a foreign language. Terms like *hypothesis*, *theory*, and *independent variable* may be new or foreign to the reader. Additionally, some of the statistical jargon can be confusing, and a novice can be overwhelmed by the complexity of the analysis. Similar to an individual learning a new language, with a minimal vocabulary and understanding the exercise practitioner should be able to read and interpret results from research evidence. With practice and years of experience, capabilities will increase to the point that the novice becomes a master.

CONSUMERS OF RESEARCH

As we begin this chapter on research evidence, it is important to note that the purpose of this chapter, the book, and the whole evidence-based philosophy is not to transform practitioners into scientists. To be able to ask questions and to give perspectives on the published science, it is important that practitioners understand the scientific process, its context, and its impact on the questions and problems to be addressed. Understanding research is an important step practitioners take in becoming a valuable, contributing member of the evidence-based practice (EBP) team. A scientist has a specific occupation that is concerned with asking and answering questions (table 5.1); scientists have dedicated academic training and experience in their field(s) of study. The questions they ask can be mechanistic, dealing with fundamental biological questions. They can be descriptive or aimed at characterizing the response of a population to exercise. They can also be clinical or applied. Applied questions relate to outcomes research that studies the effects of a new protocol, device, or nutritional supplement in a population.

Scientists in each of these categories strive to generate results and disseminate their findings, but one may ask who the consumers of this research are (Amonette, English, Spiering, & Kraemer, 2013). Often, other scientists are the primary consumers. Scientists publish papers in peer-reviewed journals to which other scientists in the same or similar disciplines subscribe. Scientists read research articles published in these journals as they develop new study ideas that either extend or replicate the findings published by

Table 5.1   Examples of Scientific Questions

Types of scientific questions	Sample questions
Mechanistic or basic questions	How do the dihydropyridine receptors interact with ryanodine in skeletal muscle contraction?
	What is the mechanotransduction mechanism for resistance exercise and bone formation?
Questions on characteristics	Does the back or front squat result in greater activation of the gluteal muscles in adults who are older?
	Do junior-level female gymnasts have greater bone mass than age-matched controls?
Practical or clinical questions	Does training using the back squat exercise improve sit-to-stand capabilities of adults who are older?
	Does 6 months of gymnastics training improve bone mineral density in young females?

other scientists. On completion of their studies, they publish in the same peer-reviewed journals (figure 5.1).

This process works well if the sole purpose of research is for other scientists to learn and generate knowledge for self-consumption. Before the advent of modern communication mediums, scientific papers were a mechanism to communicate findings to scientific colleagues; they were not necessarily intended for public consumption. The problem with this model is that the ultimate recipients of knowledge generated via science should be the clients, athletes, and patients seeking to improve their health, fitness, or performance. If information is propagated only through the scientific community, it will never reach the population that needs it. Exercise practitioners who develop protocols, provide advice on supplementation, and buy devices that are used for clients, patients, and athletes should be the primary consumers of science.

In order to be an effective consumer of science, one does not need to be an expert scientist, although a fundamental understanding of research and basic terminology is crucial to the accurate interpretation of data. When problems arise, exercise practitioners should be skilled to the extent that they can ask an answerable question, find evidence, read and interpret the evidence, rank the evidence, and then make a decision to incorporate or exclude it from practice (figure 5.2). More difficult or complicated questions related to research can be directed toward exercise and sport scientists who

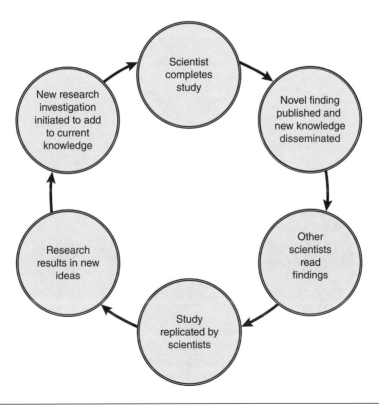

Figure 5.1 A self-edifying model of scientific discovery in which research evidence is disseminated only within the scientific community.

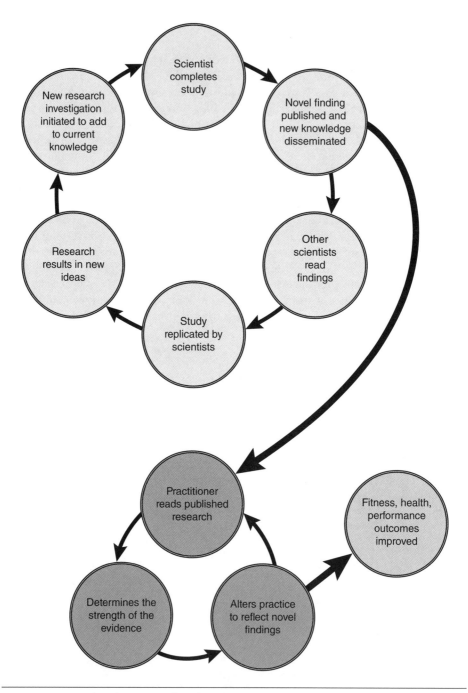

Figure 5.2 Results when scientific information is propagated to practitioners and implemented into practice. This process is ongoing—practitioners should continue to read published research and reevalute their practice as needed.

can provide a more rigorous evaluation of the data. When practitioners develop these basic skills, they become the bridge from the scientist to the beneficiaries (i.e., clients, patients, and athletes).

WHAT IS RESEARCH?

This book strongly emphasizes research and science because of their value in generating information and contributing to the body of human knowledge. The previous chapter discussed various forms of research design. Moving back one level, it is important to ask the question: What is research? A quote attributed to the great physicist Albert Einstein provides a partial answer to this question: "If we knew what it was we were doing, it would not be called research, would it?" Perhaps a more inclusive definition of research is provided by the National Institutes of Health (NIH, 2013):

> A systematic, intensive study intended to increase knowledge or understanding of the subject studied, a systematic study specifically directed toward applying new knowledge to meet a recognized need, or a systematic application of knowledge to the production of useful materials, devices, and systems or methods, including design, development, and improvement of prototypes and new processes to meet specific requirements.

The NIH definition highlights a few important components of research. First, research is a "systematic" and "intensive" method. In other words, the process of research is purposeful. When scientists endeavor to answer a question, they do so using a systematic method. Framing the study with independent and dependent variables, they systematically exclude or control for confounding factors and minimize the potential for bias. The systematic, intensive method is designed for the purpose of "increasing knowledge." The increase in knowledge could be at the mechanistic level or could be an "application of knowledge." In most cases, the exercise practitioner is concerned with the application of knowledge, not the knowledge alone. For example, an exercise practitioner working with a client who has multiple sclerosis would be interested in the pathophysiology or the symptoms associated with the disease. Such information may provide background knowledge that improves exercise prescription. However, the more important question to the exercise practitioner is likely based on outcome(s): Does exercise reduce or increase the symptoms of the disease? If so, are there some types or intensities of exercise that are better for improving outcomes?

The NIH definition suggests that the practical component of research is useful for the "production of useful materials, devices, and systems or methods." Research is profoundly important for practice. It is through research that new devices are developed, improving outcomes in clients, patients, and athletes. New systems of exercise training and methods of exercise application are developed through the systematic process of research, all leading to improvements in practice.

COMPONENTS OF A RESEARCH PAPER

As stated earlier, the implementation of research findings requires that they be disseminated to the exercise practitioner and that the practitioner be able to read and interpret the findings. Although research literature can seem like a foreign language, scientific manuscripts are ordered in a particular way. In a scientific paper, the authors are telling a story; their primary purpose is to communicate a new idea or concept expressed by their data (Knight & Ingersoll, 1996b). First, the story defines a problem and explains what is currently known about a phenomenon. Then it explains the research question that the authors developed to address the problem (i.e., what is unknown), the research methods used to acquire the data to answer the question, and the statistical techniques used to analyze the data. This is followed by a summary of the data generated from the project and a discussion of those data in the context of other published research or of how they fit into the context of current understanding of a phenomenon (Kraemer, Fleck, & Deschenes, 2016; Ohwovoriole, 2011). Although each journal has its own particular style, there are many commonalities between journals (Tipton, 1991).

Abstract

Research articles begin with an abstract. This is a summary of the research findings and typically includes an overview of each section of the paper. An abstract is limited by journal specifications and may vary in length from 75 to 350 words (Knight & Ingersoll, 1996a). The brevity of an abstract necessitates that only the most important information be included. Abstracts generally follow the flow of a manuscript and may contain information about the background, methods, results, discussion, and conclusions (Alexandrov & Hennerici, 2007). Some journals require a structured abstract in which each section of the manuscript is summarized separately. An example of a structured abstract is provided in figure 5.3. In contrast to a structured abstract, unstructured abstracts flow in the same general manner but without delineated sections (figure 5.4).

Another key difference between journals and their specifications for abstracts is related to the results. Some journals specify that the abstract must contain data; that is, the authors must include means, change scores, correlations, and so on. They may also contain P values for the statistical differences. Here is an example of results summarized for an abstract with the actual data and P values included:

Peak aerobic fitness improved ($P = 0.025$) from 45 ± 2.5 mL$\cdot$kg$^{-1}\cdot$min^{-1} to 55 ± 2.5 mL$\cdot$kg$^{-1}\cdot$min^{-1} following the training intervention. In contrast, the nonexercising control group did not improve aerobic fitness ($P = 0.77$) from baseline (44 ± 3.7 mL$\cdot$kg$^{-1}\cdot$min^{-1} to 46 ± 5.5 mL$\cdot$kg$^{-1}\cdot$min^{-1}) following 12 weeks of training.

Other journals require only summary statements concerning the analysis. In these abstracts, there may be no actual data or statistical statements in the summary of results. Here is an example of a summary of results that does not include data:

The exercise intervention improved aerobic fitness following 12 weeks of training, but there was no change in the nonexercising control group.

The purpose of the abstract is simply to provide a summary of the study. The evidence-based practitioner should never read the abstract alone to make definitive judgments about a training intervention, protocol, or the effectiveness of a novel device or supplement. The abstract should provide enough information to allow readers to know whether the data are relevant to the question posed in the first step of EBP and whether they should spend the time to acquire and read the entire research article.

J Motor Learn Dev, 2015, 3(1), 39-52.

Exploring Associations Between Motor Skill Assessments in Children With, Without, and At-Risk for Developmental Coordination Disorder.

Valentini, N., Getchell, N., Logan, S.W., Liang, L.Y., Golden, D., Rudisill, M.E., & Robinson, L.E.

Abstract

Background: We compared children with, at-risk for, or without developmental coordination disorder (DCD) on the Test of Gross Motor Development (TGMD-2) and the Movement Assessment Battery for Children (MABC) through (a) correlations, (b) gender and age comparisons, (c) cross tab analyses, and (d) factor analyses.

Method: Children (N = 424; age range: 4–10 years) from southern Brazil completed the TGMD-2 and MABC and placed into groups (DCD: ≤ 5th%, n = 58; at-risk: > 5th to ≤ 15th%, n = 133; typically developing (TD) >16th%, n = 233).

Results: The strongest correlation was between total performance on the TGMD-2 and MABC (r = .37). No gender differences were found for performance on the MABC while boys performed better than girls on the TGMD-2. Cross tab analyses indicated a high level of agreement for children who performed in the lowest percentiles on each assessment. Factor analyses suggested that, for both the TD and at-risk groups, three factors loaded on the motor assessments. In contrast, the DCD group loaded on a sport skill, general skill, and a manipulative skill factor, accounting for 42.3% of the variance.

Conclusions: Evidence suggests that children who perform very poorly on one assessment are likely to perform poorly on the other. Children with DCD may have sports-related skill deficiencies.

Figure 5.3 Example of a structured abstract with clear section delineations (Valentini et al., 2015).

Reprinted from Valentini et al. 2015.

Introduction or Background

The introduction or background section of a paper immediately follows the abstract in most articles. As the name implies, it contains background or introductory information related to the research question. The section is a set of paragraphs in which the investigators develop the backdrop of the problem, frame the question to be studied, and address any paradoxes or controversies related to the research problem in the literature (Sifft & Kraemer, 1982). Typically, this section contains a brief review of the most relevant literature. Well-written introduction sections are narrowly focused and succinct and logically lead the reader to the study purpose.

Many times the background contains a "timeline" for how the current research question arose, highlighting a series of studies leading up to the current investigation. Most questions addressed in research are follow-up questions to other research. Scientists

Int Sport Coach J, 2015, 2(1), 29-38.

Sustained Participation in Youth Sports Related to Coach-Athlete Relationship and Coach-Created Motivational Climate.

Rottensteiner, C., Konttinen, N., & Laakso, L.

Abstract

The main purpose of this study was to examine the links of coach-athlete relationship (CAR) and perceived coach-created motivational climate to persistence in youth sport. A total of 1692 persistent and 543 withdrawn football, ice hockey, and basketball players, aged 15–16 years, completed the Coach-Athlete Relationship Questionnaire and the Perceived Motivational Climate Sport Questionnaire. Results indicated that persistent players reported higher scores in CAR and task-climate than withdrawn players. Persistent players also represented higher competition level, higher amount of training, and more years of involvement in sport than withdrawn players. Cluster analysis identified three profiles: 1) High CAR, high task climate, and moderate ego climate, 2) Moderate CAR, moderate task climate, and moderate ego climate, and 3) Low CAR, low task climate, and high ego climate. Differences between profiles were found in terms of relative proportion of continuing players, competition level, and amount of training. In all, Profile 1 appeared to be the most beneficial from the perspective of sport persistence. The present findings lend support for the view that coach-athlete relationship and motivational climate together can have implications for young athletes' maintenance in organized sports.

Figure 5.4 Example of abstract without clear section delimitations (Rottensteiner, Konttinen, & Laakso, 2015).

Reprinted from Rottensteiner, Konttinen, and Laakso 2015.

read a series of articles and realize, based on these papers, that there is a gap in the information. Therefore, the new paper fills a gap in knowledge. Alternatively the aim may be to try to replicate the findings of others. The introduction or background section also presents debatable topics related to the study purpose, introducing alternative theories and related literature.

Over time, information regarding a subject accumulates (the traditional "accumulation theory of science") (figure 5.5). Although this is not the only or most complete theory of science, it is a helpful concept for understanding the goal of the background section of a paper. According to this theory, facts from published research pile up over time, and each additional fact adds to the pile. As more and more facts are added, scientists move closer to fully understanding a phenomenon. Using this theory, the background section contains the authors' interpretation of one small aspect of the pile of facts that have already accumulated, how this new research will add to the pile, and more importantly, how it will add to an understanding of the phenomenon.

The introduction or background section is not an exhaustive review of the literature. It is typically limited by the journal's specifications for length and perhaps the number of allowable references. Thus, when reading the background of a paper, one should

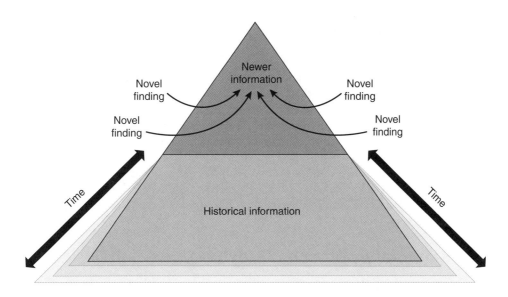

Figure 5.5 Accumulation theory of science. Although not completely accurate, the accumulation model reflects to some extent the structure of the introduction section of a paper.

remember that this is only the authors' interpretation of the literature and that the sequence of studies leading up to the current investigation may not be fully represented. That is, there may be other papers not cited by the authors that also contribute to an understanding of the phenomenon.

Important as well is a clear understanding of the research question the authors were investigating. From this, a hypothesis is developed to guide the direction of the study. Although some scientific papers are exploratory or descriptive, most scientific manuscripts contain a conceptual framework that leads to a research hypothesis. A conceptual framework is a series of related concepts that are integrated to form the basis of a theory (Jewell, 2011). The conceptual framework is generated from previous research findings and may be expressed schematically or in written form; it is the investigators' working idea of how a phenomenon operates (Jewell, 2011). Based on this conceptual framework, the investigators form a research hypothesis that is a testable component of the current theory; thus, it may also be used to refute a common theory (Jewell, 2011). The null hypothesis states that there is no effect of the independent variable on the dependent variables and therefore no observed difference between the control and experimental groups. The statistical methods will dictate either acceptance (confirming there was no difference between groups or over time) or rejection of the null hypothesis, supporting a relationship in the data.

Hypothesis statements may be included in the final paragraph of the introduction or at the beginning of the methods section. One of the key requirements for a hypothesis in a scientific manuscript is that it must be testable. Hypothesis statements are essential; they frame the methods, results, discussion, and conclusion sections, and each division of the paper relates back to the study hypothesis.

Materials and Methods

The methods section of a paper describes how the researchers tested the hypothesis, that is, the methods by which the data were collected to answer the research question. Well-written methods sections describe the study protocol in explicit detail. Protocols should be described such that another investigator could replicate them. Detail is critical because replication is an important aspect of science. The laws of probability state that research findings can occur by chance (e.g., false-positive results). Replication confirms previous findings and provides stronger evidence that the initial findings were accurate. Lack of detail in a methods section can mislead other investigators who are trying to replicate a study and may ultimately lead to false research conclusions. The methods (or materials and methods) section is often subdivided into overview, subjects, and procedures subsections.

Experimental Approach, Study Overview, or Design

One of the most important sections within the methods deals with the experimental approach to the problem which provides an overview of the study protocol or design. This section contains a detailed description of how the subjects were selected and assigned to various groups for testing. After reading the protocol, one should understand the study type: randomized controlled trial, nonrandomized trial, prospective cohort, and so on. This information will elucidate potential limitations, confounding factors, and sources of bias that could influence the findings. This section should also include information about blinding and randomization, which helps readers to determine the strength of the evidence. After reading this section, it should be clear how the investigators tested their hypothesis, and a skilled reader can evaluate whether the design was appropriate.

The overview section within the methods may also include an outline of the overall study procedures. Although the testing procedures are not described in detail, there may be information concerning when and where the testing occurred. There may be general information regarding subject conditions at testing (e.g., fasted, standardized diet, free from exercise for 24 h). The location of this section differs depending on the journal; it may precede or follow the subjects section.

Subjects, Participants, or Patients

The subjects section of a paper contains important information about the characteristics of the participants (Sift, 1984). There may be a short description of how the participants were recruited, including any remuneration for participation. The subjects' characteristics and clinical data are generally included in this section of the paper (Jewell, 2011). There may also be information about their physical state or training level. Examples of subject characteristics often found in this section include the following:

Demographic Information

- Age
- Sex
- Race or ethnicity

- Sport or activity
- Competitive level
- Educational level
- Socioeconomic status

Clinical Information

- Height
- Weight
- Body mass index, percent fat, lean body mass
- Diagnosis and date of diagnosis, age of onset
- Comorbidities
- Level of function
- Medications
- Number or types of previous injuries or surgeries
- Physical activity level
- Exercise behaviors (modes, durations, frequencies, and so on)

In clinical exercise literature, volunteers may be recruited from a nursing home, assisted living facility, or hospital. This provides clues about the volunteers' level of physical function. In applied exercise literature, the volunteers may be members of a sports team, college students, or community-dwelling participants that are active or sedentary. Inclusion and exclusion criteria are generally specified in the subjects section, and as an evidence-based practitioner searching for evidence that may affect one individual, it is important to read the subjects section carefully to ensure that the population closely matches your client, patient, or athlete. A concluding statement concerning ethics committee approval for human subjects, informed consent statements, and consent methodologies for testing of minors or other vulnerable populations is always provided in this section.

Specific Procedures

In contrast to the overview of the study procedures provided at the beginning of the methods section, an exhaustive description of the testing protocol is given later in the methods. The description of the specific procedures details the protocol of each test to the extent that it could be replicated in near exactness by another investigator. Well-written methods sections present the methods in the order in which they were actually performed. In exercise science literature, such elements as the warm-up protocol before testing, the timing, and test termination criteria are specified. Training studies also describe the protocol used for exercise in the intervention group. Information is provided concerning the exercises used and the frequency, intensity, and volume of training performed by the participants. This may also include monitoring procedures. Some studies are "self-monitored"; that is, the protocol is provided to the subjects and they complete it on their own, self-reporting their adherence. This has the potential to undermine the internal validity of the study and should be considered when one is deciding on the strength of the evidence. In contrast, some training studies ensure that all sessions are monitored by a qualified or certified exercise trainer. This allows for monitoring of adherence and provides greater confidence in the outcome of the study.

The accuracy of the written study procedures is important to scientists and practitioners. As discussed earlier, a complete description of the procedures allows for replication of the study by other scientists. This can confirm or potentially refute the findings. It also allows exercise practitioners to implement the exact protocol or slightly modify it in practice. Exercise practitioners may choose to use previously validated testing procedures in their own batteries to optimize the validity and reliability of their testing. They may also desire to base their own program on a protocol supported by scientific evidence to increase the probability of a successful outcome. Unless the procedures are clearly defined, this becomes difficult for the practitioner.

Another important aspect of the methods section of a paper is the description of any potential confounding variables. In exercise science research, information related to nutrition, hydration levels, sleep, and standardization of testing conditions is crucial to the interpretation of the findings. Without this information, it is impossible to determine if the outcome measures were influenced by the independent variable alone or another, confounding factor.

Statistical Analysis

The statistical analysis section describes the mathematical methodology used to test the study hypothesis. Statistics are descriptive or probabilistic techniques that determine differences or relationships within data. There are hundreds of statistical tests; discussion of these is far beyond the scope of this book. However, novice readers working in sport performance, exercise training, or testing environments should understand several important concepts (Sifft, 1983, 1986, 1990a, 1990b).

The statistical section typically leads with a description of the tests used, which are tied to the study hypothesis. Each statistical test has a set of assumptions or necessary conditions to ensure the accuracy and appropriateness of the test. This may include information such as normality of distribution, independence of observations, and equal variance between groups. The authors should state how the statistical assumptions for the specific tests were met; unfortunately, not all authors report these data. It may be important to discuss this information with a science team member who is skilled in statistical analyses.

"Significance" in classical statistical analysis refers to the probability of reaching an erroneous conclusion. Two types of error are accounted for in testing the null hypothesis. A type I error is typically known as a "false positive," meaning that there is an error in rejecting a null hypothesis (which states that there is no difference), when in fact the null hypothesis is actually true and there was no difference (more accurately, no statistically significant difference). This often occurs when the sample size is large and therefore the statistical power is so great that statistically significant differences can be observed despite only small differences between the means. For example, a study finds that an exercise program was beneficial to produce strength increases when in fact the program was not an effective stimulus. The arbitrary cut-point for statistical significance is typically set at $P \leq 0.05$, meaning that, if the investigators find a difference, there is only a 5% probability that the finding was a false positive. The actual P values, or type I error rate, are reported for all major outcomes in the results section of a paper.

A type II error occurs when investigators report no difference (accept the null hypothesis) due to treatment when in fact there was; they accept the null hypothesis when they should have rejected it. Different from the type I error, this error is commonly made when sample sizes are small or experimental conditions are not tightly controlled and investigators accept the null, stating that no differences were observed between the group means. For most exercise science studies, this is the bigger threat, saying that no differences were found when in fact there were differences. The probability of a type II error is usually set at 20%. In other words, if there is no statistically significant difference, the authors accept a 20% chance that they are wrong. Statistical power is mathematically related to type II error (also called beta, β) and is calculated as $1 - \beta$ (e.g., $1 - 0.20 = 0.80$ or 80% power). Statistical power is rarely reported in the exercise science literature but can be easily calculated using statistical software or publically available tools via the internet. Power hinges on the difference in means, variance, and sample size and is often less than 80%. Because power is rarely reported and often not even calculated by investigators, null findings in journals may be the product of poorly powered studies (e.g., a 30%, 40%, or 50% type II error probability). The default probabilities that are used for type I and II errors (5% and 20%, respectively) imply a fourfold greater cost or severity for making a type I error; however, unlike the situation in medicine, where an ineffective treatment may be economically costly if not physically harmful, ineffective exercise interventions are often just a waste of time (e.g., minimal physiological adaptations for training time invested). Thus, some statisticians have suggested that the consequences of type I and II errors be evaluated on a case-by-case basis and that alpha (the significance level or type I error rate) and beta (type II error rate) be adjusted accordingly (Ploutz-Snyder, Fiedler, & Feiveson, 2014).

Statistical difference is important, but the entire interpretation of a study's findings should not hinge on the reported P values or statistical power. Instead, readers should examine the statistical significance in the context of the precision of measurement and the study protocol and procedures. Following is a short list with simplified definitions of important statistical terminology and tests.

Common Statistical Terminology

Sample size: The number of participants in a study *(N)* or in a group *(n)*.

Mean: The average score of a variable or group.

Median: The middle score of a variable or group.

Range: The distribution of data defined by the top and bottom scores.

Variance: The spread of scores relative to the mean.

Standard deviation: The most commonly reported calculation of variance in a research article. It is computed as the square root of the variance.

Standard error: Measurement of variance computed as the standard deviation divided by the square root of the sample size *(n)*.

Distribution: How data are spread out or lie in relation to one another. A "normal" distribution of data lies on a bell-shaped curve.

Alpha value (type I error rate): The probability of stating that there is a difference between groups when in fact there was not. Alpha or the P value cut-point for statistical difference is typically defined as 5% or 0.05, respectively.

Beta value (type II error rate): The probability of stating that there is not a difference when in fact there was. Type II error rate is typically defined as 0.20 (20%) but is rarely reported in papers.

Statistical power: The difference of 1 – beta (e.g., if beta is 0.20, statistical power is 0.80 or 80%).

Statistical assumptions: Conditions that must be met to ensure the accuracy and appropriateness of a statistical test.

Common Statistical Tests

Effect size: A simple way of defining the difference between two groups of data.

t-test: The most common test used to describe differences in two groups. If data are independent observations (e.g., different people), an unpaired t-test is used. If the data are dependent (i.e., same people tested over time), a paired t-test is used.

Chi square (χ^2): A statistical hypothesis test used to determine if data drawn from a population are consistent with a hypothesized distribution of data.

Relative risk (RR): The probability of an event occurring in an exposed compared to a nonexposed population.

Odds ratio (OR): The odds of an event occurring in someone exposed versus nonexposed.

Analysis of variance (ANOVA): A statistical test used to determine differences in more than two groups. If the groups are dependent (i.e., same people), an ANOVA with repeated measures is used. An ANOVA computes an F ratio, which indicates a difference in groups.

Multifactorial ANOVA: A statistical test used to determine differences when there is more than one independent variable.

Analysis of covariance (ANCOVA): A method that controls for a group of covariates or random variables within a sample (e.g., sex, body mass index, age).

Post hoc test: An analytical test used to determine differences between groups after a global test indicates significance. For example, if an ANOVA F ratio is significant, a post hoc test such as Tukey's honestly significant difference (HSD), Bonferonni procedure, or Newman-Keuls method is used to determine pairwise group differences.

Correlations: The relationship or association between two variables. Variables can be positivity correlated; i.e., when one variable increases, so does the other. They can also be negatively correlated; i.e., when one variable increases, the other decreases.

Regression analysis: A statistical method used to determine relationships between variables. A correlation is a simple regression analysis between two variables. Multiple regression computes combined and independent correlations between many variables. Regression techniques can be linear or nonlinear.

Results

The results section contains the summarized data resulting from the study procedures (Starck & Fleck, 1982). This section of the paper specifies the findings from the study methods related to the hypothesis. The data may be in text, figures, charts, or tables.

Although there may be some redundancy between the written component of the results and the figures, charts, and tables, each item typically contains novel information, and it is thus important to read and interpret each. Along with conveying the data in a summarized form, the results section also discusses the statistical significance of group comparisons. Although it is uncommon to do so, authors should also report the statistical power in the results section of the paper for nonsignificant differences (Moher, Dulberg, & Wells, 1994). Results sections should contain only the data, with no accompanying explanation of why something is or is not different. It is simply the data that are presented; arguments and explanations as to "why" are left to the discussion section of the paper.

Discussion

The discussion section immediately follows the results and is typically the longest section. The purpose of the discussion is to frame the findings of the current research investigation in light of previously published research. If the primary purpose of the study was replication, the discussion section may consider in detail how the data confirm or refute the findings of previous investigators. If the findings are at odds with others', this section may present reasons why the results differ from those of previously published works.

Novel findings are also discussed in the context of previously published works. The discussion may reveal how novel findings of the current investigation add to the body of knowledge on a given subject. Principally, the discussion is the authors' synthesis of their results with previous findings and explanation of how their work adds to what is already known about a phenomenon.

Research evidence always has limitations, weaknesses, or flaws. These weaknesses may be the result of study design, subject recruitment, statistical power, lack of blinding, or confounding factors, among others. Even when a research scientist works diligently to control for outside influences, human subjects research always has minor inadequacies. For example, in exercise training studies, a research team may closely monitor the exercise protocol for an intervention group to ensure that they closely adhere to the prescribed intensities and volumes of training, but they cannot control what happens outside of the laboratory. Typically, information such as diet, medication use, and additional physical activity are self-reported. These data are prone to error and may obfuscate true intervention effects (Suchanek, Poledne, & Hubacek, 2011; Trabulsi et al., 2006). There are no perfect studies, and any limitations in a study should be integrated into the speculations or generalizations put forth within the discussion. This paragraph highlights potential flaws in the research investigation and how the research team attempted to ameliorate them.

Clinical or Practical Applications

When reading a journal article, the exercise practitioner's goal is to determine, "What does this research mean for practice?" Some journals require authors to include a clinical or practical applications section. This section explains how the results may influence daily practice. Sometimes the data from a research investigation can be rather disconnected from typical industry practice techniques. Thus, this section of the paper is the authors' opportunity to make the case for the real-world applicability of their findings.

Conclusions

The final section in a research manuscript (other than references) is the conclusion; in many journals, the conclusion is simply the final paragraph of the discussion. Key findings from the research investigation are once again stated and tied back to the introduction section. Conclusion sections are short, generally no more than one or two paragraphs. Included in this section are often a few key questions that the novel research findings have raised and an indication of future research directions.

References

The reference section of a paper is last, immediately after the acknowledgments. Many journals limit the number of references, although this number varies widely. In a narrative review or a consensus statement, journals allow more references because the nature of a review is to examine relevant literature. Formatting differs by journal, but the primary identifiers of a work (i.e., authors, title, journal, volume, year, and page numbers) are always included; this allows readers to find the cited papers and to examine them in more detail.

STEPS TO CRITICALLY READING RESEARCH

Reading and interpreting research may be one of the most important tasks of an evidence-based practitioner. It is important to develop the skills necessary to identify the key findings and to disregard the superfluous information. Although critical appraisal is a topic left to a later chapter, next we present some practical guidelines for reading research evidence (figure 5.6).

Read the Title and Abstract

The title of a research article can provide an important clue to the most important findings in the manuscript. But you should never read only the title and abstract. Some journal titles explicitly state the major finding of the paper. In a 1989 publication in *Archives of Neurology*, an article titled "Focused Stroke Rehabilitation Programs Improve Outcome" appears on pages 700-701 (Reding & McDowell, 1989) (figure 5.7). From the title alone, without reading the paper or the abstract, the reader learns that the investigators determined that a stroke rehabilitation program was effective in improving clinical outcomes. However, the very next article in the same volume of *Archives of Neurology* is titled "Focused Stroke Rehabilitation Programs Do Not Improve Outcome" (Dobkin, 1989) (figure 5.7). Which article is correct?

When the evidence-based practitioner relies on the title or abstract alone, he can make significant mistakes in the interpretation of research evidence. Also, examples such as the one just presented clearly show that research can be contradictory. The practitioner should read the title and abstract to determine whether the paper is relevant to the evidence-based question. If it is, he should proceed to a more careful reading and analysis to determine if the study methods and results support the findings stated in the abstract and title.

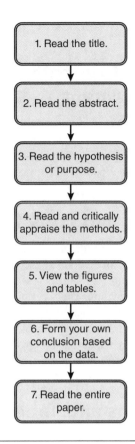

Figure 5.6 Practical method for reading a research paper and identifying and disregarding the authors' potential personal bias in interpretation of the data.

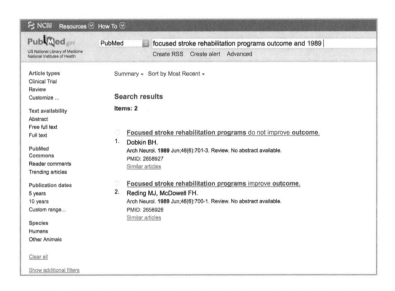

Figure 5.7 Two studies with directly contradictory titles that appeared consecutively in 1989 in *Archives of Neurology*.

Reprinted from PubMed.gov.

Read the Purpose, Construct, Theory, or Hypothesis Statement

After reading the title and abstract to determine the relevance of the study to the evidence-based question, the practitioner can skip to the final statement in the introduction that presents the study purpose, construct, theory, or hypothesis statement. This should provide the information needed to determine how the study should be designed to best answer the research question.

Read the Methods and Look for Flaws

The methods section is the most important part of a paper to read carefully and critically. It is from the methods section that an evidence-based practitioner can determine many of the potential biases and confounding factors associated with the study. This is also the section of the paper that determines the level or strength of evidence (chapter 8) provided by the research. As authors critically appraise their own research, they should carefully consider the study design, testing protocols, devices used, training and intervention protocols, statistical analyses, and study group allocations. If the authors are unclear in their presentation of any component of the methods, this should be a "red flag" to the reader and could potentially be a fatal flaw in the design.

Examine the Graphs and Tables

Graphs and tables are simple tools that authors use to draw out and present the most important results. In a manuscript, tables and graphs should be able to stand alone. The evidence-based practitioner should be able to read the caption and evaluate a figure or table without reading the results section and determine the major findings. While studying the graphs and tables, the reader should pay special attention to the variation and potential outliers. Large standard deviations or errors indicate widely divergent responses from participants; also, one or two responses can drive the summary statistic and create a false finding, or a result that is not representative of the population (figure 5.8). Similarly, outliers in correlations can leverage the regression line and disproportionately influence the association (figure 5.9). Statistical significance, which is typically provided on a table or graph, should be interpreted within the context of a critical examination of the data.

Arrive at Your Own Conclusions Before Reading the Results Section

After critically examining the graphs and tables, read the results section of the paper. Note the statistical statements, but carefully study the means and standard deviations. In addition to looking for statistical significance, ask yourself if the reported differences would result in a meaningful difference in practice (a "clinically meaningful" difference). Also, ask yourself if the study results support or refute the study hypothesis. Are the findings amenable to practical implementation with your clients, patients, and athletes? Form your own conclusions about the data before reading the authors'.

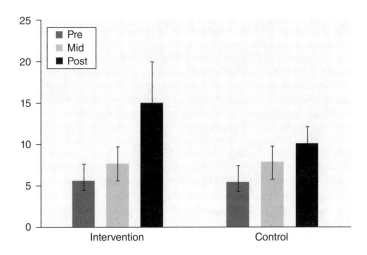

Figure 5.8 Theoretical difference in means showing changes in scores between two groups during pre-, mid-, and posttesting. Notice the large standard deviation in post-testing for the intervention group influencing the pairwise and between group differences. This could be due to outliers that are not reflective of true population responses.

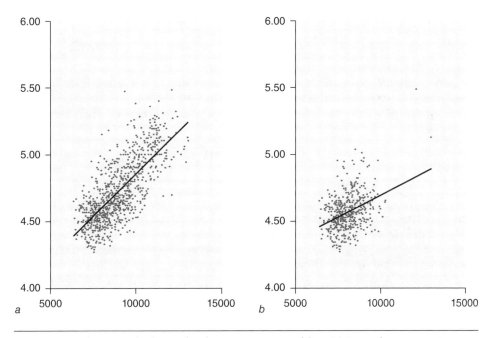

Figure 5.9 Theoretical relationship between two variables. *(a)* A graph representing a strong relationship. *(b)* A poor relationship that is being leveraged by two outlier data points.

Go Back and Read the Entire Paper

After carefully reading, critically appraising, and reaching your own conclusions related to the data, as an evidence-based practitioner you should then go back and read the entire paper, including the introduction and discussion sections. This will help to provide historical context to the paper and also show how the results fit within the context of other published data. Reading these sections last is recommended because they contain the authors' opinions regarding the data. It may be advantageous to form your own opinion, based on the data alone, before reading the authors' opinion. While reading the entire paper, practitioners can generate a list of referenced articles that merit further investigation, then acquire and critically assess them using the same process.

Consider the Findings in Light of Other Published Works

No matter how well a study has been designed, it is important to never form strong opinions based on a single study. Always look for alternate explanations and for supporting and contradicting work. When possible, use multiple papers to form viewpoints and make practice-based decisions. It is crucial to understand that a single paper is part of a larger story being told by the complete body of literature. A single paper could contain a random finding (statistical error) or could have been influenced by an unreported confounding variable. Replication is crucial to ensuring the legitimacy of a finding. If the paper is the first to have been published in a particular area or is investigating a new phenomenon, it is possible that future studies may strongly support the findings; it is also possible that future research may never confirm the results. Thus, a study that has not been replicated, no matter how strong the design, should be cautiously interpreted. Even after you have decided to implement research evidence in practice, continue to search for evidence and do not be afraid to change when better evidence emerges.

CONCLUSION

Research evidence is a critical cog in the evidence-based philosophy of decision making. Not all exercise practitioners are scientists, but all must be good consumers of research. Those new to reading research can feel overwhelmed and confused since they are in effect reading a foreign language. It can help novice readers navigate this foreign world to know that scientific manuscripts are structured in a consistent format that is designed to " tell a story" of how the research question arose, describe the methods used to answer the question, explain the results obtained, and discuss the findings in the context of other research. Following the reading tips provided in this chapter will help evidence-based practitioners critically appraise the literature and reach their own conclusions concerning the data. Reading research is a skill that develops with practice. Evidence-based practitioners who want to improve the outcomes of their clients, patients, and athletes should immerse themselves in the culture and language of research and expect the process to become easier and more fruitful with time.

PART II

The Six Steps of Evidence-Based Practice

Anyone who has ever assembled a new piece of furniture knows the difficulty and frustration of this seemingly simple task. A new television stand when fully assembled may be 1 meter tall × 1 meter wide × 0.5 meter deep, but it comes in a box that is one-fourth its final size. The box can be full of nonspecific wooden boards, cams, bolts, pegs, small tools, and other assembly materials. When all of these components are scattered across the living room floor, there is only one good place to begin—the instruction manual. A well-organized and clearly written instruction manual can make the assembly task easy, providing simple step-by-step instructions that, if followed, will in a few hours yield a functional new piece of furniture.

The purpose of part II of this book is to provide the reader with a step-by-step instruction manual for evidence-based practice. We present an in-depth view of the five classical steps employed in the field of medicine and then describe a new additional step for the exercise science field. At the end of this section, readers should clearly understand how to ask an answerable question; use search tools such as PubMed and Google Scholar to find evidence; rank or define the strength of the evidence; implement the evidence in an athlete, client, or patient; and then confirm that the implementation was beneficial. Readers should also understand the importance of reevaluating questions, as new scientific evidence emerges constantly, further supporting or refuting exercise program theories, devices, and nutritional supplements.

Chapter 6

DEVELOPING A QUESTION

Learning Objectives

1. Identify the types of evidence-based questions.
2. Provide examples of scenarios in exercise science in which questions about evidence-based practice arise.
3. Discuss the five components of an evidence-based question.
4. Provide examples for how phrasing or rephrasing a question may ultimately affect the answer to the question.
5. Provide examples of well-worded and poorly worded evidence-based questions.

Many exercise professionals complete a long course of formal education. After primary and secondary study during childhood and adolescence, students go to a university for a 4-year undergraduate degree that is often followed by several years of graduate studies. Some college programs allow students to complete internships; this results in real-life experiences that add to academic knowledge. Mentors pass their experience on to the students with whom they work to ensure that they are well prepared to enter the exercise science field.

At the end of their formal education, students typically seek employment. Excited to begin their working career and to use the knowledge gained over 16+ years of education, graduates enter the work force. Invariably, within the first week or even the first day, new graduates come to a startling realization. Despite their rigorous formal education and internship experience, they do not know everything there is to know. In fact, experienced practitioners will echo the thought of the great philosopher Socrates: "True knowledge exists in knowing that you know nothing." On a routine basis, exercise practitioners are presented with scenarios or questions for which they do not have the answer. In addition to novel situations, new treatment approaches and prescriptive techniques are continually coming into use, raising questions about their efficacy. It is always prudent to understand *why* certain techniques are used and to question whether or not a new approach is the best option for the client, patient, or athlete. The answer may increase our confidence about implementing the technique, compel us to investigate other options, or perhaps convince us to abandon a technique due to lack of evidence or evidence demonstrating its ineffectiveness. In any case, the key to long-term learning and effectiveness as a practitioner is the ability to ask critical questions.

QUESTION DEVELOPMENT AREAS

Learning to ask good clinical or practical questions is a key component to becoming an effective evidence-based exercise scientist or practitioner. The answer to a question can be profoundly affected by the way it is asked or phrased. In medicine, it has been suggested that questions arise from at least six different problem-solving areas: intervention or therapeutic techniques, etiology, diagnosis or testing, prognosis, meaning, and economics (Sackett, Straus, Richardson, Rosenberg, & Haynes, 2000). Of these six areas, three are relevant to exercise science: interventions, testing, and economics. These problem-solving areas drive practitioners to ask questions concerning exercise philosophies, program designs, exercise for special populations, exercise devices, and nutritional supplements.

Program Interventions or Exercise Techniques

Exercise interventions (i.e., program designs) and techniques make up perhaps the most common problem-solving area with which an exercise scientist or practitioner is confronted. We may be motivated to ask if a certain set or repetition scheme is really the most effective way to increase strength or power in an individual. We may also seek to determine if an exercise routine, device, or method is appropriate for a particular population. For example, is high-intensity (load) strength training effective for improving strength and power in people who are elderly? Questions may also arise concerning the safety of techniques. It may be that a technique is effective but not safe for a particular population. Additionally, one might ask what the potential for improvements in muscular size and performance is for a given patient, client, or athlete. The answer would in part dictate goals and types of training programs used. Thus, beyond just the efficacy of a program, secondary and tertiary questions abound when we are dealing with exercise training in an individual.

Occasionally, new interventional techniques, philosophies, or devices surface in the exercise science discipline, and an evidence-based practitioner must be prepared to ask and answer questions to determine whether to implement them in practice. The following are questions about a few recent trends that warrant investigation:

- Are high intensity and volume training programs an effective and safe method to improve physical fitness in untrained middle-aged individuals?
- Does running in minimalist shoes improve running efficiency?
- Is high-intensity interval training more effective than steady-state aerobic conditioning to improve peak aerobic capacity in healthy young individuals?
- Is stability or "core" training more effective than traditional heavy, periodized resistance exercise to develop trunk muscle strength in trained individuals?

All these questions pertain to relatively new concepts that have emerged in exercise science. Some have received widespread acceptance in the field, but is there strong evidence to support their effectiveness or safety?

Testing Techniques

Questions may also arise concerning the appropriateness of testing, assessment, or screening techniques. Each of these is extremely important for the prescription of exercise. When developing an exercise program, most practitioners perform some type of initial testing. The initial testing allows the practitioner to determine a baseline level of fitness, track progress over time, and assess individual strengths and weaknesses in order to develop a more focused, customized program. Typically, the tests assess some construct associated with functional performance. If the prescription is for a sport athlete, ideally the construct is associated with performance, and improvement in the construct is associated with more successful play. However, function is not relegated only to sport; performance may also be associated with more common tasks. If you are writing a prescription for an individual who is elderly, perhaps the functional task targeted for improvement is stair climbing capability or the ability to rise from a chair or toilet. In this case, the initial assessment should include tests shown to predict stair climbing or chair rise performance, and the subsequent training or rehabilitation program should improve performance in the test and task. It is essential that tests be selected in response to a question and based on supporting research evidence.

Testing may also be used to screen for increased risk of disease or injury. Thus, questions may be developed that concern the association between the score on a certain test and disease. If approached by a client who is apparently healthy and is interested in improving general health and fitness but who has a family history of cardiovascular disease, it is prudent for the exercise practitioner to search the literature for exercise tests that predict increased probability of cardiac events. It may also be important to develop questions and search for literature concerning the safety of tests and to choose assessment tools that are safe for the population of interest. In other words, if two tests have data suggesting that they are equally effective at screening for disease but one test is inherently riskier, choose the safer test.

Economics

Another factor that influences the development of an evidence-based question is budgetary concerns or economics. Interestingly, this has been a point of contention in the medical community (Law & MacDermid, 2008; Vos, Houtepen, & Horstman, 2002). One of the by-products of the evidence-based medicine movement is the use of evidence to make decisions concerning funding of certain treatments or interventions. In some cases, insurance companies have demanded evidence to support the use of treatments and have stopped covering some that have little or no evidence to support their effectiveness. This has led to resentment within the medical community in that it minimizes the value of physicians' professional experience and diminishes their role in treatment selection. Some physicians argue that although there may not be a scientific paper to support a treatment's efficacy, they have observed its effectiveness many times in practice.

While we acknowledge the potential dangers of this economically driven approach to evidence-based practice (EBP), it is important to note the potential for positive economic effects as well. Law and MacDermid and others have suggested that there are at least five economic factors from which evidence-based questions can be derived: cost consequence, cost minimization, cost-effectiveness, cost utility, and cost-benefit (Law & MacDermid, 2008). Three have obvious implications for exercise science practitioners.

Cost consequence analysis is a technique that compares the economic impact of two different treatment methods or the fiscal consequences of treating versus not treating. This sort of analysis may motivate some individuals to seek preventive or early intervention treatment. Consider, for example, someone who has been diagnosed with insulin resistance, often a precursor to diabetes. This person would be seeking the best course of action with respect to her health but would in most cases also have financial concerns and limitations. A cost consequence analysis could be performed on the following question:

What is the economic impact of a medical regimen to control blood glucose versus working with an exercise practitioner to make behavioral changes to improve blood glucose tolerance and insulin sensitivity?

A second type of economic analysis has been termed cost minimization (Law & MacDermid, 2008). Cost minimization analysis compares two or more treatment options with similar reported outcomes to determine the least costly intervention. Suppose you are looking to purchase a new piece of equipment to measure peak aerobic capacity in a clinic. After performing a rigorous product search, you narrow your choices to two metabolic carts that fit your requirements. After searching the literature, you find that there are validation studies supporting the accuracy of both devices. Assuming that there are no differences in maintenance costs and so on, a cost minimization analysis would lead you to purchase the less expensive device.

Cost-effectiveness is perhaps the most important and relevant issue for exercise practitioners. In most professional organizations, budgetary resources are limited and managers are forced to make decisions based on the cost-effectiveness of a purchase or treatment. If a new device or treatment option is expensive, it must provide a commensurate impact for the client, athlete, or patient. Purchasing an expensive piece of equipment warrants the development of an evidence-based question and then a search for information concerning effectiveness. If the device is effective but not significantly more effective than options already available to the organization, the purchase should receive careful consideration. Another possibility is that a treatment technique or device may not be practical in that its use requires additional resources, such as significant professional oversight. An organization should consider maintenance costs, staff training, staff supervisory requirements, and so on; high costs for these factors increase the necessity that a device produce superior results for clients, patients, and athletes. To state this simply, devices that have high costs should provide remarkable results. Ultimately, research evidence should drive these important decisions.

TYPES OF QUESTIONS

In the evidence-based literature, two broad types of questions have been described: background and foreground or direct questions. The primary determinant of the type of question to be asked is the type of information sought. The way the question is phrased will increase or decrease the quantity and relevance of information obtained in a search. In both cases, it is important to phrase the question in an objective manner so as not to bias the search or influence the type of information the search returns.

Background Questions

Background questions are broad and are often used to obtain general information on a topic or subject. Exercise practitioners may ask a background question when they simply do not know where to start. These types of questions are typically mechanistic or epidemiological. In other words, they try to answer who, what, when, where, why, and how. Suppose an exercise practitioner is approached by a cancer patient who asks him to develop an exercise regimen. The patient is currently undergoing treatment for leukemia, and her oncologist has explained that the disease, chemotherapy, and resultant changes in nutrition and activity patterns may result in cachexia. The physician suggested hiring an exercise practitioner to design and implement an exercise program during treatment. The exercise practitioner has heard of cancer cachexia, but his knowledge of the condition is minimal. He may develop background questions to help him gain a rudimentary understanding of the condition and to better formulate direct questions. These are examples of background questions that could be asked:

- **What is cancer cachexia?**
- **What are the physiological consequences of cancer cachexia?**
- **How is cancer cachexia diagnosed?**
- **What are the biological mechanisms underlying cachexia that lead to muscle strength and power loss?**
- **Why are some patients cachectic in response to treatment and others not?**
- **At what stage of treatment do patients often become cachectic?**

These questions do not focus on a treatment or the effectiveness of an intervention, but the answers may help the practitioner to better focus interventional questions. The background search on these questions is likely to suggest that cancer cachexia results in a loss of muscle mass, strength, and power and an increase in fatigue. This may influence the evidence-based practitioner to investigate treatment options known to improve these physiological constructs in apparently healthy individuals.

The answers to background questions may be found in peer-reviewed and non–peer-reviewed sources. Textbooks are often solid resources for general, stable information. They may also be helpful in identifying biological mechanisms. However, when searching for information, practitioners should keep in mind the publication date of a textbook and always search for newer peer-reviewed information when available. Expert opinion or consultation is another valuable source for answering background questions. In the current example, perhaps the best source of general information concerning cancer cachexia would be consultation with the referring physician.

Foreground or Direct Questions

In contrast to background questions, foreground or direct questions address a focused aspect of treatment, intervention, or testing (Amonette, English, & Ottenbacher, 2010). In addition, they typically possess a quantifiable component and lead to a decision with respect to an intervention. A direct question typically includes four components that can be remembered by the acronym PICO: population or patient, intervention, comparison, and outcome. A fifth component, time, is sometimes added to the question (and acronym). Here is an example of a direct question:

In a middle-aged patient with cancer cachexia, is resistance exercise more effective than sedentary behavior to maintain skeletal muscle strength over a 12-week or greater treatment period?

These are the PICO-T components of this question:

Population—middle-aged patient with cancer cachexia
Intervention—resistance exercise
Comparison—control or typical sedentary behavior
Outcome—muscle strength
Time—12-week or greater treatment period

COMPONENTS OF A FOREGROUND QUESTION

The phrasing or definition of each component can lead to a dramatically different answer; therefore, it is important to carefully consider each one when formulating a question. Table 6.1 gives an overview of each of the five components of a foreground question. Sloppily or inaccurately defining these components can lead to erroneous conclusions. In the next sections, we review each of these components in more detail.

Population or Patient

The population, patient, or client can significantly influence the answer to a question and needs to be carefully considered when one is phrasing a question. Because most evidence-based questions are written specific to a single individual or a group of indi-

Table 6.1   Importance of the Five Components of a Foreground Question

Component	Importance
Population	Different populations may respond differently to a stimulus. Ideally, the research population should match the practitioner's client, patient, or athlete population.
Intervention	When possible, the practitioner should search for evidence using the exact intervention. Although some generalization is possible, moving too far from the target intervention can lead to incorrect answers.
Comparison	Compare to the gold standard or standard-of-care treatment when possible. Attempt to use only a pure control comparison when the standard of care is to do nothing.
Outcome	Carefully define your outcome measure to ensure that it tests the construct most important to functional outcomes in your client, patient, or athlete.
Time	Time may be important when measurement is limited by the assessment tool or when an intervention works but is slow to cause change in a particular test.

viduals, it is important that the question targets research populations that are similar. Although research conducted in different populations may be of some use, the results should be interpreted with caution because population characteristics (e.g., age, sex, training status) can have a dramatic effect on outcomes.

Chronological Age

The age of the client is perhaps the most obvious characteristic to match. Age can significantly influence the acute effects of exercise, including altered hormonal (Hakkinen, Pakarinen, Newton, & Kraemer, 1998; Kraemer et al., 1998), immune (Ceddia et al., 1999; Shinkai, Konishi, & Shephard, 1998), cellular signaling (Haddad & Adams, 2006), and motor responses, among others. Each of these factors may contribute to differing adaptation rates in older compared to younger individuals (Kosek, Kim, Petrella, Cross, & Bamman, 2006) and may necessitate altering the exercise dose based on age (Bickel, Cross, & Bamman, 2011). Additionally, individuals who are older may be more deconditioned than younger people. Thus, a given exercise stimulus applied to both younger and older individuals may result in similar responses, but the magnitude or rate of response may be noticeably different. Consider the following question:

In individuals who are elderly, does heavy resistance training significantly improve lower body strength compared to sedentary, nonexercising controls?

A cursory review of the literature demonstrates that heavy resistance training can have a profound effect on lower body strength in people who are elderly (Hunter, McCarthy, & Bamman, 2004). Evidence consistently supports lower body strength changes from 50% to 100% in 10 to 12 weeks of training three times per week in individuals who are older (Charette et al., 1991; Ferketich, Kirby, & Alway, 1998; Frontera, Meredith, O'Reilly, Knuttgen, & Evans, 1988). One study reports a 257% increase in strength over a 10-week period (Singh et al., 1999). If the population in the foreground question were changed, the search would return different results. For example, if we changed the population to young, college-aged individuals, the research would strongly support resistance exercise as a profound stimulus to improve lower body strength, but the magnitudes of change would be smaller. Therefore, it is important to closely define the population in order to best assess the effectiveness of a stimulus and to more completely answer the direct question.

Exercise Age or Experience

Another important variable to define in the population is the training age or training experience of the population. Similar to chronological age, experience can significantly affect the physiological response to a stimulus. This is an often cited criticism of EBP in some disciplines. Sport scientists often argue that much of the published literature on exercise adaptations uses young, recreationally trained college-aged students and that the results from these studies do not necessarily reflect the responses of elite athletes (Stone, Sands, & Stone, 2004). Indeed, training experience can significantly affect the response to exercise (figure 6.1). If an individual has been consistently training for a period of time, the magnitude of response to a stimulus may be significantly less than for a novice. Moreover, it could be that different training volumes, intensities, and frequencies result in differential responses in novices versus experienced exercisers. When

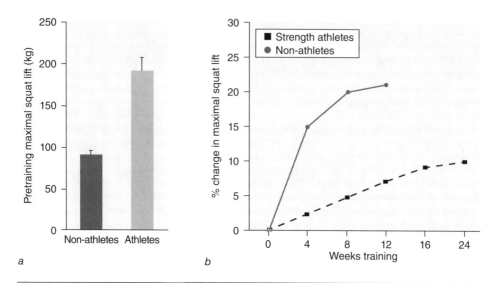

Figure 6.1 Differential responses between strength-trained persons and nonathletes.

Reprinted, by permission, from K. Hakkinen, 1985, "Factors influencing trainability of short term and prolonged training," *NSCA Journal* 7(2): 32-37.

possible, it is important to closely match the training experience of the population in question to the population represented in the research so as to provide the strongest, most relevant evidence. If research evidence conducted in the target population is not available, findings from other populations can be implemented, but close monitoring is warranted to determine their effectiveness in the target population.

Sex

Sex is another factor that can affect training outcomes. As with training experience and age, some physiological responses are different in men compared to women. These differences may affect the expected outcomes of a training stimulus. One example in the medical discipline lies within pharmaceutical research. The lack of sex-specific research has been an issue with regard to evaluation of drugs and the appropriate doses for men versus women. For some drugs, the ideal dose may be different in men and women, but there is a lack of information on these dissimilarities in the published research. Some contend that this could lead to negative side effects from inappropriate dosing.

A more relevant example in exercise science may lie in the "dosing" of resistance exercise between men and women. Although the general structure of a resistance exercise program should not be substantially different between men and women, some differential physiological responses to training stimuli suggest that small differences might be beneficial to obtain adequate overload stimuli. For example, it is well known that men and women have different resting levels of testosterone and that the elevation of testosterone in response to a training stimulus differs between sexes. Because testosterone promotes skeletal muscle anabolism and aids recovery from intense muscular work, it could be that there are subtle differences in responses between males and females to the same exercise protocol. Thus, it is important to search for sex-specific research when it is available in the literature.

In addition to differential physiological responses, women and men exhibit different orthopedic injury rates, muscle firing patterns, and biomechanics (Cowley, Ford, Myer, Kernozek, & Hewett, 2006; T. E. Hewett, Ford, & Myer, 2006; Hewett, Myer, & Ford, 2006; Myer et al., 2009; Renstrom et al., 2008). Thus, the findings of a study using men may not be appropriately extrapolated to women; when possible, evidence-based practitioners should look for sex-specific studies to answer questions regarding their client population.

Disease Condition or Disability

A very important factor to consider when developing a direct question is the presence of disease or disabilities. The pathology of certain diseases may influence exercise responses and potentially blunt positive effects or magnify risks. Proper definition of the disease condition may be especially important when one is considering the activity-specific risks of an intervention. In other words, a stimulus that is safe for an apparently healthy individual could be dangerous for an individual with diabetes. Suppose you are designing a program for someone who is morbidly obese. His physician has diagnosed him as borderline hypertensive and prediabetic and strongly supports the initiation of a comprehensive weight loss program. You pose the question:

Is overground running a safe and effective method to promote weight loss?

Unless the population has been defined first, the answer to this question would be an overwhelming "yes!" Running overground is an activity that requires high energy expenditure and, when performed regularly with mild hypocaloric dietary intake, results in weight loss. Overground running is also relatively safe. Although it is possible for a runner to trip, fall, or develop an overuse injury, for example, given a proper progression, the risks are relatively small and the benefits are large. However, if the population of interest is persons who are morbidly obese, risks arise that would not have been evident for normal-weight or slightly overweight individuals. Due to excess mass, there is an increased mechanical stress on joints that may increase the potential for injury (Felson & Chaisson, 1997). Additionally, obesity impairs thermoregulation, so running in a hot environment may be problematic (Bar-Or, Lundegren, & Buskirk, 1969). Ventilation is more difficult for people who are obese (DeLorey, Wyrick, & Babb, 2005), and they may be more prone to losing their balance and less able to correct a fall when running (Corbeil, Simoneau, Rancourt, Tremblay, & Teasdale, 2001). Hypertensive individuals may have an increased risk of a cardiac event. Each of these factors may influence the exercise practitioner to initially implement a safer method of caloric expenditure.

Although this example may be obvious to most exercise practitioners, other disease or disability states and their associated risks may be more subtle. Carefully screening clients and defining risks must be a priority for the safety of the client. This will also help to ensure that the evidence-based practitioner is searching for the most relevant information to improve the health and fitness of the client.

Intervention

Searching for information on the efficacy of an intervention is the primary reason most practitioners engage in the evidence-based process. When applying an exercise stimulus, intervention, or protocol to an individual, practitioners should ensure that it has a high probability of success. Thus, defining the intervention is the key step in developing a direct question; sometimes this is easy, as the emergence of a novel intervention is often the catalyst for an evidence-based search. Exercise practitioners may have a method, exercise mode, or programmatic philosophy that they regularly employ. They then learn of a new intervention through a colleague, seminar, professional journal, or catalogue and ask themselves whether this new intervention could be helpful to their clients, athletes, or patients. Thus, a question is formed and evidence is sought to support or refute use of the novel intervention.

Defining the intervention can be frustrating at times. Often, novel interventions have little research evidence to support their effectiveness. This is usually evident within product development. Ideally, a product manufacturer would develop a concept; build a prototype; use independent labs to test the effectiveness of the product; redesign based on findings; test the product again; and then begin mass production, marketing, and sales. Because originality and being first to market are extremely important for sales, this process is almost never followed. Instead, a product appears on the shelves, people begin using it, the scientific community takes notice, and then experiments are performed that either validate or invalidate the product. This can be frustrating to practitioners searching for sound information on which to base their decisions.

When defining the intervention, it is important to be as specific as possible but flexible when there is no evidence regarding the intervention. When searching for evidence for a product, first look for research studies completed on that specific product. If such studies are not available, look for research on similar interventions or products. Although you may not be able to perfectly extrapolate the results from these studies, they will provide some initial information to guide decision making.

The intervention in a direct question may also be a protocol or protocol technique. Questions can be formed to help define the appropriate exercise intensity, volume, duration, frequency, or any of the acute programming variables associated with exercise prescription. Although position stands, textbooks, and expert opinion are good starting points for determining these variables, more specific questions in defined populations must be developed to better assess the effectiveness of an intervention in individual clients, athletes, or patient populations.

As a side note, it is important for evidence-based practitioners to assess the evidence supporting techniques that are routinely used in practice. Exercise practitioners can be caught in a professional rut, using techniques and protocols simply because they are familiar. Searching for evidence to determine a protocol's effectiveness may confirm or increase one's commitment to using a technique; it may also cause one to reconsider its use or to search for a better alternative.

Comparison

Along with defining the intervention, it is equally important to carefully choose a comparison intervention. The ultimate goal of the direct question is to determine the effectiveness of an intervention relative to another treatment option. When one is investigating a novel intervention, the comparison point is generally the current gold standard treatment. The comparison intervention can significantly influence the answer to the direct question. Selecting the wrong comparison intervention can inflate the effectiveness of the intervention in question and potentially lead the evidence-based practitioner to make poor exercise treatment decisions. Consider the following foreground question:

In adults who are older, is whole-body vibration more effective than sedentary behavior to improve lower body strength?

To answer this question, the evidence-based practitioner would search for research that implemented a whole-body vibration exercise protocol in adults who were older. Ideally, the study would include a group randomly assigned to receive whole-body vibration and a matched group randomly assigned to receive no treatment and to continue sedentary behavior. Indeed, there is evidence supporting the use of whole-body vibration to improve strength in adults who are older compared to sedentary behavior. Does this mean that whole-body vibration is the best treatment to improve strength in adults who are older? What if the question were rephrased such that the comparison intervention was heavy resistance exercise?

In adults who are older, is whole-body vibration more effective than resistance exercise to improve lower body strength?

To answer this question, the evidence-based practitioner would search for research comparing strength outcomes from whole-body vibration and heavy resistance exercise training. The ideal comparison study would include subjects randomized to receive each treatment with consistent outcome measurements before and after the study. One could use a study by Roelants and colleagues comparing the effects of 24 weeks of whole-body vibration training with traditional resistance exercise in 89 women between 58 and 74 years of age (Roelants, Delecluse, & Verschueren, 2004). The women were randomly assigned to either a vertical vibration group, a resistance exercise group, or a control group that did not exercise. Muscle strength was assessed using isometric and dynamic knee extension force production tests. Static strength was measured at 50° of knee flexion; dynamic strength was measured at 100°/s using an isokinetic dynamometer. These tests were completed at 0, 12, and 24 weeks of training. The vibration group completed static and dynamic exercises on the plate with oscillatory amplitudes ranging from approximately 2 to 4 mm of peak vertical plate displacement and frequency ranging from 35 to 40 Hz. Resistance exercise subjects completed a progressive weight training program using machines in a community center. Over the entire study period, the subjects completed two or three sets of 8 to 20 repetitions using leg extension and leg press exercises. No periodization protocol was reported by the investigators. The control group did not engage in any exercise over the 24-week period but continued to participate in their normal activities.

The Roelants study was designed such that it answered both direct questions: In adults who are older, is whole-body vibration more effective than sedentary behavior and traditional resistance exercise to improve lower body strength? The study showed that the control group had no increase in strength during the 24-week study. However, the whole-body vibration and resistance exercise groups both improved strength over the 24 weeks, and there was no difference between groups in the magnitude of change. In other words, whole-body vibration is more effective than no exercise to improve strength, but it is not more effective than traditional resistance exercise.

It is important to note that the findings from this study are specific to the protocols and population used. One might argue that resistance exercise at an 8- to 20-repetition maximum (RM) is not the most effective method to improve strength; in fact, this is certainly the case. Might there have been an even greater effect from resistance exercise training if the investigators had used heavier loads or periodized their resistance training plan? Also, because the investigators did not match subjects for baseline strength or activity levels before randomization, these factors may have confounded the results.

Often the perfect comparison study does not exist. Instead, the practitioner may find studies comparing the novel intervention to control and a different study comparing the gold standard condition to control. This is frequently the case when one searches the literature and is among the criticisms often raised by opponents of the evidence-based approach: There just is not enough good clinical research to make truly informed decisions. In rebuttal, great clinical or practical research on a topic often exists; it just is not all in one convenient package. When evidence-based practitioners compare the findings of two different studies, they should also carefully compare the methodologies. Practitioners should search for studies with the same outcome measures, study durations, frequency of training, subject age and sex, and so on to ensure a valid comparison between studies.

Perhaps the easiest way to compare two studies is to simply look at the mean percent change between two studies. One must be cautious when doing this as it does not account for variability in the data. The preferred method to mathematically compare two studies that employ similar outcome measures and are closely matched in conditions is to calculate an effect size. Although the concept of effect size emerged from the social psychology literature and some argue it may not perfectly apply to biological sciences, it is an easy and convenient way to mathematically view differences in groups while accounting for variability. There are several ways to calculate effect size, and one of the simplest yet most effective is Cohen's d. Cohen's d accounts for the magnitude of change between a pre- and postmeasurement relative to the variability in the entire sample. It can be calculated by determining the difference in means divided by the combined standard deviation. The following criteria have been suggested for small, medium, and large effect sizes: small effect (0.20-0.49), medium effect (0.50-0.79), large effect ($\geq$ 0.80) (Riffenburgh, 2012).

Suppose we are comparing the findings of three different studies that used different interventions. The studies all used the same outcome measure, isometric knee extension peak torque, and studied similar subjects over similar time periods. The studies report the following findings: intervention A (pre = 118 ± 15 N·m ; post = 128 ± 23 N·m), intervention B (pre = 135 ± 23 N·m ; post = 155 ± 22 N·m), and intervention C (pre = 175 ± 45 N·m; post = 205 ± 58 N·m). A glance at the data shows that interventions A, B, and C resulted in 10, 20, and 30 N·m changes, respectively. Intervention C must be most effective, correct? If we account for the variability using effect size, we reach a different conclusion (figure 6.2).

Intervention A: $d = (118 - 128) / 19 = 0.53$

Intervention B: $d = (135 - 155) / 22.5 = 0.89$

Intervention C: $d = (175 - 205) / 51.5 = 0.58$

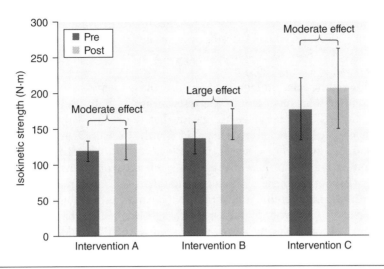

Figure 6.2   Effect size can be used to compare the magnitude of difference between studies.

When variability is accounted for, we discover that interventions A and C have moderate effect sizes, but treatment B has a large effect size. Even though intervention C shows the greater absolute increase in mean peak torque, intervention B is more effective than intervention C. If the three studies were otherwise similar in experimental conditions, an exercise practitioner might select B as the preferred intervention option.

Outcome

When one is framing a direct question, the outcome measure that the intervention is proposed to improve should be clearly defined from the beginning. The intervention and comparison should both be expected to affect the outcome measure so as not to skew or soften the comparison, as previously discussed. The evidence-based practitioner should also ensure that the proposed outcome measure has good validity and reliability. In other words, if you are trying to determine if an intervention is effective to improve power, ensure that the outcome defined in your direct question is a good test to assess the construct of power.

Selecting the wrong outcome measure to assess the effectiveness of an intervention can lead to erroneous results, misinformation, and a poor answer to the direct question. Suppose you are trying to answer the following question:

> **In adults who are older, does high velocity compared to slower, heavy resistance training result in greater improvements in power?**

An important consideration in the proposed question is how you define an improvement in power. What is the outcome measure to objectively demonstrate that one method is better than the other? Suppose you search the literature for any and all studies that compare high-velocity and heavy resistance training regardless of their measurement of power. You will find studies that report changes in power using a Wingate cycle test (Miszko et al., 2003), pneumatic-based power measurements systems (Fielding et al., 2002), free weights (Bottaro, Machado, Nogueira, Scales, & Veloso, 2007), and isokinetics (Henwood & Taaffe, 2005).

Depending on the selected outcome measure, the evidence-based practitioner may reach different conclusions because the tests measure slightly different constructs. Careful attention should be given to ensure that the selected outcome measure is likely to be improved by the intervention and that the measure is appropriate for your client, patient, or athlete. For example, if you believe that an improvement in average power output during a Wingate cycle test is not reflective of a functional performance improvement in your older clients, then you should not select it as the outcome measure to determine if the intervention is effective; that is, studies using this outcome measure should be excluded from your literature searches. If you believe that another test (e.g., power output generated during a back squat at 30% to 70% 1RM) is more valid for your population, search for studies using this outcome measure. The direct question could be rewritten as follows:

> **In adults who are older, does high-velocity compared to slower, heavy resistance training result in greater improvements in power as measured by an explosive back squat test at 30% to 70% 1RM?**

Time

As the final component of a direct question, time is optional and is not used in all situations. If we are assessing the effectiveness of a screening test, it would not be reasonable to use time as a component of the question. Time is important when comparing an intervention between studies that may be time sensitive and when using outcome measures that require a certain amount of time before there is a measurable change.

When comparing effectiveness between two interventions, especially from two different studies, it is important to ensure that the studies applied the interventions over the same amount of time. If one intervention was longer than the other, this could create a potentially invalid comparison and result in an erroneous conclusion. This is particularly important in investigating interventions that stimulate greater adaptations over time. That being said, practitioners should be careful not to eliminate a study simply because the intervention was applied over a different period of time than in other studies. Many times, outcome measures are assessed at multiple times throughout a study. It may be possible to match intervals within two studies even when the total intervention time is different.

Another purpose for defining the time in a direct question is to ensure that enough time is provided for between-group differences to manifest themselves (figure 6.3a). If we examine resistance training interventions, we find that there is little difference in strength outcomes for studies that are 4 to 6 weeks in length. Over this short period of time, any novel stimulus is effective to improve strength, and while there may be differences between stimuli, the differences are not great enough to show statistical differences (figure 6.3b). However, over longer periods of time, differences become apparent. Adaptations arising from short-term programs lasting 1 to 3 months may still be in dynamic states of change and have not yet attained their genetic maximum. Additionally, one cannot discount the role of psychology and motivation in any training program. Thus, comparisons need to address a host of different variables to correctly

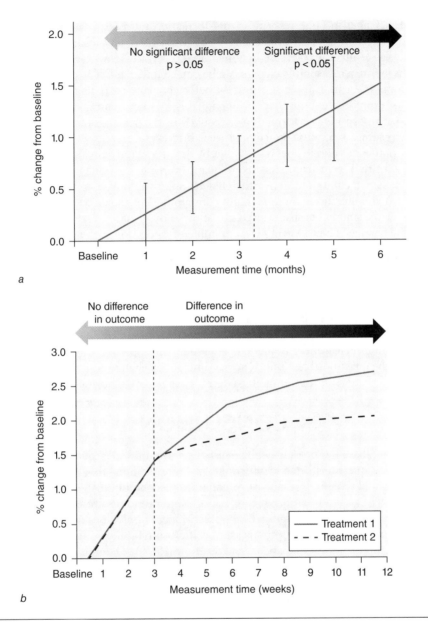

Figure 6.3 Selecting studies that are too short in duration can lead to (a) an incorrect assessment concerning the effectiveness of a treatment, especially when one is relying on statistical difference alone; or (b) an incorrect assessment concerning differences between treatments.

judge the efficacy of a program. In addition, a program may have responders and non-responders; thus, the comparison of group means may be deceptive.

A final reason to define time in a direct or foreground question that is closely related to the previous reason is that some outcome measures take time in order to show any measurable difference. Quantification of changes in bone mineral density is a good example. Changes in bone resorption and formation rates occur acutely in response to appropriate stimuli. After even a single exercise session, physiological processes are initiated that result in an increase in bone mineral density. However, it may take up to 4 months before a measurable change can be detected with a DXA (dual-energy X-ray absorptiometry) scan. Is it really 4 months before bone mineral density is increased as a result of exercise? Assume that an individual is in a net neutral state at baseline in which bone resorption and formation are equal (i.e., the person is not gaining or losing bone). An exercise stimulus applied for 1 week increases formation without a change in resorption. If bone volume does not change, by definition, the individual increased bone mineral density. The instruments used to quantify changes in bone mineral density are just not sensitive enough to detect the change. So, if we looked at short-term studies using DXA as a measurement, one might come to the conclusion that an intervention is not effective simply because we did not allow enough time to see a physiologically detectable change. The following is an example of a question with a time component:

In individuals who are elderly, is heavy resistance training more effective than running to improve bone mineral density in the lumbar spine over a 6-month period as determined by DXA scan?

The time component is very important to avoid mistakes in interpretation of the literature. Although its definition in an EBP question may not always be necessary, each scenario should be carefully evaluated, and where appropriate, timing should be factored into the question. The importance of timing may change as new instrumentation is developed that is sensitive enough to detect changes over shorter periods of time.

CONCLUSION

The evidence-based approach begins with the formation of a question. In exercise science, these questions often arise from novel interventions that appear in practice or in testing and screening scenarios, or come from economic analyses concerning the effectiveness of an intervention or test. Background questions are the basis of understanding and are often necessary when a practitioner simply does not know where to start. Foreground or direct questions are specific and typically contain a well-defined population, intervention, comparison, and outcome (PICO). Defining a time period may be necessary depending on the nature of the intervention. Evidence-based practitioners should carefully consider each component of their questions, as poorly defining any component can result in misleading evidence and erroneous conclusions.

Chapter 7

SEARCHING FOR EVIDENCE

Learning Objectives

1. Understand how to search for preliminary, primary, and secondary sources of evidence.
2. Review the importance of peer-reviewed evidence.
3. Gain familiarity with various databases of peer-reviewed research.
4. Develop an understanding of search strategies, particularly using PubMed.

After a focused question is developed, we must search for evidence to answer the question. The best evidence is useless if it is undiscovered; thus, a thorough, rigorous search is essential. That search is the focal point of this chapter. Many large databases are widely available, each holding enormous amounts of research evidence. Thus, the first vital component of a successful search is the proper use of keywords; one broad keyword will often yield hundreds or even thousands of results whereas too many (or overly specific) keywords will return very few studies. The second indispensable ingredient of a successful search is to conduct it in the proper place; the right keywords will yield few useful results if one searches in the wrong database.

An important note: throughout this chapter, readers will find examples in the text and figures of searches conducted by the authors at the time this textbook was written. Readers who replicate these searches will very likely obtain different results (e.g., more articles than reported in this textbook). This well illustrates the dynamic nature of the scientific literature and the importance of remaining current with the latest findings.

SEARCHING FOR PRELIMINARY INFORMATION

It is fair to say that without a strong working knowledge of a given area of inquiry, it is extremely difficult to effectively discriminate the findings of the search process. In the absence of a solid foundational understanding, it is likely that an evidence-based search will take much more time and effort and that it will encounter many more dead ends. Thus, it may be necessary to conduct a preliminary search for background information to facilitate a successful foreground or direct search. The search for peer-reviewed scientific articles that follows is the primary search. Later in the process, after online databases have been used to conduct an evidence-based search, additional valuable evidence can be discovered via the reference lists of articles found in the primary search; we call this a secondary search (figure 7.1).

Textbooks

After developing a question, a good place to begin searching for evidence is in a textbook; this is particularly true if the question involves or requires an understanding of basic physiology or training practice. Textbooks are amalgamations of broad knowledge on a variety of subjects. Their strength is the collection and presentation of vast bodies of scientific knowledge distilled into a neat and easily accessed package.

Textbooks in general, however, have several weaknesses. One, the treatment of any given subject is relatively generic and brief; that is, major concepts are presented, but fine details and subtle aspects are often not discussed. The second and potentially more serious drawback of a textbook is its temporal limitation. Textbooks take many months if not several years to write; they represent a synthesis of the complete body of scientific knowledge on many topics. The writing period is followed by months or years of

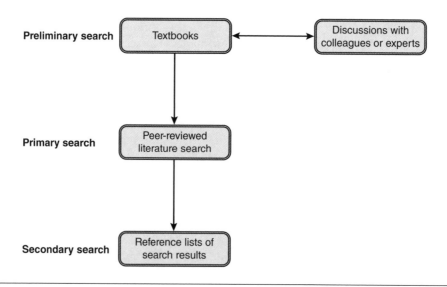

Figure 7.1 Practical order of progression in the search for evidence.

editing, typesetting, proofing, printing, and distribution. Although some effort may be made to incorporate late-breaking scientific advances, by the time a "brand-new" text-book reaches a college bookstore, the information that it contains is probably at least 2 to 3 years old, although top publishers now aim to publish materials within a year of submission (Oxman, Sackett, & Guyatt, 1993). For well-established physiological and training principles, this temporal lag is of little concern, as several years of additional scientific advances are unlikely to undo foundational understandings (Amonette, English, & Ottenbacher, 2010). However, in instances when rapid advances do occur, the textbook will present a limited or, occasionally, incorrect picture of a scientific understanding. For example, most people have thought of growth hormone (GH) as a single entity (i.e., the single 191-amino acid isoform) (Nindl, Kraemer, Marx, Tuckow, & Hymer, 2003). It now appears that aggregates (GH bonded to each other) may play more important roles in human physiology than previously thought. Unknown to many in the study of exercise endocrinology is that the aggregate forms make up the highest molar concentration of GH in the blood (Kraemer et al., 2010; Thomas et al., 2013). So, although textbooks are excellent "one-stop shops" to gain a general grasp of a particular topic, they are not good sources of recent or finely detailed information, as in the example of growth hormone. Thus, if a search begins and ends with a textbook, one's answer is likely to be limited, lacking in detail, and quite possibly inaccurate due to new scientific knowledge that was not available when the textbook was written.

Discussions With Experts

Along with reading through textbooks, another valuable step is to consult colleagues and superiors (e.g., past or current professors) and solicit their knowledge on the subject (McAlister, Graham, Karr, & Laupacis, 1999; Oxman et al., 1993). This is particularly helpful if one's knowledge base in the area of interest is very small; instead of starting from ground zero, one can look to people with experience in the area to provide a thorough background understanding, including common misconceptions and points of unresolved controversy. Of course, such discussions are only a prologue to an evidence-based search, but they are valuable for deepening one's understanding of relevant issues and unknowns in the area of interest.

Understand that all investigators have their own biases based on their research, which often frames their beliefs on a topic; this is an acceptable approach if factual and data driven. Too often in the exercise and fitness world, scientists with one area of expertise overstep this, getting into other areas where they have no legitimate credibility based on prior work. However, the mantle of their degree, prominence or expertise in another area, or their peripheral involvement with small parts of projects, allows for professional or scientific "mission creep" into areas in which they are not trained (e.g., an expert in skeletal muscle protein metabolism expounding on endocrine topics and research). This usually results in a very limited view or understanding of the topic. However, that being said, most scientists are transparent with regard to their knowledge, understanding their strengths and limitations within their research discipline and area of expertise.

A note about this idea of baseline knowledge that an evidence-based practitioner brings to a given search is worth adding here. This book is devoted in its entirety to a description of the necessary steps in the process of an evidence-based search, with the understanding and assumption that even a novice in exercise science could, via an appropriately conducted evidence-based search, gain a thorough, cutting-edge understanding of any given area or question in the field. However, the evidence-based approach is expedited (and in all likelihood heightened) by a thorough fundamental understanding of the area of interest. Thus, it is reasonable to expect that as one's professional knowledge base deepens and widens, evidence-based querying can be conducted more quickly and result in more thorough, mature conclusions (Sackett, 1997).

FINDING PEER-REVIEWED SOURCES

After one has gained a fundamental understanding of a given topic, the best place to search for research evidence is in an appropriate database (Coomarasamy & Khan, 2004). Databases house many thousands, if not millions, of research articles—all easily searchable. Some databases are particularly large and are available to the public (e.g., PubMed, Google Scholar); others are much more specialized in scope and may require a subscription (e.g., SPORTDiscus), which many university libraries provide to students and faculty through library fees that are included with tuition (Law & MacDermid, 2008; Lu, 2011). The advantage of a database like SPORTDiscus is that its holdings are specific to sport and exercise; thus, search results are likely to be more focused, reducing the effort involved in parsing them. The disadvantage of more focused, limited holdings is that the database does not index many journals and thus may not return an article relevant to the question at hand (Falagas, Pitsouni, Malietzis, & Pappas, 2008). Table 7.1 presents a list of search engines useful to the field of exercise science.

PubMed is one of the largest and most accessible databases available, operated jointly by the National Center for Biotechnology Information (NCBI), the U.S. National Library of Medicine (NLM), and the National Institutes of Health (NIH). Its holdings cover a broad spectrum of biomedical and other sciences; thus it is used here as an example of how to search a database. PubMed provides unlimited access to article abstracts; but aside from older papers and some open-access journals, full-text articles require a paid subscription. Fortunately, most universities have organizational subscriptions to a litany of journals as well as interlibrary loan agreements with other institutions to ensure access to journals to which they do not subscribe. Thus, those with access to a university library can obtain most articles either instantly or within several days if they use interlibrary loan. For individual practitioners or those working in settings that do not have institutional journal subscriptions, it will likely be necessary to obtain many articles via a university-based colleague.

SEARCHING A RESEARCH DATABASE

A popular maxim in research is "You didn't do it unless you published it," which emphasizes the importance of disseminating one's research. A major portion of scientists', researchers', and university professors' performance evaluation is linked to how many peer-reviewed papers they publish; this is the ultimate measure not only of the successful completion of a research project, but also of the work's soundness and subsequent acceptance into the pantheon of knowledge—the scientific literature. Of course, beyond a seemingly narcissistic wish to demonstrate to the scientific community the stellar work one has done, a much more utilitarian purpose is to enable others (e.g., clients, athletes, patients, other scientists) to benefit from one's work. This is, of course, at the heart of scientific endeavor: to add one's work to the ever-growing body of knowledge

Table 7.1 Common Databases Used to Search for Peer-Reviewed Evidence

Database	Description
Medline	Likely the most comprehensive database to search for medically based journals. Has robust, expanded search features to facilitate directed searches.
PubMed	Public access to the Medline database. Users can freely search and read abstracts. Contains some links to free online journal articles, also contains specific features to allow searches for randomized controlled trials or systematic reviews.
Google Scholar	Publicly available database containing peer-reviewed information. Contains links to many free online journals and abstracts. Also provides links to government reports.
SPORTDiscus	Database specifically for exercise and sport science. Contains many of the same articles as other medical databases, but also searches journals specific to the exercise discipline. Some articles are non–peer reviewed.
CINAHL	Allied health sciences database that contains some unique features and journals; popular within the field of nursing.
Web of Science	One of the most comprehensive databases for allied health sciences. Contains links to conference proceedings; user can search for the number of times a particular article has been cited in peer-reviewed literature.
Cochrane Library	Contains links to systematic reviews for a variety of allied health sciences by discipline. Also allows users to search specifically for funded trials.

for the purpose of improving the knowledge of the population. To facilitate this, it is essential for research articles to be easy to locate and for those needing to find research evidence to be able to do so.

Keywords

The primary technique used to search research is the use of keywords. When scientists submit a manuscript to a journal for publication, they must also provide 5 to 10 keywords—words that characterize the general nature of the work, the methodologies employed, or the populations studied—to facilitate others' efforts to find the article (International Committee, 1997). Most journals include the author-selected keywords in the published paper, often at the end of the abstract or as a header or footer. For example, if one searches "strength training" on PubMed, more than 20,000 results will be returned; these studies represent the full spectrum of published studies with any pertinence to strength training. Obviously, for an evidence-based practitioner, a more refined search is needed, as it would take weeks to sort through these thousands of studies. For instance, within the "strength training" search, one paper has the additional keywords "physical inactivity," "insulin resistance," "obesity," "diabetes," "high-intensity exercise" and "resistance training," while another has the very different keywords "postactivation potentiation," "muscle fatigue," "warm-up," "twitch contraction," "quadriceps femoris," and "maximal voluntary concentric torque"; clearly, these are very different papers although they have "strength training" in common. This underscores the importance of a focused search. You can waste a great deal of time sifting through dozens (or thousands!) of articles that you could easily have eliminated from the search result simply by adding additional search terms (keywords) to narrow or restrict the search.

Typically, some of the keywords for an article are broad and others are narrow. In the second paper just mentioned, keywords like "postactivation potentiation" and "twitch contraction" are relatively specific to a particular line of neuromuscular research, whereas "muscle fatigue" and "warm-up" are much more general terms and could apply to a far wider body of research. For authors, a keyword selection strategy such as this is ideal as it ensures that a paper will appear in a broad, general search and will also come up in a narrower, more focused search. Thus, a search often incorporates both a broad keyword and then one or two narrower keywords; this will usually yield a search result of highly pertinent and useful articles (Rosenberg & Donald, 1995). For instance, a search with the terms "strength training AND postactivation potentiation" yields 33 articles, a highly manageable (and relevant) return.

PubMed MeSH Terms

PubMed employs what are called MeSH (Medical Subject Headings) terms to classify studies. When a paper is published by a journal that is indexed in PubMed, an indexer assigns that paper 10 to 15 MeSH terms to categorize it. When one searches for articles using MeSH search terms, all papers indexed within that particular category are returned. MeSH terms are similar to hash tags used on Twitter in that both are created and used for the purpose of centralizing or unifying discussion or media on a particular topic. Using these terms in a search can be useful because papers indexed or stored within a category have a common theme and are often interrelated. Finding and using the correct MeSH search terms is like locating the correct shelf in the library, full of books related to your question.

It is important to understand that MeSH terms are standardized terms; they differ from keywords, which are assigned by the author and can be anything (Haynes, McKibbon, Wilczynski, Walter, & Werre, 2005; Haynes, Wilczynski, McKibbon, Walker, & Sinclair, 1994). The MeSH term list is updated each year to include new and emerging medical terms; in 2013, there were almost 27,000 MeSH descriptor terms with over 213,000 entry terms that funnel searches toward one of the descriptor terms. A list of MeSH terms and their subordinate entry terms can be found on the PubMed site, but it is perhaps even easier to begin typing in a search term and see what other similar terms appear via autofill. For instance, if one types "strength" into the PubMed search window, almost two dozen variations appear (e.g., "strength training," "muscle strength," "strength testing"); this provides a list of keywords or MeSH terms (or both) that expand on the initial word. If a search word or term does not return any results, another alternative is to include "[Text Word]" after the word or term. This directs the search engine to search the titles and abstracts for that term in addition to searching the keywords and MeSH terms. This is helpful in cases in which the term is unusual.

If "[Text Word]" is used after a MeSH term or a common keyword, it will return an avalanche of articles.

Advanced Search Strategies

The search techniques described in the previous section are often adequate to obtain relevant evidence. However, more advanced strategies are sometimes necessary.

MeSH Search

To illustrate such advanced methods for a targeted search, we will borrow one of the questions posed in chapter 6:

> **In individuals who are elderly, does heavy resistance training significantly improve lower body strength compared to sedentary, nonexercising conditions?**

Instead of immediately searching the PubMed database, from the PubMed homepage, the user selects "MeSH" from the dropdown menu to the left of the search box and uses "resistance training" as the first search term (figure 7.2). This original query is shown in the lower window depicted in figure 7.2; the contents of this window will not change with the addition of new search terms, as it displays only the initial search. The MeSH search for "resistance training" reveals that it is a MeSH term introduced in 2009 with a variety of corollary terms (e.g., Strength Training, Weight-Lifting Exercise Program). We can also see the various subheadings under "Resistance Training" that will allow us to build a more targeted search. We can select "methods" (indicating that we wish to search for articles that have "resistance training" in the methods section) and click "Add to search builder" (the term will be added to the upper window depicted in figure 7.2) and then click "Search PubMed." This returns over 1000 articles; we need to add in the other important components of our search to narrow it a bit (Ebbert, Dupras, & Erwin, 2003). Thus, we add "AND elderly" to the search box at the top of the results page and click "Search" again. This time, 400 articles match the search criteria. A repeat of this search with the addition of "AND strength" yields 285 articles. In the event that these directions do not appear to match the PubMed website, please note that the US National Library of Medicine that maintains the PubMed site may cosmetically alter it after this textbook is published.

Although "elderly" is one of the search terms, that does not necessarily mean that this is the population evaluated in the study; it may just be a topic that is discussed in the article. Thus, on the screen showing search results, we should next employ a PubMed filter from the left margin (figure 7.3). Click "Show additional filters" and ensure that "Ages" is selected. Next, click on "Ages" and "Aged: 65+ years." Finally, we must actually click on "Aged 65+ years" and ensure that a check mark appears and that "Filters activated: Aged: 65+ years" is displayed at the top of the search results. This filter has further reduced our results to 180 articles. Adding a filter for "Article type: Clinical Trial" trims the results to 114 articles. This is still a large number of articles, but an amount that can be sifted through. Because we are interested in healthy populations, it is easy to eliminate some articles with just a glance at the title (e.g., studies that evaluate exercise responses in unhealthy populations such as those with chronic obstructive pulmonary disease [COPD] or asthma).

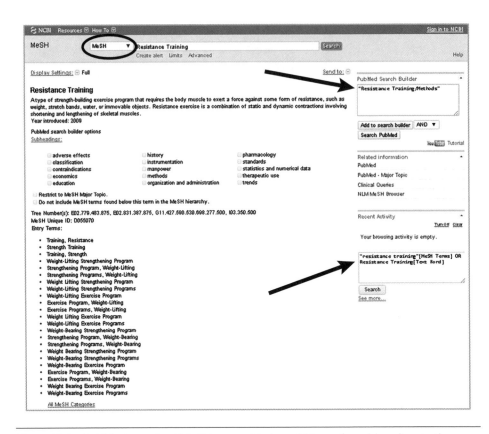

Figure 7.2 PubMed search of MeSH term with the addition of a subheading to the PubMed search builder.

Reprinted from Pubmed.gov.

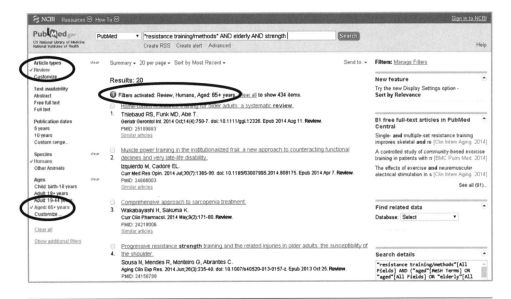

Figure 7.3 Results of PubMed search with several filters activated.

Reprinted from Pubmed.gov.

We can change "strength" in our search to "lower body strength"; this decreases the search return to 15 articles. However, this might be somewhat limited even though our interest is in lower body training. Adding (or removing) search terms and filters should be done in the results screen; we do not need to return to the MeSH search builder that we employed at the outset of this example unless we wish to add another subheading to the search.

Another option is to remove the "Clinical Trials" filter and add a filter for "Review"; this results in 12 review papers. Review papers can be particularly helpful in providing a general survey of a particular topic—all within one article. However, there are disadvantages of review articles: (1) If the article is older (e.g., if the review is 10 years old), scientific advances may have significantly changed understanding in the area of interest; and (2) although review articles should be an impartial synthesis of the literature, they can omit reference to some papers through lack of care or due to difficulty in assimilating them with other findings (reviews are rarely exhaustive). One major exception to this is systematic reviews. As the name implies, systematic reviews are a thorough, exhaustive search of the literature in a particular area. Systematic reviews carefully specify their search strategy and their inclusion and exclusion criteria: The studies included in the paper are not selected based on the whim of the authors or their fit to a particular line of thinking but instead in accordance with a set of well-described, objective criteria. Thus, if one can find a well-conducted systematic review addressing a topic of interest that evaluated quality evidence, these are usually exceptional resources that thoroughly and objectively evaluate an area of research.

Clinical Queries

From its homepage, PubMed also provides the option to make Clinical Queries (figure 7.4), a targeted search of clinical trials (Anders & Evans, 2010; Doig & Simpson, 2003). This feature is particularly helpful for medical questions but is also useful for questions within the exercise science field.

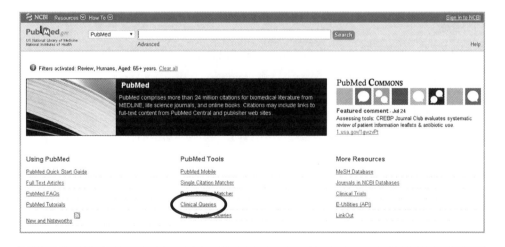

Figure 7.4 Clinical Queries option in PubMed.
Reprinted from Pubmed.gov.

A Clinical Query on "resistance training" yields results in three categories: Clinical Studies, Systematic Reviews, and Medical Genetics (figure 7.5). Under "Clinical Studies," several different categories can be selected to focus the results of the Clinical Studies search: Etiology, Diagnosis, Therapy, Prognosis, and Clinical prediction guides. Searchers must also select "Broad" or "Narrow" scope. Selecting "Therapy" and "Narrow" returns over 2000 articles for the Clinical Studies category and over 500 systematic reviews.

If we click "See all" for Clinical Studies and enact several of the same filters that were previously used (Clinical Trial, Humans, Aged: 65+ years), we reduce the number of articles to ~750. Addition of the terms "AND heavy" or "AND high intensity" reduces this to 100 articles or less. This illustrates the two sides of a database search: Too many articles are impossible to reasonably evaluate, while too few articles provides insufficient information and maybe more importantly, the potential for bias, that is, the exclusion of important research.

Cross-Listing

Another valuable feature of PubMed and many other databases is cross-listing. When the abstract of an article is selected for viewing, the database automatically lists five related articles (figure 7.6). It also provides an option to "see all," a feature that reveals a

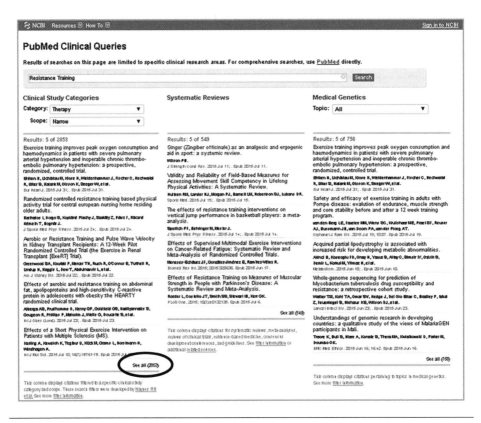

Figure 7.5 Results of PubMed Clinical Queries.

Reprinted from Pubmed.gov.

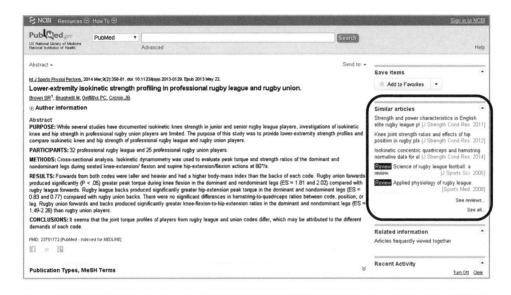

Figure 7.6 An example of the cross-listing feature inherent to PubMed. In the right-hand corner of the webpage, articles related to the selected abstract are listed and available for search.

Reprinted from Pubmed.gov.

larger number of related articles. This can be a particularly useful method for searching because the articles that are cross-listed are linked in unique ways, and sometimes a useful article will be cross-listed with several other highly relevant articles.

A Quick Word on Google Scholar

PubMed is probably the most often searched database for peer-reviewed evidence in exercise science. However, there are other publicly available sites that are useful for exercise-related searches. One in particular, Google Scholar (www.google.com/scholar), is an emerging tool that has strengths and weaknesses (Anders & Evans, 2010; Falagas et al., 2008; Kulkarni, Aziz, Shams, & Busse, 2009; Neuhaus, Neuhaus, Asher, & Wrede, 2006). A subdatabase of the search engine Google, Google Scholar searches return broader results than PubMed. An NIH committee reviews health-related journals and recommends journals to the NLM for indexing in PubMed. The NLM uses the recommendation of the NIH committee along with other criteria for selection. This ensures that higher-quality health-related journals appear in PubMed, but it also excludes non-"health-related" journals and many quality articles published in nonselected journals. Although using Google Scholar reveals many journals common to PubMed, it also reveals some articles not found in the PubMed database. Because PubMed safeguards the quality of the journal, the exercise practitioner must carefully appraise the articles found in Google Scholar—in fact, all articles should be carefully appraised regardless of the search engine used to discover them.

Google Scholar works like PubMed or any other search engine; the user enters keywords, and articles are returned (figure 7.7). It allows the user to include or exclude citations and patents; select a custom range of dates; and search by author, journal, and

Figure 7.7 An example of a Google Scholar search.

Reprinted from Google Scholar.

a number of other variables. In the right-hand portion of the screen, articles available in full text are listed and available to click and view. Clicking on the link to the articles takes the user directly to the journal website where the abstract is usually free for viewing. If the journal is open access, the full-text article will also be available.

CONDUCTING A SECONDARY SEARCH FROM REFERENCE LISTS

A final and often invaluable place to search for relevant peer-reviewed research is the reference lists of articles that have already been acquired; this is the secondary search. This process, repeated iteratively, will almost always lead to better and better evidence. You may have an article that addresses a part of the evidence-based question; for example, it evaluates aerobic exercise instead of heavy resistance exercise in adults who are older. It is very likely that the aerobic exercise article will have one or more references to studies on resistance exercise in persons who are elderly. The exercise practitioner can look for these cross-references in the reference list itself, or sometimes it is easier to find them in the introduction or discussion of the paper. The introduction discusses existing knowledge and sets the stage for the current study; the discussion integrates the findings of the study with those of other investigations. Thus, both are good places to browse for mentions of other related work.

Perhaps the best type of paper for cross-referencing is a systematic review. Systematic reviews can be invaluable—a virtual one-stop shop if you are fortunate enough to find a systematic review on the topic of interest. Let's return to our previous Clinical Query; this time we will select "See all" of the Systematic Review results. Over 500 articles are returned, so we need to filter and narrow the results. Adding an Age filter of "65+ years" reduces the number of articles to ~100. The addition of "AND high intensity" to the search term decreases the results to 17 articles. From this, we can easily select several systematic reviews that are exactly what we are searching for.

If a systematic review is well-conducted and highly relevant to the evidence-based question, it is likely that a minimal search will be sufficient to reveal any articles omitted by the authors. Even if the systematic review was recently published, there may have been a 6-month to 1-year lag time between the search conducted by the authors and the publication date. Thus, a practitioner searching for evidence should conduct a thorough search for articles starting 1 to 2 years before the systematic review was published. The practitioner could use the same search terms used by the investigators or add additional terms to the search.

On a side but important note, along with reading the recent research it can be important to read and understand the historical context of a concept or problem. Many problems surfaced decades ago, and reading and understanding the history of the literature can help the practitioner appreciate the route to the current state of the research problem (for example see Kraemer et al., 2010). It may also help to refocus the question or stimulate a unique perspective in new research.

CONCLUSION

Searching for evidence is the initial step to take after one develops an evidence-based question. Since the information uncovered in this step is key to decision making, it is important that the exercise practitioner carefully and thoroughly perform this step. Evidence ranges from information derived from preliminary sources such as textbooks to primary, peer-reviewed sources found in journals. PubMed is the most commonly used publicly available search engine. Learning the essential features of the database along with developing strong keywords can lead to a more successful and fruitful search. Google Scholar is a second publicly available database that may reveal articles not found in PubMed. Relevant evidence can also be found via a secondary search of the reference lists of papers discovered in the primary search. The next chapter discusses how the practitioner assesses the quality or strength of the discovered evidence.

Chapter 8

EVALUATING THE EVIDENCE

Learning Objectives

1. Discuss the importance of critically appraising the literature.
2. Understand the strengths and weaknesses of the different types of evidence.
3. Define important terms such as bias, confounding factors, and blinding.
4. Identify critical appraisal tools that are publicly available for use in grading evidence.
5. Become familiar with two common grading systems associated with evidence-based practice literature, including the model used by the American College of Sports Medicine.

When uncertainty in practice arises, the evidence-based practitioner develops a question and performs a thorough search of the literature to uncover relevant evidence. The evidence may range from expert opinion in the form of a conference presentation or personal communication to randomized controlled trials or a systematic review with homogeneous findings. After finding the evidence, the practitioner must determine if it is valuable and worth using and whether it supports or refutes the protocol, device, or supplement in question. This can be a difficult task because a search may uncover an abundance of information that is either directly or indirectly related to the evidence-based question. The next step—perhaps the most important in the evidence-based approach—is to critically appraise and grade the evidence. As alluded to earlier in the book, there is always some evidence to support a device, protocol, test, supplement, or programming theory. If there were not, the theory base for use would be purely random. Evaluating the evidence in light of the well-known fundamental laws of physics, physiology, and training methodology, the evidence-based exercise practitioner must now ask how strong the evidence is and whether its strength is sufficient to warrant incorporation into practice.

WHY IS IT IMPORTANT TO RANK THE EVIDENCE?

In a criminal investigation, detectives search for clues to elucidate the events surrounding a crime. Using this discovered evidence, they attempt to recreate past events that ultimately lead to an individual who is responsible for the criminal behavior. The evidence ranges in strength and quality and can be broadly divided into two categories: circumstantial and direct evidence. Circumstantial evidence, as the name implies, includes a series of facts or circumstances that a lawyer uses to create a narrative—a version of events that argues for the guilt (prosecutor) or innocence (defense) of the defendant. This type of evidence is indirect and relies heavily on judgment—ultimately the judgment of the jury to determine whether they believe the narrative presented by the lawyer. Direct evidence, on the other hand, is factual information that directly supports or refutes an occurrence. It can include testimony from an eyewitness who saw an individual at a crime scene or forensic evidence such as fingerprints or DNA. When a judge or juror is making a final decision regarding the guilt of a defendant, not all evidence is equal. A lawyer may create a logical sequence of events suggesting that an individual has never been to the scene of a crime, but if blood is found at the scene containing DNA that matches the defendant, then the circumstantial evidence is trumped by the direct evidence. Even within direct or forensic evidence there are levels of certainty. Fingerprint matching, once the gold standard of forensic evidence, is prone to error and could be disproved in court by DNA matching, a more precise technique. In a trial, evidence is presented for and against the defendant. Even when a defendant is exonerated, there is some evidence to suggest that this person was the perpetrator. If there were not, the person would not have been charged with the crime. It is the strength, quality, and eventually the jury's critical appraisal of all of the evidence that determine the fate of the defendant.

Similar to criminal evidence, expert opinion, observational research, and experimental research have different levels of strength (figure 8.1). However, even within the different types of evidence, aspects of study design or other factors can suggest that the evidence is stronger or weaker. In comparing multiple pieces of evidence, understanding how to read, evaluate, appraise, and potentially eliminate them from consideration are important skills for the evidence-based exercise practitioner. Utilizing the EBP approach, trained exercise practitioners evaluate and weigh the strengths of evidence in the context of their question, determining how or if the information should be utilized or implemented in their clients, athletes, or patients.

A legitimate concern to this approach is the evaluator's level of understanding of the topic, including the background to the question. This highlights the need for honesty and sincerity in evaluating one's own or a team member's ability to provide an accurate assessment of information. Even journals soliciting reviews from highly trained researchers sometimes receive responses from reviewers saying, "Sorry, this is outside of my area of expertise." Thus, deep evaluation of evidence, especially research evidence, requires a certain degree of expertise in the area and an understanding of the different permutations and implications of research findings and the complete body of knowledge that surrounds the question. This further strengthens a prevailing theme of this book—the need for a team of evidence-based practitioners to evaluate information.

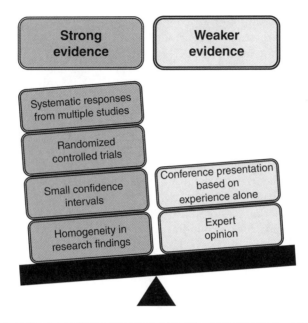

Figure 8.1   After retrieving evidence, the exercise practitioner must determine the quality or strength of the evidence. When one is making the final decision to include or exclude the evidence, it is the strength, not the quantity, of evidence that most profoundly influences the decision.

Nonetheless, we believe that with a fundamental understanding of research design and the critical aspects of study design, any exercise scientist or practitioner can use research to influence their practice. With a long-term commitment to reading the breadth of research in a particular area, one may even become an expert. In this process, it is important that we understand our own limitations in background and experience regardless of where we are in our lifelong pursuit of greater understanding. Learning how to evaluate a scientific paper in this chapter will not make a practitioner an expert overnight; this requires education and years of professional experience. Nevertheless, beginning the journey to becoming a skilled evidence-based practitioner is imperative, and even novice evidence-based practitioners can make valuable contributions within a team setting.

Thus, the purpose of this chapter is to teach readers how to appropriately grade or judge information in their particular area of expertise, ultimately arriving at less biased answers to the questions that surround practice.

CRITICALLY APPRAISING THE LITERATURE

Critically appraising the literature begins with thoroughly reading the evidence and classifying it. The first way to classify an article is to determine if it is based on expert opinion (e.g., narrative review), observational research, or experimental research. After determining the general type of evidence, one can more thoroughly evaluate the study design. Within each study design are common biases and confounding factors that may be inadvertently introduced or uncontrolled by the investigators.

A confounding factor, from a research design perspective, is a variable other than the independent variable that is associated with or affects dependent variable or outcome (Riffenburgh, 2012). If not controlled, a confounding variable can lead to an erroneous conclusion. Suppose you are reading a study that tested the effects of a particular interval exercise protocol compared to a steady-state program on heart rate responses. It would be important for the authors to specify that they controlled for the consumption of stimulants before testing. If, by chance, a large number of people in one group but not in the other consumed a stimulant such as caffeine before exercise, an increase in heart rate could be due to the caffeine, not the exercise protocol. In this case, caffeine consumption would have confounded the result, leading to an inaccurate conclusion from the data if not accounted for in the design and controls used in the investigation. Other examples abound for confounding variables that are uncontrolled or not mentioned in a research paper. These include nutritional behaviors, sleep, motivation in the testing session, menstrual status, birth control method in women, hydration status, and learning effects.

Bias is a systematic error in the design or execution of a study that leads to an erroneous result (Gordis, 2009). An abundance of potential biases have been presented in the literature (Sackett, 1979); many are discussed in this chapter along with individual study designs. Biases can be introduced purposefully or inadvertently. Either way, they reduce the strength of the study and call its findings into question. An example of bias in exercise science is dietary recall bias. Suppose you read a study examining the efficacy of an exercise program to promote weight loss in adolescent girls; the study evaluates three different groups: girls who are obese, overweight, and normal-weight. Each performs exercise that is identical in intensity, frequency, and duration. Dietary intake is documented by a 3-day dietary recall performed once per week for the duration of the study. At the end of the study, the results suggest that the normal-weight and overweight groups lost a significant amount of weight, but not the obese group. The 3-day dietary recalls show that the obese group actually consumed less calories than the other groups. Can it be concluded from this investigation that diet and exercise are ineffective to promote weight loss in adolescents who are obese? Probably not, as the 3-day dietary recall represents a potentially fatal bias. Young adolescents who are obese and who are concerned about their weight tend to underreport dietary intake (Ventura, Loken, Mitchell, Smiciklas-Wright, & Birch, 2006). It is likely that the investigators in this hypothetical example introduced bias into the study by choosing the dietary recall method to document energy intake.

Bias and confounding factors are significant problems associated with all forms of evidence. Each evidence type is prone to certain errors, and knowing these errors can inform a better appraisal of the research. Although the peer review process should catch many biases and confounding factors, questionable studies are still published in well-respected journals. Journal reviewers are typically experts in the area of a paper they are reviewing, but they may not be aware of peripheral topics that affect the study design (e.g., a muscle metabolism expert may not be familiar with sleep dynamics and stress effects). Thus, studies with uncontrolled bias or confounding factors often appear in the research literature despite a rigorous peer review process.

Critically reading an article does not denote criticism of the author or the investigative team. Many biases and confounding factors are impossible to control. For example, how does one create a placebo for "chewing tobacco," an almost impossible problem under typical experimental conditions (Ksir, Shank, Kraemer, & Noble, 1986)? It is imperative for the investigative team to admit and understand this limitation, designing a study in which the effects are obviously attributed to the independent variable. This process helps to minimize the possibility that evidence-based exercise practitioners will incorporate an erroneous finding into the training plan of a client, patient, or athlete, thus blunting program effectiveness.

In the next sections, we review the strengths and weaknesses of the evidence types introduced in chapter 4.

Expert Opinion

Expert opinion is a valuable source of evidence; in many cases it may be the strongest form of available evidence supporting the efficacy of a given protocol, device, test, product, or supplement. This evidence type may be obtained through personal conversations with a practitioner or scientist, conference presentations (which may be peer reviewed), and narrative reviews or Internet blogs, among others. Expert opinion is not peer reviewed and can be prone to significant personal bias. Nonetheless, evidence based on expert opinion has several strengths.

Strengths

Valuable information can be learned from experts, much of which cannot be acquired via formal scientific means. Practical information, such as sequencing of training protocols and "how-to" implementation questions, is best obtained through expert opinion—specifically from individuals with vast experience in the discipline. Exercise practitioners often practice on the edge of science. In other words, to a small extent, they tend to incorporate new devices, protocols, and techniques that may not yet be completely supported by science. Evidence-based practitioners, of course, minimize the number of these elements in their programs; these elements certainly do not form the base or foundation of practice. They are incorporated minimally in an attempt to gain an edge, to provide that "little extra." Because science can be a slow, meticulous process, practitioners may begin to incorporate novel interventions that will be supported by science in the future. Thus, they actually provide evidence for the efficacy of a device before science does.

Limitations

The limitation of expert opinion is that it is only as strong as the evidence on which it is based. If expert opinion is based on a strong body of unbiased, well-designed, replicated scientific experiments, it is as strong as the research evidence itself. However, if expert opinion is based on a small number of personal observations without careful control or a reference group, then the evidence is weak.

Expert opinion is frequently based on mechanistic evidence, or a basic understanding of biology and physics. Basic science forms the fundamental understanding of how human physiology will respond to a stimulus—it is the foundation on which clinical research practices. In the pharmaceutical industry, a drug cannot proceed to clinical trials without a thorough understanding of the pharmacokinetics and the drug's mechanism of action. These drugs are thoroughly tested in animal models before they are ever tested in humans. Basic science testing serves to eliminate potentially harmful drugs and helps scientists arrive at more precise dosages that may minimize adverse events in clinical trials. Likewise, basic science and mechanistic evidence are on the front line of research in exercise science. Our understanding of the exercise responses of skeletal and cardiac muscle, bone, hormones, the immune system, and bioenergetics helps us to determine parameters for training studies. The acute responses of these systems have often been extensively studied before a training study is begun. This improves efficiency, as the resources needed to perform a strong exercise intervention training study can be significant. These basic science studies help to determine which protocols are worth testing.

However, basic science studies are very specific in their design; they typically investigate the response of one system and may not always measure the integrated physiological response. If an exercise intervention results in an acute response in muscle protein synthesis or markers of bone formation, this does not necessarily mean that the protocol, if repeated every other day for 6 months, will stimulate significant hypertrophy or an increase in bone mass. Likewise, if an acute biomechanical experiment suggests that a significant amount of torque is placed on a joint using a novel device, this does not necessarily translate to a significant improvement in strength or hypertrophy with chronic use. Both examples support the notion of probable improvement, but strength gains, hypertrophy, and bone mass improvements are the total result of complex integrated physiological responses. These acute protocols warrant further investigation; often, if the basic science experiments were well-designed and controlled, the practical experiments will confirm the chronic training response. However, the basic experiments or the recommendations based on a basic experiment should not be the final piece of evidence to inform practice decisions.

Case Series and Case Studies

Case studies can be a valuable form of evidence. In medicine, they often describe novel occurrences such as a unique treatment response or presentation of a disease. A case series is a series of case studies. In a case series, a physician or scientist may report the cases of several patients, forming the basis for more rigorous scientific investigations in the future.

Strengths

Similar to expert opinion, the primary strength of a case study or case series investigation is that sometimes it is the only evidence available. It may provide the basis for an initial protocol or a starting point for practitioners interested in using a novel intervention in their patients. Case studies or case series may be particularly valuable in populations in which research evidence is limited. For example, there is limited information regarding exercise interventions to improve cardiorespiratory fitness in patients with traumatic brain injury. In 2008, Mossberg and colleagues reported that a training protocol using a body weight support treadmill improved aerobic capacity in two patients with traumatic brain injury (Mossberg, Orlander, & Norcross, 2008). This study provided a protocol and evidence in two patients that exercising on a treadmill with a support system may be beneficial. The exercise intervention worked in two patients; this does not mean that it will work in all. However, the study is a starting point for therapists searching for exercise strategies and forms a basis for future experimental research.

When reading and evaluating a case study, it is important to note whether the cases were identified retrospectively or prospectively (Young & Solomon, 2009). If the cases were identified retrospectively, one may not be able to identify the level of adherence to the study intervention, as it is typically self-reported (see later discussion of retrospective cohort study). In a prospective cohort, the investigators may have great control over the intervention; in some cases, such as in the study by Mossberg and colleagues, the researchers are able to observe and implement the intervention over a given period of time. Thus, they have complete control over the exercise volume, intensity, mode, and duration. Such evidence provides stronger support for the intervention.

Limitations

There are two major limitations to case studies and case series: small sample size and the lack of a control group. Because case studies and case series report on a very small sample (as few as one participant), the results cannot be extrapolated to the entire population. Cases are often extraordinary patients, which is one of the primary reasons the journal publishes the paper. An unusual case may not represent your client, patient, or athlete, whose responses to the treatment may differ significantly. Although multiple cases (case series) improve the probability that the responses seen in the article will be more representative of your client, this is not always so. Thus, the results of these studies must be interpreted with caution, and one must be careful when informing practice with this evidence.

The second limitation to case studies and case series is the lack of a control or placebo group. There is no reference point against which to judge the results. The evidence-based exercise practitioner cannot know if the patient would have improved without treatment or perhaps would have improved to a greater extent with another treatment. There is also limited control associated with a case study; thus, the exercise practitioner cannot fully determine whether the improvement was due to the novel intervention, a confounding factor, or chance. Consider the following example. In 2000, Dickerman and colleagues reported the lumbar spine bone mineral density in a unique athlete—the world record holder in the back squat lift (Dickerman, Pertusi, & Smith, 2000) (figure 8.2). They reported that the bone mineral density in this individual was the highest ever measured. The obvious conclusion is that the compressive forces during heavy squats result in enormous improvements in lumbar spine bone mineral density. Although the authors found this proposition plausible, there are several other possible explanations. It could be that the individual was genetically predisposed to higher bone mineral density. Perhaps it was not that squats caused the high bone mineral density, but instead that he was able to tolerate such a load because he was genetically predisposed to have higher bone mass. It is possible that he took a drug or supplement that caused the accumulation of bone. It may be that he consumed a unique diet and high amounts of calcium and vitamin D. Of course, the high compressive forces are still the most plausible explanation, but this explanation is confounded by all of the other possible explanations. The result is also biased by the lack of a control group. The intention of the case study authors was not to suggest that exercise practitioners change their practice based on findings in one individual. Instead, the goal was to report a unique finding and stimulate thought in scientists and practitioners alike. Similar to expert opinion, case studies and case series should never inform the foundational aspects of practice unless there is no other evidence on which to base one's practice.

Int J Sport Physiol Perform, 2013, 8, 582-592

Natural bodybuilding competition preparation and recovery: A 12-month case study

Rossow, L.M., Fukuda D.H., Fahs C.A., Loenneke J.P., Stout J.R.

Abstract

Bodybuilding is a sport in which competitors are judged on muscular appearance. This case study tracked a drug-free male bodybuilder (age 26–27 y) for the 6 mo before and after a competition.

Purpose: The aim of this study was to provide the most comprehensive physiological profile of bodybuilding competition preparation and recovery ever compiled.

Methods: Cardiovascular parameters, body composition, strength, aerobic capacity, critical power, mood state, resting energy expenditure, and hormonal and other blood parameters were evaluated.

Results: Heart rate decreased from 53 to 27 beats/min during preparation and increased to 46 beats/min within 1 mo after competition. Brachial blood pressure dropped from 132/69 to 104/56 mmHg during preparation and returned to 116/64 mmHg at 6 mo after competition. Percent body fat declined from 14.8% to 4.5% during preparation and returned to 14.6% during recovery. Strength decreased during preparation and did not fully recover during 6 months of recovery. Testosterone declined from 9.22 to 2.27 ng/mL during preparation and returned back to the baseline level, 9.91 ng/mL, after competition. Total mood disturbance increased from 6 to 43 units during preparation and recovered to 4 units 6 mo after competition.

Conclusions: This case study provides a thorough documentation of the physiological changes that occurred during natural bodybuilding competition and recovery.

Figure 8.2 A fascinating case study published in the *International Journal of Sport Physiology and Performance* describing a single subject response to preparation and recovery for a bodybuilding competition.

Reprinted from Rossow et al. 2013.

Case–Control Studies

Case–control studies are a common evidence type in medicine and therapeutic sciences. They are of particular interest in medicine and therapy because research can focus on a particular disease or disability commonly seen in practice. It is impossible in many instances to randomly select the cases; thus a convenience sample of cases (i.e., anyone who will volunteer) is selected and then matched to controls who are recruited and enrolled.

When reading case–control evidence, one should pay careful attention to the cases; they are in fact the most important component of the study (Young & Solomon, 2009). The cases should be explained in detail, with complete information on how the disease or disability presented, the cause of injury if applicable, the years since injury, current medical and therapeutic interventions, and any other information that sheds light on the current or past physical state of the patient. Omission of such information can mislead any determination of whether the case is relevant to your client, patient, or athlete.

Because the key feature of a case–control investigation is the matching of control subjects to the cases, the exercise practitioner should ensure that the case and controls are as similar as possible (Young & Solomon, 2009). As noted in chapter 4, matching characteristics often include sex, age, height, weight, and body mass index. Previous exercise history may also be an important characteristic in the exercise science literature. Typically, the authors of the investigation provide the specifications that were used to match the cases to the controls. For example, an author might state that the subjects were matched for age ± 2 years. When critically appraising, the reader should ensure that the ranges provided are sensible and appropriate. Further, the practitioner should ensure that both the cases and controls were drawn from the "same population." It may be important in some studies to match for race, ethnicity, socioeconomic status, and other variables with the potential to confound or bias the results; if a study uses cases who are wounded soldiers, logically the controls should also be drawn from the military.

Outcome measures or dependent variables should be measured identically in the cases and controls (Young & Solomon, 2009). This may be difficult in some populations because cases may have an impairment or disability that requires modification of the protocol. If modification is necessary, readers should consider whether the modification could have affected the outcome that is being examined in the study. Finally, readers should determine whether the study measures are subjective or objective. Subjective study measurements can result in significant bias. Biases are present in any study population but may be exaggerated in clinical populations. For example, if an investigator wanted to determine the effect of previous exercise habits on aerobic fitness in patients who suffered a stroke compared to uninjured controls, a case–control study would be an appropriate design. However, both the cases and the controls would have to recall previous exercise habits, which can be prone to bias. The recall difficulties might be amplified by the brain injury resulting from the stroke. Therefore, when subjective measurements are implemented, the reader should carefully consider potential sources of error.

Strengths

The primary strength of a case–control study is its external validity to the appropriate patient population. When investigators develop a research question relative to a particular disease, it makes sense that they would draw their subject samples directly from the pool of patients with the disease. When performed correctly, the matching or pairing techniques for the controls can improve the strength of the evidence; case and control subjects should be matched for as many relevant characteristics as possible. The goal is to set up the experiment such that the defining difference between the two groups is the disease alone. This is not completely possible unless the case and control subjects are monozygotic twins; and indeed, some case–control research matches monozygotic twins with (case) and without disease (control) (Pittaluga, Casini, & Parisi, 2004) to eliminate the powerful influence of genetics on health. These studies are rare, but genetically matched cases and controls dramatically strengthen this research design.

Most case–control studies are population specific. That is, the investigator identifies cases and then draws controls from the same population. Daly and colleagues studied the sprinting biomechanics of elite male Gaelic Games athletes to determine if muscle activation patterns or joint kinematics were associated with hamstring injuries (Daly, McCarthy Persson, Twycross-Lewis, Woledge, & Morrissey, 2015). They appropriately used a case–control study to investigate their research question. First, they identified Gaelic sport athletes with hamstring injuries returning to sport; the cases were matched to control subjects who were also Gaelic sport athletes but with no previous hamstring injuries. Muscle activation patterns and joint kinematics were different between the athletes of the same sport, with and without previous injury, during a running test. The authors determined that there were deficits in muscle activity patterns in the injured athletes and that they had altered knee and hip biomechanics that could place increased strain on the hamstrings. Thus, they suggest that these differences are associated with hamstring injuries in Gaelic Games athletes.

Limitations

Along with strengths, there are limitations to case–control studies. Sackett has identified over 50 potential areas of bias associated with this study design (Sackett, 1979); here we discuss two of the most relevant to exercise science. The first and perhaps most important limitation is selection bias. Selection bias can relate to either group, but it arises primarily as a limitation of the study design. In case–control research, neither group is randomly selected from the population. Instead, a convenience sample of cases presenting at a clinic, hospital, or lab or through available medical records is used. Then, control subjects are found that match the characteristics of the case subjects. This may be accomplished by searching medical records and finding individuals with similar physical characteristics or drawing a sample from the community, purposefully selecting individuals with similar physical characteristics. Either way, random selection is almost always violated in both groups for the sake of the design.

A second important limitation is misclassification bias. Misclassification bias occurs when individuals are admitted into the case group when in fact they are disease or injury free or are admitted into the control group when they in fact have the disease. The degree of misclassification depends on the nature of the disease or injury. For example, if someone is studying the effect of exercise in patients with breast cancer, it is unlikely that subjects will be misclassified. Cases would likely be identified through mammograms and then confirmed by biopsy of the tumor; thus they would be clearly diagnosed with breast cancer. Likewise, controls would be confirmed to be breast cancer free via mammograms. The potential source of error in this type of study would be if a tumor were to manifest in a control subject after the initial group assignment. In this instance, a control participant would be incorrectly identified as cancer free. Misclassification bias can be a significant issue in sport athletes. The incidence of certain injuries, such as concussions, may be underreported in high-velocity contact sports (e.g., rugby, Australian rules football, American football). Thus a player may have suffered one or many "mild" concussions and not reported the injury to avoid missing game time or losing his spot on the roster, or he simply may not have understood the severity of the injury. If participants are used in a case–control study investigating outcomes related to concussion, they may be misclassified into the control group since the injury was never reported.

A second avenue for misclassification bias, and perhaps a more frequent occurrence, is the diagnosis of disease by medical judgment. Individuals are said to be hypertensive if their resting blood pressure is equal to or greater than 140/90 mmHg (James et al., 2014). Practically speaking, hypertension is diagnosed when people present with a blood pressure greater than the diagnostic threshold. They are then asked to return to the office 1 to 4 weeks later for confirmation; and if they are again above the threshold, they are classified as hypertensive (Weber et al., 2014). In contrast, people are not suspected of having hypertension if they present consistently with a blood pressure below the diagnostic threshold. Hypertension is an important risk factor because it can be an indicator of underlying conditions such as cardiovascular disease and chronic kidney disease (Levey et al., 2003; Turner et al., 1998). Suppose an investigator was interested in cardiovascular outcomes and exercise and performed a case–control study on individuals who were hypertensive. The inclusion criteria for the case group might include hypertension diagnosed and confirmed by a physician, whereas the normotensive control group would be free of a hypertension diagnosis. There are several opportunities for potential misdiagnosis in this design. First, the diagnosis of hypertension, although seemingly objective, can be somewhat arbitrary. For example, an individual who is 140/90 would be diagnosed as hypertensive whereas someone who is 135/88 would not receive the diagnosis. It is possible that both have the same underlying problems causing an increase in blood pressure but due to a difference of a few millimeters of mercury (possibly because of human or machine measurement error), one is classified with the disease and the other is not. Second, in a well-known phenomenon, "white coat syndrome," some individuals present with higher blood pressure in the physician's office due to underlying fear or apprehension (Ohkubo et al., 2005; Owens, Atkins, & O'Brien, 1999). Such a person who is normotensive could be diagnosed as hypertensive and misclassified in a study. In both examples, erroneous conclusions could be drawn from the data due to the misclassification of hypertensive patients.

The major limitation of the study by Daly and colleagues (2015) discussed previously is the lack of baseline measurement in the injured Gaelic sport athletes. Since the study was retrospective, the reader must take into account that the altered muscle activity patterns, knee and hip biomechanics, may have surfaced after injury and therefore could not have caused the original injury. Therefore, practitioners must address and carefully weigh critical factors before making a decision to use or exclude results from a case–control study. A critical appraisal tool for grading case–control studies is provided in table 8.1. This table was developed by the STROBE (STrengthening the Reporting

Table 8.1 Checklist of Items That Should be Included in Reports of Case–Control Studies

	Item no.	Recommendation
Title and abstract	1	(a) Indicate the study's design with a commonly used term in the title or the abstract
		(b) Provide in the abstract an informative and balanced summary of what was done and what was found
INTRODUCTION		
Background/rationale	2	Explain the scientific background and rationale for the investigation being reported
Objectives	3	State specific objectives, including any pre-specified hypotheses

(continued)

TABLE 8.1 *(continued)*

	Item no.	Recommendation
		METHODS
Study design	4	Present key elements of study design early in the paper
Setting	5	Describe the setting, locations, and relevant dates, including periods of recruitment, exposure, follow-up, and data collection
Participants	6	*(a)* Give the eligibility criteria and the sources and methods of case ascertainment and control selection. Give the rationale for the choice of cases and controls
		(b) For matched studies, give matching criteria and the number of controls per case
Variables	7	Clearly define all outcomes, exposures, predictors, potential confounders, and effect modifiers. Give diagnostic criteria, if applicable
Data sources/ measurement	8*	For each variable of interest, give sources of data and details of methods of assessment (measurement). Describe comparability of assessment methods if there is more than one group
Bias	9	Describe any efforts to address potential sources of bias
Study size	10	Explain how the study size was arrived at
Quantitative variables	11	Explain how quantitative variables were handled in the analyses. If applicable, describe which groupings were chosen and why
Statistical methods	12	*(a)* Describe all statistical methods, including those used to control for confounding
		(b) Describe any methods used to examine subgroups and interactions
		(c) Explain how missing data were addressed
		(d) If applicable, explain how matching of cases and controls was addressed
		(e) Describe any sensitivity analyses
		RESULTS
Participants	13*	*(a)* Report numbers of individuals at each stage of study—e.g., numbers potentially eligible, examined for eligibility, confirmed eligible, included in the study, completing follow-up, and analyzed
		(b) Give reasons for nonparticipation at each stage
		(c) Consider use of a flow diagram

	Item no.	Recommendation
Descriptive data	14*	(a) Give characteristics of study participants (e.g., demographic, clinical, social) and information on exposures and potential confounders
		(b) Indicate number of participants with missing data for each variable of interest
Outcome data	15*	Report numbers in each exposure category, or summary measures of exposure
Main results	16	(a) Give unadjusted estimates and, if applicable, confounder-adjusted estimates and their precision (e.g., 95% confidence interval). Make clear which confounders were adjusted for and why they were included
		(b) Report category boundaries when continuous variables were categorized
		(c) If relevant, consider translating estimates of relative risk into absolute risk for a meaningful time period
Other analyses	17	Report other analyses done—e.g., analyses of subgroups and interactions, and sensitivity analyses
DISCUSSION		
Key results	18	Summarize key results with reference to study objectives
Limitations	19	Discuss limitations of the study, taking into account sources of potential bias or imprecision. Discuss both direction and magnitude of any potential bias
Interpretation	20	Give a cautious overall interpretation of results considering objectives, limitations, multiplicity of analyses, results from similar studies, and other relevant evidence
Generalizability	21	Discuss the generalizability (external validity) of the study results
OTHER INFORMATION		
Funding	22	Give the source of funding and the role of the funders for the present study and, if applicable, for the original study on which the present article is based

*Give information separately for cases and controls.

Note: An Explanation and Elaboration article discusses each checklist item and gives methodological background and published examples of transparent reporting. The STROBE checklist is best used in conjunction with this article (available free on the websites of PLoS Medicine at www.plosmedicine.org, Annals of Internal Medicine at http://annals.org, and Epidemiology at www.epidem.com). Information on the STROBE Initiative is available at www.strobe-statement.org.

Reprinted, by permission, from STROBE, 2009, STROBE statement: Strengthening the report of observational studies in epidemiology.

of OBservational studies in Epidemiology) initiative and can be used as a checklist to determine the inclusion or exclusion of essential elements in a case–control study.

Cohort Studies

Cohort studies are the gold standard of epidemiologic research (Gordis, 2009). Most cohort studies seek to determine relationships or associations between exposure and disease. In many cases, this type of evidence is the strongest form available. For example, if one is interested in studying the relationship between chemical exposure (e.g., benzene) and cancer risk, a randomized controlled trial in humans would not be ethical or possible. Instead, a cohort design in which an investigator studies the relationship between occupational exposure (or lack thereof) and disease risk is ideal (Glass et al., 2003). Although this is a strong design, an exercise practitioner must carefully appraise cohort studies to ensure proper control of confounding factors and minimization of bias.

As discussed in chapter 4, cohort studies can be either prospective or retrospective. Prospective studies start with groups of individuals and follow them forward in time; retrospective studies start with a population and ask questions about their past. Both have strengths and limitations, and the exercise practitioner should identify the type of cohort study before proceeding with a critical appraisal. Secondly, it is important that the practitioner determine if the cohort represents a true subset of a particular population (Young & Solomon, 2009). It is essential that both the exposed and unexposed groups be drawn from the same population. Drawing from a single population helps to eliminate some of the biases between groups, although there are always confounding factors in a cohort study. When reading and appraising, the exercise practitioner should ensure that all of the potential confounders associated with the disease, injury, or condition are clearly identified (Young & Solomon, 2009). It is also important that the measurements be precise and consistent across groups . In many cases, the confounders can be controlled through statistical modeling. When confounders are not controlled, this should alert the reader to the potential for invalid findings.

Because cohort studies typically use large samples, it is common for subjects to drop out of the research project. Well-written cohort studies identify the number of subjects in both groups that were lost to follow-up in either the methods or the results section (Young & Solomon, 2009). When possible, the reason for loss should be identified since it could have a direct effect on the interpretation of the results. A final important consideration in critical appraisal of prospective cohorts is the study length (Young & Solomon, 2009). Certain diseases progress slowly over time, while others remain latent only to progress rapidly many years after exposure. If the length of follow-up in a cohort study is too short, an investigator may incorrectly conclude that disease is unrelated to exposure when in fact it was related but had a later onset. In the context of other research projects, the exercise practitioner should carefully consider whether the study length was sufficient to justify the authors' conclusions.

Strengths

The primary strength of a cohort study as it relates to exercise science is the sample size. Perhaps the best example of a cohort study is the Framingham Heart Study. Initiated in 1948, the Framingham study investigated the effects of lifestyle factors and certain biomarkers on cardiovascular disease. Now in its third generation, the study has enrolled nearly 20,000 participants, and it is not an exaggeration to state that most findings on cardiovascular risk factors and lifestyle are due in part to work with this cohort. The tremendous success of the project lies in its volume and duration. Obviously the study was very well-designed, but many of the findings are the result of a long-term commitment (>65 years and counting) in a large sample of subjects that has grown to include children and grandchildren of the original cohort. It is simply not possible to conduct a comparable randomized controlled trial and in the case of the Framingham study, a randomized controlled trial would not have been the appropriate design to address the study hypotheses.

Another strength of a cohort study design is the opportunity to observe natural outcomes. Because the independent variable occurs naturally, it is not influenced by the investigator. Thus, the outcomes may better reflect the occurrence of the independent variable in the general population. Multiple outcomes can also be tested with a single experimental design. Cohort studies generate large databases of information that can be retrospectively analyzed to answer research questions other than the original. As the prevalence of electronic medical recording continues to grow, it is likely that the number and quality of cohort study designs will increase.

Limitations

There are several limitations to the cohort design. One important weakness is the lack of control of the independent variable. Because the exposure occurs naturally in the population, its level is not controlled, and this can lead to errors in the interpretation of results. In studies of exercise and physical activity using cohort designs, participants are often dichotomously classified into exercising and sedentary groups, and there are obviously degrees of each. Depending on how these groups are defined, a participant may have an active, physically demanding job but not engage in vigorous exercise and thus be classified as sedentary. A person reporting a history of exercise might train vigorously for 30 min four times per week but have an office job and sit for 8 to 10 h per day. Different levels of exercise intensity, volume, and duration as well as different exercise modes are also unlikely to be accounted for in this design; that is, an individual who exercises three times per week for 30 min a day may be placed in the same group as a triathlete who exercises 2 to 3 h per day 6 days a week. Placing these individuals in the same group may result in underestimation or overestimation of the impact of exercise.

Cohort studies rely heavily on self-reporting, which, as we have noted before, may lead to errors or bias. People often have difficulty remembering daily activities, and recall error may be increased depending on the time interval since the event. With exercise specifically, individuals may over- or underreport their activity, and answers may also be influenced by the way the questions are presented (Hutto, Sharpe, Granner, Addy, & Hooker, 2008; Tsai, Chee, & Im, 2006).

Possibly the greatest limitation of the cohort study design is that it cannot establish causality. The results simply demonstrate an association between two or more variables; the association or correlation may be strong or weak. In fact, some of the most important health-related findings are based on association, not causation. The association between lung cancer and smoking was established through epidemiological cohort research. When the seminal report from the Surgeon General was published in 1964, there was no randomized controlled trial supporting a causal link between smoking and poor health outcomes because such a study would be unethical (Bayne-Jones et al., 1964). However, there was an abundance of observational data supporting relationships between smoking and poor health outcomes. Even with the large amount of data published since 1964, one still cannot say that smoking causes lung cancer, only that there is a very strong association between the two. The association between the variables can be strengthened if the eight criteria for causation described by Hill (1965) are met:

1. Strength: Strength of the association indicates that when there is exposure, there is a high probability of development of the disease.

2. Consistency: Consistency indicates that the finding is replicable. It has been demonstrated not only in a single population, but also in multiple populations at different times and in different places.

3. Specificity: There is a much higher incidence of the particular disease or condition within a particular exposed population; the incidence is significantly lower in a different, unexposed population.

4. Temporality: The exposure must be present before the onset of the disease or the condition.

5. Biological dose response: Risk for the disease or condition increases or decreases in relation to the magnitude of exposure.

6. Biological plausibility: Given the current understanding of human biology, it is conceivable that exposure could cause an increased risk for disease or the condition.

7. Coherence: The potential cause–effect relationship should be consistent with other published research.

8. Experiment: If exposure is decreased, the incidence of the disease or condition is reduced.

To move from mere association toward causation, the exercise practitioner should examine Hill's criteria. Failure to satisfy a number of the criteria listed could indicate the presence of confounding variables. A critical appraisal tool for cohort studies is provided in table 8.2. This list, developed by the STROBE initiative, determines the inclusion or exclusion of essential elements in a study.

Randomized Controlled Trials

Cohort research is the trademark study design for epidemiological questions, but a randomized controlled trial is the ideal methodology to test most practice-based scientific hypotheses and, for exercise practitioners, to answer evidence-based questions. It is the preferred method, when feasible, because it allows for careful control of the independent variable in the tested population. The randomized controlled trial study design also allows the researcher to compare to a reference (control) group that receives a placebo treatment or no treatment at all. In order to reduce bias in the measurement of the dependent variables, group allocation can be concealed from the investigators until the conclusion of the project—this would be a single-blind study. To further control for bias, some research questions allow the participants to be blinded to group or treatment allocation. When both the investigators and the participants are blinded to group assignment, the study is termed a double-blind randomized controlled trial.

The first important item to address is the project's randomization procedure (Young & Solomon, 2009). Ideally, participants are selected at random from the population.

Table 8.2   Checklist of Items That Should be Included in Reports of Cohort Studies

	Item no.	Recommendation
Title and abstract	1	*(a)* Indicate the study's design with a commonly used term in the title or the abstract
		(b) Provide in the abstract an informative and balanced summary of what was done and what was found
INTRODUCTION		
Background/ rationale	2	Explain the scientific background and rationale for the investigation being reported
Objectives	3	State specific objectives, including any prespecified hypotheses
METHODS		
Study design	4	Present key elements of study design early in the paper
Setting	5	Describe the setting, locations, and relevant dates, including periods of recruitment, exposure, follow-up, and data collection
Participants	6	*(a)* Give the eligibility criteria, and the sources and methods of selection of participants. Describe methods of follow-up
		(b) For matched studies, give matching criteria and number of exposed and unexposed
Variables	7	Clearly define all outcomes, exposures, predictors, potential confounders, and effect modifiers. Give diagnostic criteria, if applicable
Data sources/ measurement	8*	For each variable of interest, give sources of data and details of methods of assessment (measurement). Describe comparability of assessment methods if there is more than one group
Bias	9	Describe any efforts to address potential sources of bias
Study size	10	Explain how the study size was arrived at
Quantitative variables	11	Explain how quantitative variables were handled in the analyses. If applicable, describe which groupings were chosen and why

(continued)

TABLE 8.2 *(continued)*

	Item no.	Recommendation
METHODS *(continued)*		
Statistical methods	12	*(a)* Describe all statistical methods, including those used to control for confounding
		(b) Describe any methods used to examine sub-groups and interactions
		(c) Explain how missing data were addressed
		(d) If applicable, explain how loss to follow-up was addressed
		(e) Describe any sensitivity analyses
RESULTS		
Participants	13*	*(a)* Report numbers of individuals at each stage of study—e.g., numbers potentially eligible, examined for eligibility, confirmed eligible, included in the study, completing follow-up, and analyzed
		(b) Give reasons for nonparticipation at each stage
		(c) Consider use of a flow diagram
Descriptive data	14*	*(a)* Give characteristics of study participants (e.g., demographic, clinical, social) and information on exposures and potential confounders
		(b) Indicate number of participants with missing data for each variable of interest
		(c) Summarize follow-up time (e.g., average and total amount)
Outcome data	15*	Report numbers of outcome events or summary measures over time
Main results	16	*(a)* Give unadjusted estimates and, if applicable, confounder-adjusted estimates and their precision (e.g., 95% confidence interval). Make clear which confounders were adjusted for and why they were included
		(b) Report category boundaries when continuous variables were categorized
		(c) If relevant, consider translating estimates of relative risk into absolute risk for a meaningful time period
Other analyses	17	Report other analyses done—e.g., analyses of sub-groups and interactions, and sensitivity analyses

	Item no.	Recommendation
DISCUSSION		
Key results	18	Summarize key results with reference to study objectives
Limitations	19	Discuss limitations of the study, taking into account sources of potential bias or imprecision. Discuss both direction and magnitude of any potential bias
Interpretation	20	Give a cautious overall interpretation of results considering objectives, limitations, multiplicity of analyses, results from similar studies, and other relevant evidence
Generalizability	21	Discuss the generalizability (external validity) of the study results
OTHER INFORMATION		
Funding	22	Give the source of funding and the role of the funders for the present study and, if applicable, for the original study on which the present article is based

*Give information separately for exposed and unexposed groups.

Note: An Explanation and Elaboration article discusses each checklist item and gives methodological background and published examples of transparent reporting. The STROBE checklist is best used in conjunction with this article (available free on the websites of *PLoS Medicine* at www.plosmedicine.org, *Annals of Internal Medicine* at http://annals.org, and *Epidemiology* at www.epidem.com). Information on the STROBE Initiative is available at www.strobe-statement.org.

Reprinted, by permission, from STROBE, 2009, *STROBE statement: Strengthening the report of observational studies in epidemiology.*

After random selection, the participants are randomly assigned to a group depending on the study design. The randomization procedure can be completed using a simple coin flip in a two-group design or via a computer-generated randomizer for multiple group allocations. If subjects are completing multiple treatments (crossover design), the order in which the treatments are presented may be balanced across all subjects but randomized to individual subjects. In other words, if there are two treatments, half of the subjects (randomly) complete treatment 1 followed by treatment 2; to balance the design, the other half of the subjects complete the two treatments in the opposite order. If the study is using a small sample size, a matched-pair randomized design may be implemented. Similar to what occurs in a case–control design, participants with similar physical characteristics are "matched" and then randomized into the treatment or placebo group. This can minimize the risk of inadvertently providing one group a competitive advantage due to some inherent physiological trait. Unfortunately, this design cannot account for genetic predisposition to training adaptations, which can be controlled for only in monozygotic twin studies. As noted earlier, very few of these types of twin studies have been published in the exercise science literature, but the few in print represent one of the strongest forms of evidence (Lee et al., 2007; Monga, Macias, Groppo, Kostelec, & Hargens, 2006; Smith et al., 2003).

A second consideration in appraising an article is to determine whether the subjects were truly blinded to the treatment (Young & Solomon, 2009). It is important, when possible, to use a sham or placebo intervention. The placebo effect is a well-known psychological phenomenon whereby an improvement from an intervention is seen even when the improvement is not biologically plausible. Thus, a sham treatment is used in medical research in which a medication and a placebo can be given in a form that appears identical. The placebo is used to account for any psychological effect that could be present but unrelated to the intervention. The same is true for nutritional supplementation studies, but it is often impossible to blind individuals to an exercise treatment. However, it is possible to blind the investigators to the group assignment, or at least to conceal the treatment from the individuals assessing the outcome measures, to avoid any potential bias in rating. Biased investigators could inadvertently or purposely sway the results of a study if they know the group assignments before completing poststudy testing. Thus, authors should specify all techniques used to conceal group assignment. Similar to what occurs in cohort studies, participants may drop out from a randomized controlled trial. Investigators should disclose the attrition rate for both groups because this information may provide useful evidence concerning the feasibility of a treatment. A large attrition rate in a treatment group could indicate that the intervention is poorly tolerated (Young & Solomon, 2009). If some data points are missing for a subject but the subject's other data are not excluded from analysis, the investigator should also describe the process used to account for the missing data.

Strengths

The greatest strength of a randomized controlled trial is the degree of control it affords over the independent variable. If the researchers are testing a novel exercise intervention or supplement, they can manipulate the precise dose. Unlike a cohort study in which the independent variable occurs naturally, in a randomized controlled trial the investigator prescribes the dose and observes its effects over time. In many exercise intervention studies, each and every exercise session is monitored, allowing the investigator to answer precise questions about intensity, volume, duration, and the many other acute programming variables prescribed by exercise practitioners. Another strength of this design is that it allows the comparison of two or more interventions. Suppose an investigator is interested in comparing a current standard-of-care protocol to a novel intervention. Using a three-group design with a control group in a single experiment, both groups could be compared to each other and then to a no-exercise control group. Numerous randomized controlled trial designs can be implemented to provide strong evidence to answer research questions.

Limitations

A randomized controlled trial is only as strong as its control and the accounting for confounding factors; unfortunately, many in the exercise science literature fall short of their potential due to serious flaws. Additionally, randomization works only with larger *n* sizes used to gain a normal distribution; thus, matching before randomization may be needed to better meet the demands of linear statistics and the assumptions made in using them. Some of the flaws are due to the nature of exercise; others are due to omissions or sloppiness in data collection. As mentioned previously, it is nearly impossible to conceal the group assignment in an exercise intervention study. Participants know that they are exercising, so they are obviously not in the nonexercising control group; thus most intervention studies in exercise are single blind at best. Another common limitation of randomized controlled trials of exercise interventions is the lack of external control (i.e., control of confounding variables). Physiological adaptations are profoundly affected by diet and recovery. If the investigator carefully controls the intervention, precisely prescribing the volume, intensity, and duration, but does not monitor diet and recovery behaviors, the study results may be questioned. Designing an elegant, well-controlled randomized controlled trial is an art that few scientists should attempt alone. The numerous areas of bias and the confounding variables may result in the need for large, diverse research teams to address such factors as specific controls, study length, and instrumentation that may influence the effectiveness of the intervention in relation to the dependent variable. Indeed, this is why many of the seminal papers in the exercise science literature have multiple authors with different areas of expertise. This team approach allows the investigators to focus on their particular area of expertise and cohesively arrive at a protocol that minimizes potential fatal flaws in study design.

Another criticism (and limitation) of a randomized controlled trial is ironically also its key strength—the control of the independent variable. In order to protect the validity of the independent variable's effects on the dependent variable, the research protocol may be controlled to the extent that it no longer reflects a real-life scenario. This is a problem in practice-based research; at times, the strength of the evidence may be compromised to ensure that the protocol reflects real-world practices. Conversely, the real-world application may be compromised for the sake of scientific design. Exercise practitioners must be aware of this limitation and determine the applicability of a study's findings to their client, patient, and athlete base. Figure 8.3 provides a critical appraisal tool for grading randomized controlled trials.

Systematic Reviews

A systematic review is not a study design but rather a synthesis of articles on a selected topic (e.g., systematic review of the effects of caffeine on endurance performance). This type of article can be a valuable time-saving tool for the evidence-based practitioner. On the other hand, systematic reviews can be biased and misleading if not properly conducted. As with all other forms of research, one must appraise the quality of the systematic review to determine its usefulness.

In contrast to a narrative review, a systematic review is a focused, objective review of articles relevant to a specific topic. When appraising a systematic review, the exercise practitioner should ensure that the authors specify their search strategy and article inclusion and exclusion criteria (Young & Solomon, 2009). The authors should also specify the number of articles found and the final number of articles used and should say why the other articles were eliminated. The search and inclusion–exclusion strategy should be performed by at least two individuals, and when there is a tie (i.e., one investigator includes, the other excludes), a third investigator should make the final decision (Young & Solomon, 2009). Studies included in the analysis should be described in sufficient detail to allow the reader to understand the most important aspects of the study, including the sample size and outcome measures. Each study should be rated through the use of some type of critical appraisal technique or form.

CHECKLIST FOR CRITICAL APPRAISAL OF INTERVENTION STUDIES INCLUDING RANDOMIZED CONTROLLED TRIALS

ARE THE RESULTS OF THE TRIAL VALID? (INTERNAL VALIDITY)

WHAT QUESTION DID THE STUDY ASK?

Patients:

Intervention:

Comparison:

Outcome(s):

Was the assignment of patients to treatments randomized?	
What is best?	Where do I find the information?
Centralized computer randomization is ideal and often used in multicenter trials. Smaller trials may use an *independent* person (e.g., the hospital pharmacy) to "police" the randomization.	The ***Methods*** should tell you how patients were allocated to groups and whether or not randomization was concealed.

This paper: Yes ☐ No ☐ Unclear ☐
Comment:

Were the groups similar at the start of the trial?	
What is best?	Where do I find the information?
If the randomization process worked (that is, achieved comparable groups) the groups should be similar. The more similar the groups the better it is. There should be some indication of whether differences between groups are statistically significant (i.e., *p* values).	The ***Results*** should have a table of "Baseline Characteristics" comparing the randomized groups on a number of variables that could affect the outcome (e.g., age, risk factors). If not, there may be a description of group similarity in the first paragraphs of the ***Results*** section.

This paper: Yes ☐ No ☐ Unclear ☐
Comment:

Aside from the allocated treatment, were groups treated equally?	
What is best?	Where do I find the information?
Apart from the intervention the patients in the different groups should be treated the same, e.g., additional treatments or tests.	Look in the ***Methods*** section for the follow-up schedule, and permitted additional treatments, and so on and in ***Results*** for actual use.

This paper: Yes ☐ No ☐ Unclear ☐
Comment:

Figure 8.3 Critical appraisal tool for grading randomized controlled trials.

Adapted from Center for Evidence-Based Medicine http://www.cebm.net/critical-appraisal/ (Therapy / RCT Critical Appraisal Sheet). Under Creative Commons license 4.0.

Were all patients who entered the trial accounted for? And were they analyzed in the groups to which they were randomized?

What is best?	Where do I find the information?
Losses to follow-up should be minimal—preferably less than 20%. However, if few patients have the outcome of interest, then even small losses to follow-up can bias the results. Patients should also be analyzed in the groups to which they were randomized—"intention-to-treat analysis."	The *Results* section should say how many patients were randomized (e.g., Baseline Characteristics table) and how many patients were actually included in the analysis. You will need to read the *Results* section to clarify the number and reason for losses to follow-up.

This paper: Yes ☐ No ☐ Unclear ☐
Comment:

Were measures objective or were the patients and clinicians kept "blind" to which treatment was being received?

What is best?	Where do I find the information?
It is ideal if the study is "double-blinded" —that is, both patients and investigators are unaware of treatment allocation. If the outcome is *objective* (e.g., death), then blinding is less critical. If the outcome is *subjective* (e.g., symptoms or function), then blinding of the outcome assessor is critical.	First, look in the *Methods* section to see if there is some mention of masking of treatments, e.g., placebos with the same appearance or sham therapy. Second, the *Methods* section should describe how the outcome was assessed and whether the assessor/s were aware of the patients' treatment.

This paper: Yes ☐ No ☐ Unclear ☐
Comment:

WHAT WERE THE RESULTS?

1. How large was the treatment effect?

Most often results are presented as dichotomous outcomes (yes or no outcomes that happen or don't happen) and can include such outcomes as cancer recurrence, myocardial infarction, and death. Consider a study in which 15% (0.15) of the control group died and 10% (0.10) of the treatment group died after 2 years of treatment. The results can be expressed in many ways as shown below.

What is the measure?	What does it mean?
Relative Risk (RR) = risk of the outcome in the treatment group / risk of the outcome in the control group.	The relative risk tells us **how many times more likely** it is that an event will occur in the treatment group relative to the control group. An **RR** of **1** means that there is no difference between the two groups thus, the treatment had **no effect**. An RR < 1 means that the treatment decreases the risk of the outcome. An RR > 1 means that the treatment increased the risk of the outcome.
In our example, the RR = 0.10/0.15 = 0.67	Since the RR < 1, the treatment decreases the risk of death.

Figure 8.3 *(continued)*

What is the measure?	What does it mean?
Absolute Risk Reduction (ARR) = risk of the outcome in the control group – risk of the outcome in the treatment group. This is also known as the **absolute risk difference.**	The absolute risk reduction tells us the absolute difference in the rates of events between the two groups and gives an indication of the baseline risk and treatment effect. An **ARR** of **0** means that there is no difference between the two groups; thus, the treatment had **no effect.**
In our example, the ARR = 0.15 – 0.10 = 0.05 or 5%.	The absolute benefit of treatment is a 5% reduction in the death rate.
Relative Risk Reduction (RRR) = absolute risk reduction / risk of the outcome in the control group. An alternative way to calculate the RRR is to subtract the RR from 1 (e.g., RRR = 1 – RR).	The relative risk reduction is the complement of the RR and is probably the most commonly reported measure of treatment effects. It tells us the reduction in the rate of the outcome in the treatment group relative to that in the control group.
In our example, the RRR = 0.05/0.15 = 0.33 or 33%. Or RRR = 1 – 0.67 = 0.33 or 33%.	The treatment reduced the risk of death by 33% relative to that occurring in the control group.
Number Needed to Treat (NNT) = inverse of the ARR and calculated as 1 / ARR.	The number needed to treat represents the number of patients we need to treat with the experimental therapy in order to prevent 1 bad outcome and incorporates the duration of treatment. Clinical significance can be determined to some extent by looking at the NNTs, but also by weighing the NNTs against any harms or adverse effects (NNHs) of therapy.
In our example, the NNT = 1/ 0.05 = 20.	We would need to treat 20 people for 2 years in order to prevent 1 death.

2. How precise was the estimate of the treatment effect?

The true risk of the outcome in the population is not known and the best we can do is estimate the true risk based on the sample of patients in the trial. This estimate is called the **point estimate.** We can gauge how close this estimate is to the true value by looking at the confidence intervals (CI) for each estimate. If the confidence interval is fairly narrow, then we can be confident that our point estimate is a precise reflection of the population value. The confidence interval also provides us with information about the statistical significance of the result. If the value corresponding to **no effect** falls outside the 95% confidence interval, then the result is statistically significant at the 0.05 level. If the confidence interval includes the value corresponding to **no effect,** then the results are not statistically significant.

WILL THE RESULTS HELP ME IN CARING FOR MY PATIENT? (EXTERNAL VALIDITY/APPLICABILITY)

The questions that you should ask before you decide to apply the results of the study to your patient:

- Is my patient so different from those in the study that the results cannot apply?
- Is the treatment feasible in my setting?
- Will the potential benefits of treatment outweigh the potential harms of treatment for my patient?

Figure 8.3 *(continued)*

Strengths

The strength of a systematic review, especially for the practitioner, is that an investigator has already completed the article search and synthesized the information. If the systematic review is appropriately designed and relevant to the evidence-based question, the practitioner can read it and then search for newer articles (i.e., articles published since the search) to make a final decision concerning acceptance or rejection of the technique or intervention. If a well-designed systematic review indicates agreement across a large number of articles, there is likely agreement on a topic. In many evidence-based models, systematic reviews are the strongest form of evidence.

Limitations

The value of a systematic review is dependent on the rigor of the search and synthesis process and the quality of the articles reviewed. If the search process is not exhaustive or the synthesis process is poor, then a systematic review may be weak and unusable for the practitioner. Authors can be and are often biased in their selection of articles. Thus, one must assess the completeness of the search and ensure that the authors are clearly using all relevant articles. Likewise, even if a systematic review is well-designed, if the available studies are of poor quality, the results should be questioned. Thus, it is imperative that practitioners not simply accept a systematic review as a strong form of evidence until they have carefully appraised its quality. A critical appraisal tool for grading systematic reviews is provided in figure 8.4.

LEVELS OF EVIDENCE

Having explored the types of evidence and various common research designs, we can now discuss levels of evidence. Levels of evidence are a hierarchical arrangement of evidence types that allow practitioners to effectively evaluate a body of evidence and make appropriate programming decisions. Human physiology is complex, and often the answers we seek are not simple or straightforward; when a question is asked in a slightly different way or about a different population, the answer may be different. When it comes to evidence, evidence-based exercise practitioners may encounter four scenarios (table 8.3):

Scenario I represents an area of knowledge (1) that is new, or (2) that is difficult or costly to conduct research in, or (3) in which necessary technologies do not yet exist to permit research efforts. Scenario II represents the ideal, that is, an area of knowledge replete with much research that is homogenous in its findings. Scenario III is the most challenging: an area in which, despite much research, no consensus has been reached and uncertainty persists due to the variance in the findings. Scenario IV represents an area with little research and one in which a consensus does not yet exist.

Whereas the levels of evidence discussed in the following sections are rigid and well defined, these four categorizations (which are not actual levels of evidence) represent qualitative divisions pertaining to the volume and consensus of scientific evidence within a particular topic of research inquiry. For instance, after a review of the literature we might conclude that the evidence for a link between smoking and lung cancer is plentiful and in strong agreement, placing this topic in scenario II. It is important to remember that the literature is constantly in flux, with investigators continually adding to the

CHECKLIST FOR CRITICAL APPRAISAL OF SYSTEMATIC REVIEWS

What question (PICO) did the systematic review address?

What is best?	Where do I find the information?
The main question being addressed should be clearly stated. The exposure, such as a therapy or diagnostic test, and the outcome(s) of interest will often be expressed in terms of a simple relationship.	The **Title, Abstract,** *or final paragraph of the* **Introduction** should clearly state the question. If you still cannot ascertain what the focused question is after reading these sections, search for another paper!

This paper: Yes ☐ No ☐ Unclear ☐
Comment:

Is it unlikely that important, relevant studies were missed?

What is best?	Where do I find the information?
The starting point for comprehensive search for all relevant studies is the major bibliographic databases (e.g., Medline, Cochrane, EMBASE) but should also include a search of reference lists from relevant studies and contact with experts, particularly to inquire about unpublished studies. The search should not be limited to English language only. The search strategy should include both MeSH terms and text words.	The **Methods** section should describe the search strategy, including the terms used, in some detail. The **Results** section will outline the number of titles and abstracts reviewed, the number of full-text studies retrieved, and the number of studies excluded together with the reasons for exclusion. This information may be presented in a figure or flow chart.

This paper: Yes ☐ No ☐ Unclear ☐
Comment:

Were the criteria used to select articles for inclusion appropriate?

What is best?	Where do I find the information?
The inclusion or exclusion of studies in a systematic review should be clearly defined a priori. The eligibility criteria used should specify the patients, interventions or exposures, and outcomes of interest. In many cases the type of study design will also be a key component of the eligibility criteria.	The **Methods** section should describe in detail the inclusion and exclusion criteria. Normally, this will include the study design.

Figure 8.4 Critical appraisal tool for grading systematic reviews.

Adapted from Center for Evidence-Based Medicine http://www.cebm.net/critical-appraisal/ (Systematic Review Critical Appraisal Sheet). Under Creative Commons license 4.0.

This paper: Yes □ No □ Unclear □
Comment:

Were the included studies sufficiently valid for the type of question asked?	
What is best?	Where do I find the information?
The article should describe how the quality of each study was assessed using predetermined quality criteria appropriate to the type of clinical question (e.g., randomization, blinding, and completeness of follow-up).	The **Methods** section should describe the assessment of quality and the criteria used. The **Results** section should provide information on the quality of the individual studies.

This paper: Yes □ No □ Unclear □
Comment:

Were the results similar from study to study?	
What is best?	Where do I find the information?
Ideally, the results of the different studies should be similar or homogeneous. If heterogeneity exists the authors may estimate whether the differences are significant (chi-square test). Possible reasons for the heterogeneity should be explored.	The **Results** section should state whether the results are heterogeneous and discuss possible reasons. The forest plot should show the results of the chi-square test for heterogeneity and discuss reasons for heterogeneity, if present.

This paper: Yes □ No □ Unclear □
Comment:

Figure 8.4 *(continued)*

Table 8.3   Volume and Consensus of Information Derived From Peer-Reviewed Articles

Agreement of the evidence	QUANTITY OF EVIDENCE	
	Many studies	Few or no studies
Homogenous	Many studies with similar findings (II)	Few studies with similar findings (I)
Divergent	Many studies but with divergent findings (III)	Few studies with divergent findings (IV)

body of knowledge on a particular subject. Nevertheless, exercise practitioners must make decisions for their clients, patients, and athletes based on the present literature. It is important when making these decisions that practitioners understand the volume and consensus as well as the strength or level of evidence supporting their answer to an evidence-based question.

Classic Levels of Evidence for Medicine (CEBM)

Evidence-based practice (EBP) must have, at its core, a grading system that includes all relevant evidence sources. The Centre for Evidence-Based Medicine (CEBM) at Oxford University has established levels of evidence that are well respected and encompass many types of study designs (Law & MacDermid, 2008) (table 8.4). The highest level of evidence is a systematic review of randomized controlled trials with homogenous findings. As discussed earlier, randomized controlled trials are the least biased, most objective form of research evidence; thus, a systematic synthesis of such studies in a given research area that produces a clear consensus result represents the strongest available evidence. Next, predictably, are randomized controlled trials themselves, specifically, randomized controlled trials with narrow confidence intervals, meaning that variation in response to the intervention was low.

There may be varying levels of evidence supporting an answer to an evidence-based question. For example, a literature search may uncover cohort studies, randomized controlled trials, and systematic reviews all relevant to the question. This may be a natural

Table 8.4 Levels of Evidence or Grading Scheme Provided by the Centre for Evidence-Based Medicine

Level of Evidence	Type of evidence
1a	Systematic review of randomized controlled trials (with homogeneity)
1b	An individual randomized controlled trial with narrow confidence intervals
2a	Systematic review of cohort studies (with homogeneity)
2b	Individual cohort studies and low-quality randomized controlled trials
2c	Outcomes research (tracking studies, discussed in chapter 10)
3a	Systematic review of case–control studies with homogeneity
3b	Individual case–control studies
4	Case series and poor-quality cohort or case–control studies
5	Expert opinion without explicit critical appraisal, or based on basic science research or basic knowledge of physiological mechanisms

progression in some study areas—that is, historical cohorts lead to randomized controlled trials and eventually a rich body of literature resulting in a systematic review. In such case, the evidence is given a ranking according to the highest form of evidence available.

Levels of Evidence for Exercise Science (NHLBI/ACSM)

The level of evidence grading scheme adopted by the American College of Sports Medicine (ACSM) for position stands was originally published by the National Heart, Lung, and Blood Institute (NHLBI) (Ratamess et al., 2009) (table 8.5). The scheme uses an A to D system for grading the strength of evidence (figure 8.5). Level A evidence in this model includes a rich body of well-designed randomized controlled trials with consistent findings. Level A is awarded only when there is substantial evidence supporting a position. Level B ranking is awarded when there are fewer randomized controlled trials, the results are inconsistent, or the population studied is different from the population in which an intervention is to be implemented. When a meta-analysis is used to combine results from multiple smaller studies, the recommendation is still level B. Observational research and poorly controlled experimental research are given a rank of C in the NHLBI scheme; outcomes research with a large number of participants would also be ranked as level C. Finally, expert opinion or panel recommendations based

Table 8.5 Levels of Evidence Presented by the National Heart, Lung, and Blood Institute

Category	Source of Evidence	Definition
A	Randomized Controlled Trials (RCT; rich body of data)	Evidence is from well-designed RCT that provide a consistent pattern of findings in the population for which the recommendation is made. Requires a substantial number of participants.
B	RCT (limited body of data)	Evidence is from intervention studies that include only a limited number of RCT, post hoc or subgroup analysis of RCT, or meta-analysis of RCT. Pertains when few randomized trials exist, they are small, and the results are somewhat inconsistent or were from a nonspecific population.
C	Nonrandomized trials, observational studies	Evidence is from outcomes of uncontrolled trials or observations.
D	Panel consensus judgment	Expert judgment is based on the panel's synthesis of evidence from experimental research or the consensus of panel members based on the clinical experience or knowledge that does not meet the above-listed criteria.

Adapted from NHLBI 1998. Available: www.ncbi.nlm.nih.gov/books/NBK2003/pdf/TOC.pdf

Figure 8.5 Depiction of the grading system developed by the National Heart, Lung, and Blood Institute and adopted by the American College of Sports Medicine for position statements.

on evidence are awarded a grade of level D. Other peer-reviewed research not fitting into the other categories is also assigned a level D grade.

Each professional discipline (e.g., medicine, physical therapy, occupational therapy, nursing) has in some sense unofficially adopted a particular level of evidence model. The model selected is related to the types of research that typically appear in that discipline. In the case of exercise science, we advocate adoption of the NHLBI model (table 8.5). The model is simplified compared to others and provides two grading levels for randomized controlled trials, an area emphasized in the exercise intervention literature. For the remainder of this text, including the case study chapters, we use this model to provide categorical grades for the levels of evidence.

The grading or ranking of evidence is the essential component of the EBP methodology. Assignment of a grade specifies the strength of the quality of information related to a question. Thus, evidence-based exercise practitioners should not state that a technique, philosophy, product, or protocol is evidenced-based. Instead, they state that level A, B, C, or D evidence exists to support or refute a practice. The use of this nomenclature definitively specifies the strength of support. To overturn an answer based on level A evidence, substantial new high-quality evidence is required. On the other hand, evidence receiving a grade of D is relatively weak; the exercise practitioner should therefore approach implementation cautiously and continue to watch the literature for new information confirming or contradicting the current stand.

CONCLUSION

Evidence-based practice is a decision-making process that hinges on information—more specifically, quality information. There is always evidence to support or refute the use of a technique, supplement, device, protocol, or testing tool. The question the evidence-based exercise practitioner must address is the strength of the information. Information derived from expert opinion may be formed from strong experimental evidence. It may also be based on biased personal opinion and be in conflict with research evidence. Evidence obtained through a rich body of quality experimental studies is less prone to bias and thus forms a stronger information base. However, the type of study alone does not necessarily suggest strength. Research can be well or poorly designed; therefore, the evidence-based practitioner must develop and practice the skills of critically appraising the literature.

Chapter 9

INCORPORATING EVIDENCE INTO PRACTICE

Learning Objectives

1. Discuss the process of determining when to use new evidence.
2. Learn to value the patient, client, or athlete and to understand why the person's input is critical to the process of evidence-based practice.
3. Present critical questions that must be answered before using newly found evidence.
4. Discuss practical ways of implementing the art and science of exercise prescription.

The Galleria dell'Accademia in Florence, Italy, houses one of the most remarkable and famous sculptures in history—"David." David was carved from 1501 to 1504 by Michelangelo. An incredible depiction of King David, it has been admired for centuries for its vivid and lifelike rendition of the human body. It is hard to believe that at one time "David" was a solid block of Carraran marble. Michelangelo, with nothing but an idea and a block of marble, created this beautiful work of art—it is astounding! Michelangelo could not have produced this statue with simply an idea—he needed tools to bring his vision to fruition. The artist likely used chisels, hammers, and other carving tools to create David; undeniably the statue could not have been formed without these instruments. However, when we discuss the statue of David, no one marvels over the tools. The history books do not credit the chisels for their masterful work in creating David—they rightfully credit the artist. Billions of people who have lived throughout history could have been given the same tools and marble and never achieved the greatness of Michelangelo's sculpture. In the end, the preeminent element of art is the artist.

A quote often attributed to the great philosopher Sir Francis Bacon is "Knowledge is power." On the surface, this sounds like a logical assertion, but on deeper investigation, it may not be completely accurate. In reality, knowledge is a tool or an instrument, and it is no more or less powerful than the user who holds it. Much as with a chisel, the power of the tool lies in the hands of the artist. The ultimate goal of the evidence-based approach is to positively affect the health, performance, or outcomes of a patient, client, or athlete. To this point, we have discussed the process of writing a question, finding evidence, and interpreting the evidence. None of these steps help or hurt our patients, clients, or athletes in any way. People benefit when practitioners choose to include or exclude the discovered evidence in their exercise prescription or training plan or make other decisions to improve health and fitness. The purpose of this chapter is to discuss how we use and implement the findings of the evidence-based practice (EBP) process. Pick up your chisel and let's go to work.

TRANSLATING RESEARCH INTO PRACTICE

As discussed earlier in this book, research design is a critical element to answering questions relevant to mechanistic, practical, and clinical science. One of the key aspects of research design is control of all potential confounding factors to eliminate the possibility that the outcomes arising from the intervention (independent variable) were influenced by an external factor. Although experimental control is a strength of research, and in particular of randomized control trials, it can also create problems when one is implementing research evidence in client, athlete, or patient populations. Suppose a clinical scientist is interested in studying the effect of interval exercise compared to steady-state exercise on hypercholesterolemia in adults who are obese. A logical study design to address such a question would be a randomized controlled trial. Consider the following hypothetical study:

> One hundred participants who were obese were selected at random and then randomly assigned to one of four groups: interval exercise, short-duration steady-state exercise, long-duration steady-state exercise, and control with no exercise. In order to eliminate potential confounding factors, the participants were free from comorbidities including hypertension, insulin resistance, diabetes, overt cardiovascular disease, and recent orthopedic injuries. Before the intervention, all groups completed a treadmill $\dot{V}O_2$ maximum test and a test of blood cholesterol; the individuals conducting both of these tests were blinded to subjects' group assignment. The interval exercise group completed a protocol consisting of 1 min of exercise at a velocity equal to 90% $\dot{V}O_2$max with a 2-min recovery between intervals. A total of 10 intervals were performed; thus the total exercise time for the intervention was 30 min. The steady-state short-duration exercise group performed 10 min of exercise at 70% $\dot{V}O_2$max, matching the exercise (work) time of the interval group. The steady-state long-duration exercise group performed 30 min of exercise at 70% $\dot{V}O_2$max, matching the total exercise and rest time of the interval exercise group. The control group did not exercise for the duration of the study. Each exercise group was prohibited from performing any other exercise, including resistance exercise. Moreover, a standardized diet was provided for all groups. The intervention was performed 3 days per week for 16 weeks; at the end of the training period, blood cholesterol and $\dot{V}O_2$max were assessed to determine improvements.

On the surface, this appears to be a sound single-blind randomized controlled trial. It is in fact well-designed and controlled, but there are some potential problems when a practitioner considers implementation. Because the scientists were interested in studying training responses to an intervention in obesity alone, they eliminated subjects with diabetes, hypertension, and other common comorbidities. This was a good decision for internal validity, but not for external validity; practitioners rarely treat individuals who are obese without any of these companion conditions (Castro, Kolka, Kim, & Bergman, 2014; Despres et al., 2008). Thus, practitioners are left to wonder whether the evidence is truly applicable to their patients or clients.

A second possible problem with the evidence is the elimination of secondary interventions. Strengths of a scientific experiment may actually be perceived as weaknesses with regard to practical implementation. In practice, multiple interventions are often implemented in concert with one another. This hypothetical experiment evaluated the effectiveness of interval versus steady-state treadmill exercise alone. An exercise practitioner is unlikely to implement the interval training protocol in isolation; instead, the protocol may be combined with resistance training, a healthy diet, and other behavioral modifications. Implementation of such multifaceted protocols may enhance the effectiveness of either of the interventions. Conversely, the addition of such interventions may also reduce the effectiveness of the protocol. Obviously, it is not scientifically plausible that resistance exercise would cause an increase in cholesterol or biologically negate the effects of the interval training protocol. However, it is possible that implementation of the secondary intervention will lead to overtraining, injury, soreness, and perhaps poor adherence to the protocol due to each of these secondary results of the combined protocol. Again, this is not likely, but it is certainly something the practitioner must consider.

Another potential problem with the implementation of scientific evidence is the possibility that a client may not be able to tolerate the protocol. For example, the initial fitness level of the client may not be sufficient to enable him to tolerate its rigors. The practitioner is then left to decide whether a modified version of the research protocol will be effective in improving blood cholesterol profiles. It is also possible that the protocol used in the scientific experiment will result in undertraining in some individuals; that is, in contrast to the previous scenario, the intensity of the stimulus may be insufficient to induce a training response. The exercise practitioner must then decide how to effectively alter the training loads in order to induce a similar response in a client or patient with a greater initial fitness level. In situations like these, it is often helpful to consult the discussion section of the research paper. Often the authors discuss in great detail the programmatic elements that they believe are responsible for the particular adaptations they observed; for instance, they may note that their subjects completed 10 intervals (with positive outcomes) in contrast to a previous study in the same population that used four intervals with negative findings. Mechanistic evidence, from either the same study or another, would further elucidate the cause(s) for the positive adaptations from a higher volume of exercise.

In medicine, one of the often cited criticisms of the evidence-based philosophy is that it leads to a cookbook approach to practice (Sackett, Rosenberg, Gray, Haynes, & Richardson, 1996). Opponents assert that evidence-based practitioners read scientific papers and then implement the protocols directly from the papers. The example we have presented illustrates that there really is no "recipe" for prescription; that is not the purpose of science. Because people are different, protocols may need to be altered to fit individual patient or client needs. Science influences the "ingredients" of the prescription. This hypothetical experiment demonstrates that interval exercise might be more effective than steady-state treadmill exercise in improving blood cholesterol of patients who are obese. It also provides some basic parameters for intensity, volume, mode, and so on for the implementation of a protocol. Practitioners must then decide whether to implement the protocol. They must also decide how they should implement interval exercise in a manner that is best suited to their client's, patient's, or athlete's individual needs and experience level.

DECIDING WHEN TO ACT ON EVIDENCE

The decision to include or exclude evidence may be influenced by several important factors. Perhaps the most influential factor is the magnitude of the treatment effect. The exercise practitioner must ask the question, How much better is the novel treatment compared to the standard of care? This question must be analyzed in the context of several important influential factors such as time, cost, and patient, client, or athlete preferences and values.

Will the Evidence Significantly Improve the Quality of Care?

The most important question in the decision to implement scientific evidence is whether a novel treatment, protocol, or device will significantly improve the quality of care (Grimshaw et al., 2006). A secondary and perhaps more difficult question is, How do you define significance? In statistics, a significant difference between two treatments is defined by a difference between two means, their variance, and the mathematic probability that the difference occurred by chance. Typically, a result from an intervention is defined as significant if the P value is less than 0.05 (<5% chance that the difference or positive finding is false) (Riffenburgh, 2012), but does $P < 0.05$ indicate a definite improved outcome?

Statistical significance does not necessarily equate to a meaningful improvement in quality of care (Kazis, Anderson, & Meenan, 1989; Wyrwich, Nienaber, Tierney, & Wolinsky, 1999). Suppose that an exercise practitioner is interested in implementing an intervention to improve weight loss in a client or patient. She has read some anecdotal evidence that a novel exercise device ("Device Y") may be a viable tool to enhance metabolism during exercise. After searching for evidence, she finds a peer-reviewed journal article in which the acute metabolic demands of a standard exercise intervention and the same exercise with Device Y are reported. The authors report that the metabolic demands of standard exercise are 14.1 ± 1.3 mL $\cdot$ kg^{-1} $\cdot$ min^{-1}; the metabolic demands of exercising with Device Y are equal to 16.8 ± 1.5 mL $\cdot$ kg^{-1} $\cdot$ min^{-1} (figure 9.1). Indeed, there is a statistical difference between exercising with and without the device ($P \leq 0.05$).

Although there is a statistical difference between standard exercise and Device Y, the evidence-based exercise practitioner must ask the question, Is the difference clinically or practically relevant? In other words, is the 2.7 mL $\cdot$ kg^{-1} $\cdot$ min^{-1} difference a significant improvement to the quality of care for the patient or client? If the goal of the individual is weight loss, it would be prudent to assess the true impact of the device on

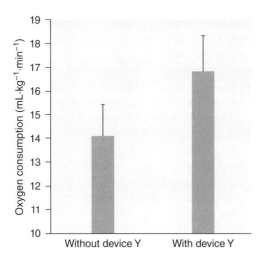

Figure 9.1 Statistical difference does not necessarily equate to practical significance. In the example, exercise with device Y results in greater oxygen consumption but may not result in practical weight loss.

caloric expenditure. In this example, suppose the client or patient weighs 90 kg. The $2.7 \text{ mL} \cdot \text{kg}^{-1} \cdot \text{min}^{-1}$ difference would be equal to 243 mL/min of oxygen consumption. This would equate to a 1.2 kcal/min increased caloric consumption. Over the course of a typical 30-min exercise bout, the individual would expend approximately 36 kcal more with Device Y compared to standard exercise. Although this example may over-simplify the problem of weight loss, if all else is equal (i.e., same caloric intake, physical activity, postexercise oxygen consumption), approximately 97 30-min bouts of exercise with Device Y would be required to obtain a 1-lb greater weight loss compared to the standard exercise intervention. The evidence-based exercise practitioner would need to carefully consider whether Device Y is worth the small practical impact it provided.

A contrasting example occurs in the world of elite athletics, where the difference between winning and losing may be tenths or hundredths of a second. Consider the competition times in the 2012 Olympics 100 m freestyle swimming event. The winning time was posted by a competitor from the United States, Nathan Adrian, who finished the race in 47.52 s; the silver medalist James Magnussen, from Australia, finished the race in 47.53 s. The race was won by one one-hundredth of a second! The time that separated the gold medalist and the eighth-place finisher, Nikita Lobintsev from Russia (48.44 s), was less than 1 s. In elite athletics, seemingly miniscule differences can be the difference between standing on the podium and finishing last in the race; therefore, interventions that produce small differences may in fact be warranted in this elite population.

Will the Patient Adhere to the Intervention?

Along with considering the impact of the evidence on the quality of care, an evidence-based practitioner must also take into account the patient's values and desires, and ultimately the probability of adherence to the treatment. The original models outlined by Sackett and colleagues suggested that a patient's preferences and values were just as important in the decision-making process as research evidence and clinical expertise (Sackett, Straus, Richardson, Rosenberg, & Haynes, 2000). A patient should have a voice in the decision making.

In medicine, the patient's preferences may have certain ethical or moral consequences (Charles, Gafni, & Whelan, 1997; Degner & Sloan, 1992; Emanuel & Emanuel, 1992). For example, a patient's religious preferences or moral beliefs may preclude his taking a drug or engaging in a particular type of therapy. Although not as likely in exercise science, it is certainly possible that a practitioner will need to consider the ethical consequences of a proposed treatment. Suppose you find evidence that a whey-based protein is effective to improve skeletal muscle protein synthesis and promote recovery after heavy resistance exercise. However, one of your clients is a strict vegan, and eating any dairy-based protein violates her personal ethics. In this case, implementation of the evidence would violate the patient's preferences and values; thus, the evidence-based practitioner must look for alternative supplements even if the evidence for them is not as strong.

In a more likely exercise science scenario, a suggested treatment may not be the patient's preference. The exercise practitioner might argue that the patient's "preferences" are not as important as the effectiveness of the treatment. In other words, "Whether you like this or not, you should do it because it is what's best." However, the behavioral psychologist would argue that people are unlikely to adhere to a treatment if it is too difficult or if they are not convinced that it is effective (Janz & Becker, 1984; Rosenstock, 1974). If two equally effective treatments are available, a practical suggestion would be that regardless of the exercise practitioner's preference, the client, athlete, or patient should make the choice, because adherence will be better if the individual believes in and is excited about the treatment. Another interesting scenario might present itself if one treatment is clearly better than the other but the patient refuses to adhere to the better treatment for personal reasons. In such a case, it would be important for the practitioner to earnestly attempt to convince the client to choose the more effective treatment. But if he still refuses and there is no harm in the less effective treatment, the practitioner would reluctantly consider acquiescing to the client's preference because in most cases with exercise, something is better than nothing.

A second practical suggestion is that evidence-based exercise practitioners develop skills in disseminating information to the "consumer" (Law & MacDermid, 2008). This is one of the major obstacles to the EBP approach. For the evidence-based philosophy to be effective, scientists must generate practical and useful information, and practitioners must read and interpret the information. If they find a new treatment option to be beneficial and cost- and time-effective, they then implement the treatment in their clients or patients. Breakdowns in the process can occur in the transfer of information from scientist to practitioner or from practitioner to patient, client, or athlete. Scientists generally disseminate evidence-based information through peer-reviewed publications or conference abstracts, proceedings, and presentations. Educators should teach all practitioners the skills needed to read, interpret, and analyze information from these sources. Such skills will help practitioners to be good consumers of science. However, it is not realistic to expect an athlete or client to be able to read and interpret scientific information. It is also not appropriate to expect an athlete, client, or patient to blindly follow the practitioner's recommendations—this violates a key premise of the evidence-based philosophy. Ideally, evidence-based practitioners should develop skills in translating and communicating the findings of peer-reviewed information in such a way that the consumer—the "end user"—can understand the rationale for treatment (Law &

MacDermid, 2008). If this information can be translated in an easily understandable way for the client, it will increase the probability of the patient's adhering to the treatment.

Is the Intervention Time-Effective?

A practical consideration when one is determining if the implementation of a treatment is warranted is whether the intervention is time-effective. Many exercise sessions are governed by a time limit although such restrictions may limit the potential effectiveness of a treatment, therapy, or training. An exercise practitioner working in a National Collegiate Athletic Association or high school environment may be constrained by a defined time limit set by the governing body for the sport. In a clinical environment, the amount of time dedicated to an intervention may be in part limited by the reimbursement time established by the insurance company. In the athletic, clinical, or commercial sectors, time is valuable, and it is prudent to ensure that the amount of time dedicated to a treatment is justified by its results.

A real-world example related to time effectiveness is seen in the emerging exercise rehabilitation technologies. A cutting-edge technology in rehabilitation for neurological diseases and injuries is robotic-assisted gait training. Devices are available that use a robotic exoskeleton that attaches to a patient's limbs to facilitate ambulation. Some devices allow the practitioner to define joint kinematics to create a precise therapeutic modality for individuals relearning the skill of walking (Morone et al., 2011). The technology is absolutely remarkable, and an evolving body of research evidence supports the benefits of such robotic technologies in rehabilitation.

While the technology and outcomes for robotic rehabilitation are exciting for the clinical exercise practitioner, the time resources required to use the devices in their current form are considerable. In many cases, three individuals are needed to position the patient in the device. Because the movement parameters are so precise, the setup must be exact to ensure proper limb movement; correctly and safely positioning the patient who has significant functional limitations requires multiple adjustments. During operation, a computer operator and a spotter are required to ensure patient safety. When the session is complete, multiple technicians are again needed to safely remove the patient. For the patients with severe disabilities who would most benefit from this technology, effectively positioning the patient for exercise may require 20 min. The patient may then be able to tolerate only 10 to 20 min of activity before another 10 to 15 min is required to remove the patient.

The essential question the evidence-based exercise practitioner must answer is whether the benefit of 10 to 20 min of robotic-assisted exercise justifies the 30 to 45 min of setup and patient removal. Also, is it worth 2 person-hours of clinician time? All this considered, is the benefit of using the device sufficiently superior to the current treatment to merit the additional time and human effort (Morone et al., 2011)? To address this, the practitioner might pose the following question: "In patients who have suffered a stroke, is robotic lower body–assisted gait rehabilitation more effective than partial body weight–supported treadmill training with therapist-facilitated leg movement for improving gait parameters?"

This question is purposefully vague with regard to the outcome measure of gait because there are not likely to be many studies that directly compare the two interventions. The best direct comparison at the time of this search was a 2009 study (Westlake & Patten, 2009). In this project, the authors investigated the effects of robotic-assisted gait training in patients with hemiparesis resulting from a stroke. Sixteen patients were randomly assigned to either robotic-assisted gait training or a control group. Both groups engaged in 12 total exercise sessions: three sessions per week for 4 weeks. The exercise sessions were approximately 30 min long, and participants completed a progressive speed protocol. The robotic training group completed all exercise sessions with assistance from the exoskeleton. The control group completed sessions identical in length, but leg and pelvis action were manipulated manually by a therapist during ambulation on the treadmill. The outcome measures were self-selected walking speed, fast walking speed, 6-min walk test (meters walked), step length ratio, and Berg Balance Scale scores, among others. All measured parameters were significantly increased with robotic-assisted training; for manual training, there were trends toward improvements but no statistically significant increases in any variable except for Berg Balance (Westlake & Patten, 2009). In a statistical conundrum that is sometimes observed, there were no between-group differences in any measured variables.

A cursory observation would suggest that robotic-assisted gait training is superior to manually assisted gait training. But the practitioner must ask whether the benefits were sufficiently greater compared to manually assisted training to warrant the greater time needed to implement the intervention. Table 9.1 compares all the tested variables. For simplicity, we focus on self-selected walking speed. It increased from 0.62 ± 0.31 m/s to 0.72 ± 0.38 m/s in the robotic-assisted group; the control group had a nonstatistical improvement in self-selected walking speed from 0.62 ± 0.28 to 0.65 ± 0.29 m/s (Westlake & Patten, 2009). Is 0.07 m/s a great enough difference to support implementing the intervention? It is clear that the results in the robotic-assisted group were better for most variables, but does this improvement justify the time and the human resources needed? In a perfect world in which time and personnel were not limited quantities, the answer would be yes, but these factors are usually key variables in decision making.

Table 9.1 Comparison of Gait Outcomes From Robotic-Assisted Gait Training and Manually Assisted Gait Training

Measure	ROBOTIC-ASSISTED TRAINING		MANUALLY ASSISTED TRAINING	
	Pretest	Posttest	Pretest	Posttest
Self-selected walking pace (m/s)	0.62 ± 0.31	0.72 ± 0.38*	0.62 ± 0.28	0.65 ± 0.29
Fast walking speed (m/s)	0.87 ± 0.55	0.96 ± 0.66*	0.72 ± 0.37	0.70 ± 0.33
6-min walk test (m)	267.3 ± 187.2	278.1 ± 176.5	234.3 ± 141.2	212.4 ± 113.5
Absolute step length ratio	0.53 ± 0.58	0.37 ± 0.46*	0.39 ± 0.37	0.34 ± 0.35
Berg Balance Scale (/56)	46.9 ± 7.5	48.3 ± 6.8*	47.0 ± 7.0	51.0 ± 5.4*

* indicates pre- post-difference ($p < 0.05$)

Adapted from K.P. Westlake and C. Patten, 2009, "Pilot study of Lokomat versus manual-assisted treadmill training for locomotor recovery post-stroke," *Journal of NeuroEngineering and Rehabilitation* 6: 18.

This is one research study; thus, it would be prudent to search for other papers with additional outcome variables. Based on this study alone, the benefits of the treatment do not appear to outweigh the costs, despite evidence of some efficacy. However, it is possible that future studies may provide more support for this emerging technology and may better substantiate its implementation. Clinicians working with patients who have neurologic pathologies and injuries should carefully monitor the literature.

Is the Intervention Cost-Effective?

Another consideration in the decision whether to implement an intervention is its cost-effectiveness. This goes beyond the simple monetary cost of the initial purchase. If the new treatment involves a piece of machinery, maintenance and operating costs must also be considered; included in these are regular repair, consumables, and the person-hours needed to maintain, clean, and repair the device. If these costs are substantial, the benefits should also be substantial compared to those of other less costly treatments.

Returning to the example of robotic-assisted gait training, the device is associated with several costs. First, there is the initial investment of purchasing the device—depending on the manufacturer, this can be quite substantial for facilities with limited budgets. Second, one must consider the time needed for operator training and the number of individuals who will need to complete training. If these individuals are not routinely using the device, they may need regular refresher sessions to maintain their competency. Most companies sell annual maintenance contracts; these costs should be factored into the purchasing decision. Practitioners must also take into account the cost of operating the device. In the case of robotic-assisted training, practitioners may need to budget two or three individuals for 30 to 45 min per session. Another consideration would be the volume of usage that the device will see; if usage volume is high, a dedicated staff may be required to operate and maintain the device. As is apparent, costs quickly escalate. These fiscal considerations must factor into practitioners' treatment decisions (figure 9.2).

For robotic-assisted gait training, there are also costs associated with the alternative treatment. Any form of gait training likely requires a treadmill, a body weight support system, and operators. Because most exercise and rehabilitation facilities that implement gait training already have a treadmill and body weight support system, the initial investment for manual gait training will probably be much less. The resources needed for gait training hardware may be better allocated to hiring additional staff. Consider the effect of using the money allocated to new equipment to hire several additional practitioners: The practitioners could implement manual gait training, conceding the small differences associated with use of the robotic-assisted gait training. In contrast to the robotic device, a practitioner is versatile and can perform exercise and rehabilitation tasks besides gait training.

This discussion has focused on an expensive intervention; what if the prospective treatment is relatively cheap? Suppose the small statistical differences observed in table 9.1 were associated with a device already available in your facility; maybe only one person is needed to operate it and setup is quick. In this scenario, it is reasonable to implement the treatment since the small differences gained by use of the device are neither time nor cost prohibitive. Patient preferences, available and needed resources, and expected outcomes associated with treatment (based on evidence) should all factor into decisions to include or exclude an intervention (figure 9.3).

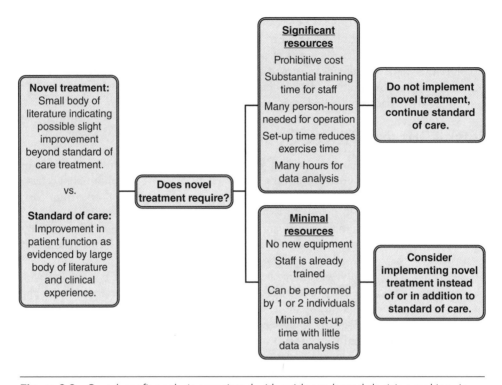

Figure 9.2 Cost–benefit analysis associated with evidence-based decision making. In the example of robotic-assisted training, the slightly greater expected improvements compared to the standard of care may not be sufficient to warrant implementation unless the facility has sufficient resources.

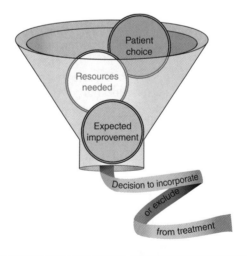

Figure 9.3 The decision to incorporate a new treatment based on evidence is dependent on the patient's preference, the resources needed, and the expected outcomes as determined by the literature.

THE INFLUENCE OF EVIDENCE

One premise behind the evidence-based philosophy is that a practitioner who engages in the six-step approach must be willing to change based on emerging or newfound information. If one is not willing to change, the process is pointless and a waste of time. Practitioners who choose to engage in the evidence-based approach must be open-minded and must believe that the incorporation of new evidence will lead to better outcomes for their clients, athletes, or patients. There are at least three possible outcomes of participating in the evidence-based process.

Confirmation of Practice

In many cases, engaging in the evidence-based approach may lead to confirmation that current practice is in line with the best practices based on evidence. This outcome is more likely for individuals with a solid foundation based on science, reputable certifications, and a strong college education. Most top-tier certifications, especially those associated with national professional societies (and particularly those with their own peer-reviewed journals), routinely and consistently update their educational materials to incorporate cutting-edge research and require continuing education credits that are research based. The foundational knowledge gained from preparation for these certifications is a great starting point for exercise science practitioners. But it should be noted that not all certifications are based on the latest or best evidence. Even for the best certifications, the information gained during preparation and testing changes over time. The top-tier certifications typically require continuing education to provide an avenue to refresh knowledge and to disseminate the latest information to certified exercise scientists.

Academic training is also a great starting point to becoming an evidence-based exercise practitioner. The best academic programs in exercise science incorporate the latest research evidence into their lectures and course materials. In many cases, the instructors at these academic institutions are also involved in the discovery and dissemination of novel exercise findings. Similar to certifications, learning from reputable individuals in the field at top academic institutions should lead to a firm foundation for practice. Several top certifications require a degree to become certified; the knowledge required to complete these certifications itself almost necessitates academic preparation. One additional benefit of formal academic training is that the top universities in the field of exercise science require students to engage in the process of EBP; teach the foundational skills of reading and interpreting research evidence; and require, support, and facilitate critical thinking. To paraphrase words often attributed to Albert Einstein, " the purpose of an education is not to memorize facts, but to train one's mind to think." Mature students realize that education is a lifelong pursuit. However, both academic training and certification are completed at a single point in time. Thus, they are sound methods for establishing competency in stable bases of information but fail to provide the dynamicity of knowledge necessary to remain on the cutting edge of practice. Even with the best education and certifications, one will make subtle changes, sometimes in contrast to what was taught in formal education, as the exercise literature evolves.

Modification or Addition to Practice

The process of EBP may lead to practice modifications; this is the most common outcome. Research or experience-based knowledge often subtly nudges practitioners in new or different directions. This type of change could be as simple as slightly altering an exercise technique for added safety, changing a set–rep scheme to be in line with new research, incorporating a novel protocol or tool that is supported by literature, or any number of other modifications to improve practice. As discussed throughout this book, the exercise industry is rapidly changing, and many emerging concepts are not supported by research; many are not even theoretically sound. Non–evidence-based practitioners may react and incorporate these concepts into practice simply because they are trendy or supported by clever marketing. Practitioners should be influenced only by new knowledge obtained through research or consistent experience, not by marketing trends.

Wholesale Change

Another possibility is that the evidence-based approach leads to wholesale changes in practice. This could be the result of dramatic new discoveries in science or years of stale, unexamined practice. From time to time, discoveries are made in research that completely change conventional thought. Within scientific theory, Thomas Kuhn described such a drastic change as a "paradigm shift." A classic example of a paradigm shift occurred in the 16th century. In the early 1500s, the prevailing thought among scientists was that the earth was the center of the universe and that all heavenly bodies rotated around the earth (Ptolemaic model). In the early to mid-1500s, Nicolaus Copernicus proposed a new model suggesting that the earth and other heavenly bodies rotated around the sun—the earth was not the center of the universe. When this idea was proposed by Copernicus, it was thought to be outlandish, revolutionary, and even heretical. It is believed that Copernicus was so afraid of the upheaval that would result from his writings that he did not allow the publication of his works until he was on his deathbed. Galileo, on the other hand, publicly defended Copernican thought and was sentenced to house arrest for heresy in the 17th century.

Most paradigm shifts in science, and exercise science in particular, do not lead to arrest or excommunication. They may be met, however, with fierce opposition by defenders of conventional thought. Consider the following question: After a proper progression, should individuals who are elderly engage in heavy or high-velocity resistance exercise? Most people who have read the literature concerning the functional effects of heavy and explosive high-velocity resistance exercise would support the use of such techniques and incorporate them into resistance programs. What if you had posed the same question in 1975? It is likely that you would have been met with fierce opposition for such a dangerous thought. There has been a paradigm shift in the conventional wisdom regarding training persons who are elderly. However, if you visit gyms or clinics and observe the exercise practices of practitioners working with people who are elderly, it is evident that many (if not most) still practice from the old paradigm. Incorporation of this new information would necessitate a large shift in practice.

Somewhat ironically, the movement to EBP may itself be a paradigm shift in the exercise science field. Although we would like to believe that most exercise practices, programs, devices, and so on are based on strong evidence, the reality is that they are not. The incorporation of many practice-based decisions is instead based on information passed down from generation to generation by people who are perceived as experts. When the root source of the information is carefully investigated, it may be shown that it is based on outdated evidence or opinion alone. Therefore, shifting the paradigm to truly investigate the evidence for practice may be a drastic philosophical change.

IMPLEMENTING RESEARCH-BASED EVIDENCE

Once the decision has been made to implement evidence supported by research, the essential question practitioners must answer is how they should incorporate the evidence. As discussed earlier, research scientists go to great lengths to carefully control for extraneous variables. Therefore, exercise practitioners cannot simply copy the protocol from the scientific experiment; instead they must find a way to incorporate or integrate interval exercise, for example, into practice. Consider the following fictitious case example.

> An exercise practitioner at a facility serving individuals who are older has a new client who is 55 years old and reports that her physician has diagnosed her with osteopenia. Specifically, the doctor reported that her hip bone mass was low. She has no other known health conditions or orthopedic risks. The practitioner knows from his academic training that bone formation can be stimulated through the compressive forces of resistance exercise and by the moments resulting from the torque the tendon generates as it pulls on the bone. After searching the literature, he finds a study that compares the effects of deep knee flexion leg press to shallow knee flexion leg press in women who are older. Forty-five women between the ages of 45 and 65 who had been diagnosed with osteopenia in the hip were randomly assigned to one of three groups: a deep knee flexion leg press group, a shallow knee flexion leg press group, and a control group. Both the deep and shallow knee flexion groups performed a periodized resistance protocol for 16 weeks of equal relative intensity and volume. The exercise was performed on a supine leg press machine in which the feet are fixed to a plate and users press their body from the plate while lying on a padded sled with two padded bars across their shoulders.

Although the two groups performed exercise at the same relative intensity, the shallow knee flexion group lifted a higher absolute load because the shallow knee flexion angle affords a greater mechanical advantage. The shallow knee flexion group descended to 45° of knee flexion, and the deep knee flexion group descended to 110°. The control group performed no exercise for the 16-week period. The outcome measures used to assess the effectiveness of the protocol included bone mineral density of the greater trochanter, lumbar spine, and calcaneus. The results are shown in figure 9.4. Both the deep and shallow knee flexion groups had significant increases in bone mineral density of the hip and lumbar spine. However, the change in bone mineral density in the hip was significantly greater in the deep knee flexion group; the lumbar spine bone mineral density change was similar between groups, but there was a trend toward greater lumbar spine change in the shallow knee flexion group (figure 9.4). There was no change in bone mineral density in the calcaneus in any group.

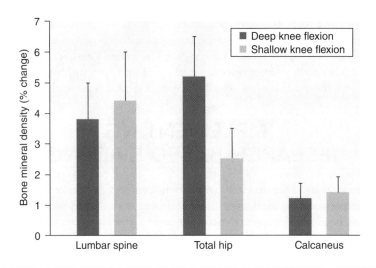

Figure 9.4 Hypothetical data from the example provided in the text.

Addressing the original question, there is level B evidence to support the use of deep knee flexion exercise to improve bone mineral density in the hip compared to heavier low knee flexion exercise. How should this information be interpreted and functionally implemented? Should the practitioner buy the exact same leg press used in the study, repeat the protocol exactly, and use only deep knee flexion leg presses? Is the trend in lumbar spine bone mineral density to be ignored because the *P* value did not reach 0.05? What if the facility has only barbells and cannot afford a leg press machine?

Each of these questions poses an interesting dilemma in the interpretation of research findings; should research be interpreted according to the letter or the spirit of the finding? The realistic interpretation of this study is that both heavy, shallow knee flexion and deep knee flexion exercises improve bone mineral density in the hip. Apparently, the joint torques in the hip are somewhat different in the deeper knee flexion exercises, which makes this exercise slightly superior. There was not a significant difference in lumbar spine bone mineral density, but a definite trend toward lower knee flexion exercises (probably due to greater compressive forces) being more beneficial for the lumbar spine. In a periodized, long-term protocol you should probably incorporate both types of exercise. In the overall training plan, because the client has lower bone mass in the hip, more exercises should focus on deep knee flexion (and greater hip flexion). However, shallow hip flexion exercises should not be eliminated because of their potential to improve lumbar spine bone mineral density. Although the study's findings are specific to the leg press, they can likely be generalized to other exercises. If you work at a facility that has only barbells, it is highly probable that a deep or shallow back squat will produce similar results.

As we have said before, exercise prescription is both an art and a science. Programs based completely on "art" or creativity will lack structure and substance. They might be fun and novel, but will they actually work? Conversely, programs based exclusively on science can be stale and monotonous. Both will probably result in initial positive changes, but at some point the results will plateau. Great practitioners have the ability to creatively develop effective programs within the lines and boundaries dictated by the scientific evidence.

CONCLUSION

After carefully reading and critically appraising information, the evidence-based practitioner must decide whether information should or should not be incorporated into practice. These decisions are based on several factors, including patient, client, or athlete preference, the necessary and available resources, and the expected differences compared to the standard of care. Each of these factors should contribute to the final decision. Engaging in the evidence-based process may confirm or strengthen your current practice, slightly modify practice, or result in wholesale philosophical changes. The likelihood of complete changes is reduced if one practices based on a solid foundation of sound academic training and top-tier certifications. However, as new evidence emerges, the evidence-based practitioner must be prepared to change if the evidence suggests a better treatment plan or expected improvement for the patient, client, or athlete. The following chapter addresses a newly proposed step: confirming the evidence by evaluating its effectiveness after implementation in practice.

Chapter 10

CONFIRMING THE EVIDENCE IN THE INDIVIDUAL

Learning Objectives

1. Discuss the principle of individuality and the potential limitations of generalizing research.
2. Provide examples of generalization of research and how it is used in exercise science.
3. Understand the importance of confirming evidence in the individual as a part of evidence-based practice.
4. Define important concepts in testing and measurement, with particular focus on reliability and validity of measurement.
5. Present ways to standardize data collection and implement longitudinal data analysis.
6. Provide a practical example of confirming the evidence in the individual.

As an evidence-based exercise practitioner, you and your team have been presented with a new client with a unique problem. After careful consideration of the case, a question was developed, carefully defining the population, intervention, comparison, and outcome. A diligent search for evidence uncovered three solid randomized controlled trials with homogeneity of findings supporting use of a new intervention; you have level B evidence to support incorporation of the intervention. After careful consideration, a practical and feasible method of incorporating the findings into your client's training plan was developed and implemented. The process is over, correct? The intervention will undoubtedly improve your client's health or fitness, resulting in positive change, correct? The research says that the intervention will work. It will work, correct?

Lurking in the back of the mind of all practitioners should be the philosophical thoughts of Karl Popper: "[No] matter how many instances of white swans we may have observed, this does not justify the conclusion that all swans are white." What if your client is the nonwhite swan? For an exercise scientist, the most frustrating principle of adaptation is the "principle of individuality." This tells us that if you apply the same stimulus, the same number of times, at the same intensity, and for the same duration in two different people, you might observe two different responses. The underlying reasons for this span the field of genetics and the adaptive mechanisms available to the individual. Finding a large number of studies with substantial sample sizes, low variability, and homogeneity of findings increases the probability of a systematic response. It does not, however, guarantee positive adaptation. Unless the client and the program perfectly match the population, independent variables, temporal timeline, training experience, nutrition, and numerous other variables, there is likely to be a somewhat different response than that elucidated by research. Even if all of these variables are identical, personal genetics will create individual responses.

The answers to the questions posed above: The intervention will probably work, will probably improve the client's health, and will most likely improve fitness. But "probably" reveals the possibility of the "nonwhite swan." In this chapter, we define a new step in the process of evidence-based practice (EBP): to confirm the evidence in the individual. The chapter also discusses basic principles of data collection, measurement, interpretation of results, and longitudinal data analysis.

GENERALIZABILITY OF RESEARCH

One of the requirements for well-designed research is that the results be generalizable. In statistics, the generalizability theory (G Theory) essentially presupposes that that there are multiple sources of error resulting from the measurement of a variable (Riffenburgh, 2012). Error in measurement may arise from many sources, and each variable or test has different sources of error. From a practical perspective, generalizability suggests that a research study finding can be extrapolated to a population with the understanding that there may be some error or variability in this extrapolation depending on the measurement.

The generalizability of research is increased by a number of variables—most importantly the precision of the test, sample size, and matching of the independent and dependent variables. Consider the following example: An exercise practitioner is contemplating recommendation of an essential amino acid drink containing 5 g of leucine in conjunction with resistance exercise to improve muscle mass in a 65-year-old woman who is her client (Phillips, 2014; Drummond et al., 2008). Before recommending the supplement, she performs a search to determine the potential effectiveness of the drink to increase muscle mass as measured using magnetic resonance imaging (MRI). The perfect experiment to determine this relationship would be as follows:

Every 65-year-old woman currently alive in every country in the world volunteered for this experiment ($N = 1,000,000+$). They all were housed at the same inpatient hospital facility where their every move was monitored for every minute of every day for the 24-week experiment. Six months before the investigation, they had all checked into the inpatient facility for a wash-out period to eliminate possible effects of prior lifestyle. At the end of the wash-out period, a single, highly skilled MRI technician performed a scan of the right leg of all 1,000,000+ participants. The MRI was performed at the same time of day for all participants. During the study, all participants performed the identical physical activity, consumed the same amount of food (relative to lean mass), slept the exact same number of hours per night, and never deviated from the protocol. The participants were equally divided and assigned to one of four groups: exercise alone, leucine alone, exercise and leucine, and control. Participants in the leucine and leucine plus exercise intervention groups were provided a nutritional supplement containing 5 g of leucine; they all consumed it at the same time each day during the 24-week study. The exercise, leucine, and exercise plus leucine groups performed a periodized resistance exercise program for the entire 24-week period. The control group did not exercise, nor did they consume extra leucine. At the end of the study, the same MRI technician performed the same MRI procedures on every participant at the same time of day as for the baseline measurement, and comparisons were made between all four groups.

If an exercise practitioner found this study and assigned 65-year-old clients to the protocol, and if the clients followed the protocol precisely, the results would assuredly be similar. However, it is clear that this study is fictitious and theoretically impossible. In reality, a large sample size in a study similar to the one described would be 30 to 50 individuals. It is unlikely that all would be 65; instead, they would all be within a narrow age range (55-65 years or 65-75 years). The experiment would be performed in a facility where participants would be recruited locally or regionally. The participants would be randomized into groups similar to those described, but the group size might be unequal, and dropout could occur in any or all groups. Unless this was a bed rest study, the participants would not live in an inpatient facility. Such parameters as their levels of physical activity, exercise modes, types, frequencies, durations, and calories would be self-reported; and along with expected measurement error, there would be recall bias and some level of variability between subjects. The MRI would be performed generally at the same time of the day, but it is unlikely that all 50 participants would be tested at the exact same time of day due to logistics. It may be possible to have the same MRI technician conduct both pre- and postintervention scans, but not necessarily. Even with all of the uncontrolled variables and sources of potential error in this experiment, to complete such a study on a sample size of 30 would be remarkable. The cost of an experiment of this magnitude with this level of control would be substantial, and an exercise practitioner looking for evidence on the efficacy of a leucine supplement would be delighted to find such strong evidence addressing the question.

The fundamental problem with this study, and with research in general, is that the researcher attempts to generalize the results from a sample of 30 to 50 individuals to an entire population, and, just as importantly, the practitioner is attempting to generalize the findings to one client. In practice, generalization occurs on a much larger scale. Tsang and Williams suggest at least four distinct types of generalization that occur at the empirical level (figure 10.1): within population, different population, different context, and different time (Tsang & Williams, 2012). We add a fifth, relevant to exercise science: laboratory-to-field generalization.

Within-Population Generalization

When one generalizes to an individual from the same population, there should be, in theory, an increased probability of a similar response. The ideal scenario in EBP is to have evidence from an investigation with a population very similar to the individual or group in which the evidence is to be applied (Tsang & Williams, 2012). The likelihood of this is greater when your question centers on a population in which research is abundant.

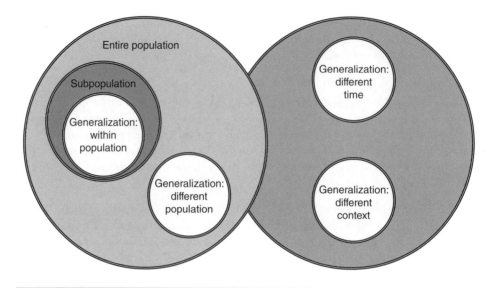

Figure 10.1 Problems may arise when generalizing research results to different populations, times, and contexts. In this figure, each shaded circle represents a sample drawn from a population and used to test a hypothesis in research.

One criticism of many exercise studies is that the subjects are often college-aged students (Stone, Sands, & Stone, 2004). This is often due to convenience. Many exercise scientists are university faculty and thus have easy access to a large number of young adults. If your evidence-based question is related to an exercise response in young adults, chances are that you can find research in this population as long as the intervention is not novel (e.g., new exercise device, emerging exercise program theory). Although there are certainly exercise studies in clinical, aging, and other particular populations, historically work in these populations has been underfunded and therefore they are understudied.

Different-Population Generalization

Often, evidence-based questions are asked about a population for which there is less published evidence (Tsang & Williams, 2012). This is an often raised criticism of sport science research (Stone et al., 2004). How many exercise intervention studies are published using Olympians or professional athletes? Many evidence-based searches by sport scientists working with elite athletes return little or no published evidence for that population. Instead, sport scientists often must rely on interventional research in sub-elite or even recreational athletes; in many cases, the responses of elite and recreational athletes are similar, but sometimes elite athletes respond differently due to their training level and genetics, among other critical factors. When generalizing across different populations, it is important to be aware of the potential for different responses. At times, it may be necessary to generalize across age groups, sexes, fitness levels, and ethnicities when evidence cannot be found for a specific population.

Context or Analogue Generalization

Contextual or analogue generalization is another common type of generalization in exercise science research (Tsang & Williams, 2012). For some conditions, access is uncommon; thus, there is a need to use an analogue comparison to obtain evidence. Analogue generalization is a common method used by scientists studying space life sciences. It is known that spaceflight, or prolonged exposure to microgravity, results in many deleterious physiological adaptations. Many of these maladaptations, such as the loss of muscle mass and function (Fitts et al., 2010), bone mineral density (Smith et al., 2014), aerobic capacity (Moore et al., 2014), and cardiorespiratory function (Perhonen et al., 2001), can be counteracted with exercise. However, the exercise devices and equipment used in spaceflight may be different from those used on Earth. Engineers and scientists have developed unique resistance exercise devices, treadmills, and cycle ergometers that have been used on the International Space Station (ISS), the Russian Mir Space Station, the Skylab space station, and crew transport vehicles. When the devices are first developed, there is no evidence to support their effectiveness in microgravity. The devices are typically built to mimic the physical properties of ground-based devices, but at some point there is a need to test the devices in a zero-gravity environment.

It is not practical to develop an exercise device, place it on a vehicle, fly it to the space station, and hope it has the same characteristics in space as on Earth. Instead, space agencies use analogues such as parabolic flights in jet aircraft to provide brief periods of zero gravity (~25 s) to study the physics and biomechanics of these exercise devices (De Witt et al., 2010). A greater difficulty lies in study of the long-term training responses associated with these devices. Because long-term exposure to microgravity can have significant and potentially lasting negative health effects, the first test of the effectiveness of an exercise device on humans is not in space. Instead, space agencies use bed rest as an analogue to spaceflight (Jost, 2008). Space agencies fund long-term studies in which inpatient subjects reside in hospital bed rest facilities. In these facilities, subjects lie at a 6° head down tilt, performing all of their activities of daily living (ADL) in this position. Scientists and engineers then develop ways to exercise using the unique spaceflight equipment in the horizontal position. Such studies provide valuable evidence supporting or potentially refuting the efficacy of exercise equipment before it is tested on astronauts in space. However, bed rest is not spaceflight. The space agencies must generalize the results from bed rest to spaceflight because spaceflight data on novel equipment are not readily available.

Contextual or analogue generalization is a common technique needed for studies of exercise equipment. Since many devices hit the market before they have been studied, it is not uncommon that a search for evidence for such equipment turns up empty. In such cases, it may be necessary to evaluate the novel equipment by proxy, using research on similar devices that do have published research. Generalization across devices is not ideal, but it is sometimes a necessary compromise when no evidence exists on the device in question.

Temporal Generalization

A fourth type of generalization proposed in the literature is time generalization, that is, the generalization of information obtained from individuals at some previous point in time (Tsang & Williams, 2012). Obviously, it is unlikely that information derived from humans at different points in time would differ. For example, if an exercise protocol was studied in 1900 and 1950 and were to be studied again in 2050, the results should be similar; humans are humans and should respond similarly. What may change, however, is the precision and accuracy of measurement; a study performed 50 years ago may not have had the instrumentation to detect relevant physiological changes. Thus, the authors could have reached erroneous conclusions because of the instruments used to measure outcomes. This suggests a need to continually search the literature for the newest evidence and to reevaluate old questions as new information may surface. It may also be necessary to interpret previously published papers in the context of new scientific information and theory.

Laboratory-to-Field Generalization

A fifth type of generalization often used in exercise science is extrapolating laboratory results and applying them to field settings. This is an interesting problem, since some tests in the laboratory do not perfectly correlate to performance on the field. One sees this in athletes who are older; they are often able to continue playing sports at high levels even when physical capabilities have declined significantly. For example, soccer players may have a diminished peak aerobic capacity at age 35 compared to 25, but they are able to continue competing at a high level and may show no signs of physical decline on the field. These players are able to adapt tactical strategies to conserve energy, becoming more efficient and requiring less energy to play the sport. This has also been shown in baseball players, who are able to maintain game-specific speed well into their 30s and even 40s despite normal loss of strength and power with aging (Coleman & Amonette, 2015). Although the laboratory may show a decline, it simply doesn't show up on the field. Thus, we must be careful when generalizing results from laboratory tests.

Another important component of laboratory-to-field generalization has to do with control of the participants. The best exercise and nutritional supplementation research studies use careful controls to ensure the accuracy of the results. The investigators often have participants stay in a hospital for 12 to 18 h before testing to ensure that they are fed identical meals and fasted and that they refrain from exercise (Dreyer et al., 2008). They then meticulously control the exercise protocols, the timing and the content of the supplement, and the postexercise recovery period. At the end of testing, they can confidently attribute their findings to the independent variable. Practitioners must then generalize these results to their own training circumstances, with some of their soccer players having been up all night studying for an exam and some having slept for 10 h. Some of the athletes consumed a well-balanced meal 1 h before arriving at the weight room; some consumed a candy bar 10 min earlier; and others have not eaten for 4 h. Thus the preexercising state of the athletes is different from the carefully controlled state of the participants in the laboratory. This does not negate the findings of the study, but suggests that when generalizing the results from the study the practitioner might expect variation.

n-OF-1 AS A MODEL FOR CONFIRMING THE EVIDENCE

Chapter 8 presents levels of evidence for grading the quality of information; levels of evidence help evidence-based practitioners quantify their strength of certainty in answering a question. The levels of evidence presented by the Centre for Evidence-Based Medicine build on the early concepts of David Sackett. In practice, the levels of evidence are based on a preponderance of evidence. When multiple well-designed randomized controlled trials consistently reveal the same result, evidence-based practitioners have increased confidence in the probability of a similar result in their clients. Theoretically, more data obtained on more subjects reduces errors in generalization as we get "closer" to sampling the entire population.

However, research is never able to sample an entire population; what should be done to address the possibility of a "nonwhite swan?" In the late 1980s, a concept complementary to that of large multicenter randomized controlled trials was proposed and was supported by many, including Gordon Guyatt (Guyatt, Heyting, et al., 1990; Guyatt et al., 1986). This practice, dubbed "n-of-1" randomized controlled trials, was developed from behavioral analysis research methodologies in which case studies in individual patients are common. The design was developed to address one of the limitations in medical research—the difficulty or temporal lag in obtaining information from large, uniform multicenter randomized controlled trials (Zucker et al., 1997). It also provides a methodology for physicians to make important clinical decisions for a patient in a timely manner. If physicians waited to prescribe treatments until many well-designed randomized controlled trials with homogeneity of findings existed, treatment options would be severely limited. More importantly, potentially useful treatments would be eliminated and patient care would suffer. Guyatt and colleagues presented the concept that individual physicians could study outcomes in their patients in a systematic way, perhaps generating strong evidence at a much faster rate than randomized controlled trials (Guyatt et al., 1988).

With use of the n-of-1 trial technique, each patient becomes a randomized controlled trial. In this methodology, if a patient presents with a certain set of characteristics, he is offered admittance into an n-of-1 trial. Standardized outcome measures are collected on the patient before treatment is administered. Studies in medicine typically compare competing drugs, low versus high dose, drug versus placebo, and treatment A versus B, for example (Guyatt, Keller, et al., 1990; Keller, Guyatt, Roberts, Adachi, & Rosenbloom, 1988). In order to blind the study, drugs are provided by a pharmacist who keeps the assignment secret. In most cases, a crossover design is employed in which a patient shifts into an alternate treatment. Physicians collect standardized outcome measures throughout the entire process; if adverse effects of a drug are manifest, the treatment is discontinued and the patient is shifted to another treatment.

The most intriguing and beneficial outcome of n-of-1 trials is the straightforward determination of the best treatment plan for the patient. At the conclusion of the "study," the physician gets the data and can then choose the best treatment for that patient. The accumulative study then continues when a new patient presents, is enrolled in the study, and is randomly assigned a treatment. The strength of the n-of-1 design is that the physician is able to objectively assess the best treatment plan, not through generalization but through actual confirmation in the patient; thus, clinical decisions can be made quickly in each patient while additional patients are enrolled, contributing to a future pooled analysis (figure 10.2). At the conclusion of the trial, the final pooled analysis with all of the patients' data included is used to make global decisions on the best treatment options for the patients. Thus one can make decisions immediately for the individual patients while waiting for the final results of the randomized controlled trial.

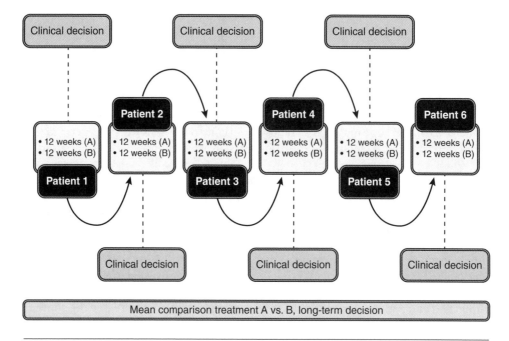

Figure 10.2　Theoretical design for an *n*-of-1 crossover randomized controlled trial. The design allows the practitioner to make the best clinical decision for the patient while collecting longitudinal group data in a systematic, unbiased manner.

CONFIRMING THE EVIDENCE WITH SYSTEMATIC TESTING

When research evidence is based on a small sample size or is being generalized to a population different from the one studied, evidence-based exercise science practitioners should test the effectiveness of a treatment in their patients. Building on the concept of the *n*-of-1 trial, we propose an additional step to the five-step evidence-based process established in the medical fields—confirming the intervention in a client, athlete, or patient. To be clear, we are not advocating that all practitioners conduct *n*-of-1 randomized controlled trials unless they are well trained as a scientist and have expertise in research methodology. However, we are advocating a systematic approach to testing and measurement within each client, athlete, or patient to ensure that selected interventions are effective at the individual level. To obtain long-term positive results, it is necessary for evidence-based practitioners to develop a testing battery that is both valid and reliable. This will allow the development of long-term tracking methodologies that facilitate direct comparisons between emerging, evidence-based interventions and older, established methodologies in individual clients, athletes, and patients and also inform a practitioner's decision about whether to continue or discard a particular treatment on an individual basis. Although incorporating the step of confirming the evidence and implementing longitudinal data tracking may require a substantial commitment, it will ensure that practitioners do not continue a treatment simply because generalization of the research suggests that it is the best option; this step will also help to identify and provide the best treatments for the "nonwhite swan."

BASIC PRINCIPLES IN TESTING AND MEASUREMENT

In order to develop a long-term tracking plan, it is necessary to understand some basic principles of testing and measurement. The assessment of a treatment, especially in a longitudinal manner, is only as good as the reliability and validity of the test and evaluation techniques. Choosing tests that are invalid with poor reliability may result in erroneous conclusions and the inclusion of suboptimal treatments or the exclusion of treatments that are effective. Thus, it is prudent to carefully consider the measurement options and not indiscriminately test a client, athlete, or patient. Testing and measurement takes precious time away from training. The evidence-based exercise practitioner should deliberate on the inclusion of each test and the evidence supporting it to arrive at a precise testing battery that will optimize time efficiency and reduce the likelihood of injury.

Scientific Research Versus Assessment

The first step in test implementation is to identify the testing rationale and determine if the purpose of the evaluation is research or whether it is client, patient, or athlete assessment. If the tests are for research, a regulatory process must be followed. Research requirements are governed by the Declaration of Helsinki (Rickham, 1964). The declaration, which was adopted by the World Medical Assembly in 1964, mandates ethical treatment of humans in research and requires biomedical scientists to protect the health of research subjects and patients. Most importantly, all research protocols must be evaluated by an unbiased ethics committee that weighs the inherent risks of the procedures against the knowledge to be gained. The ethics committee, often called an institutional review board (IRB) or committee for the protection of human subjects (CPHS), must also evaluate the informed consent process, which ensures that investigators explain the risks and benefits of the research to potential subjects in simple, easily understandable language.

Ethics committee approval is essential to the publication of data. Unfortunately, such boards are autonomous governing bodies at each institution and while they may follow guidelines for best practices and state and federal law, they are their own "entity" with no recourse for appeals by individual investigators. Public Responsibility in Medicine and Research (PRIM&R) is an organization that "attempts to advance the highest ethical standards in the conduct of biomedical, behavioral, and social science research" (www.primr.org). Some universities have seen politics and "mission creep" in the roles of IRBs due to the lack of oversight. This can affect both the investigator's work and subject safety. Thus, while the ethics board review process may not be perfect, it is essential to research work.

Scientists must also safeguard the privacy of the data and, particularly with any public dissemination (e.g., conference presentation or journal publication), ensure that no personally identifying information is presented; the potential for publication must also be explained to participants.

Testing and measurement not intended for publication do not need to be evaluated by an ethics committee. However, evidence-based exercise practitioners conducting evaluation tests should carefully consider the ethical aspects of an assessment. Similar to the situation with researchers, whether one is testing an individual or a sport team, the potential benefits should outweigh the risks. The risks and benefits should be clearly presented to clients, athletes, and patients; it is still a good idea to obtain written informed consent before testing. Although the primary intention of routine data collection may not be publication, retrospective research projects may arise from historical data collection. In training domains, such as collegiate athletics, it is prudent to seek ethics committee approval for testing to allow future retrospective analysis. Teaming with a sport or exercise scientist at the university who is familiar with the ethics committee minimizes the somewhat cumbersome process of proposal submissions and solidifies a relationship with a colleague who may help with data collection and interpretation.

Another consideration when one is preparing to test an individual or a team is the quality of data management and the safeguarding of information. Sometimes storage and archiving of data is just as important as the reliability and validity of measurement. Creation of a robust and organized database can facilitate future queries and can increase the usefulness of the information. Organizations that collect or have access to large amounts of quality data but without well-organized storage are unable to realize the full potential of their data.

Data should be safeguarded for privacy. If a medically anomalous result is obtained, an evidence-based practitioner is obligated to share this information with a client, athlete, or patient and refer him to a physician. Most information obtained in nonclinical settings is not medically sensitive, and transparent presentation of the information may be warranted. For example, a strength and conditioning coach may want to post the results of a 1-repetition maximum (1RM) test on a leader board to create competition. Such presentation is not unethical or illegal, but the coach should be transparent with the athletes about her intentions. This also may be articulated in the informed consent process, along with the procedures, risks, and benefits of testing.

Validity of Measurement

In order for a test to be useful, it must be valid. To state this simply, validity is the ability of a test to properly assess what it purports to assess (e.g., a particular physiologic capability). One should address at least five types of validity when selecting a testing battery: construct validity, face validity, convergent validity, predictive validity, and discriminant validity (Brown, Khamoui, & Jo, 2013). Table 10.1 provides an overview of these types of validity, with a basic example of each. In the next sections, each is discussed in more detail.

Construct Validity

Construct validity is the ability of a test to properly assess what it claims to measure, or in the case of exercise, a particular physical ability (Brown et al., 2013). Consider the following example: An evidence-based exercise practitioner is developing a test battery for a volleyball player. The practitioner has determined that a key physical attribute in volleyball performance is lower body power. There are a number of tests that can be used to measure or calculate power with varying degrees of validity: Wingate cycle ergometer test, Margaria-Kalamen stair climb test, isokinetic knee extension test, and broad or vertical jump. All of these tests measure some form of power.

1. Wingate cycle ergometer test: Performance of this test requires an individual to pedal a cycle ergometer with a resistance of 0.075 kp/kg (kiloponds per kilogram) of body weight at maximum effort for 30 s. The test provides peak instantaneous cycling

Table 10.1 Types of Validity Used to Determine the Appropriateness of a Test for an Individual or a Population

Term	Simple definition	Example
Validity	The proposed test is an effective measure of the desired outcome.	Treadmill $\dot{V}O_2$max test is a valid measurement of aerobic capacity for distance runners.
Construct validity	The proposed test effectively measures the desired construct.	Squat 1RM test is an effective measure of dynamic lower body triple extension strength.
Face validity	The proposed test appears to the rater and the client, athlete, or patient to be an effective measurement of the construct.	Vertical jump is a good measure of basketball-specific power.
Convergent validity	This is the ability of two tests to effectively measure the same construct.	Treadmill $\dot{V}O_2$max test and 1.5-mile run time are both good measurements of aerobic capacity.
Predictive validity	The test effectively predicts success or lack of success in a functional outcome.	Vertical jump height and body weight are predictive of speed in a 40 m sprint.
Discriminant validity	The test is able to distinguish between two different constructs.	Vertical jump and a Wingate cycle test both measure power, but two different constructs (i.e., vertical power production vs. cycling or anaerobic power).

power, average power, total work, and a fatigue index determined by the power decline over the 30-s test. Although power is measured, the primary construct assessed by the Wingate test is anaerobic capacity. The test is also sensitive to cycling experience; for example, an elite cyclist will inherently perform better than an untrained cyclist.

2. Margaria-Kalamen stair climb test: This test requires a scale, a staircase, and timing mats or a stopwatch. The athlete stands at the base of a staircase and then runs up the stairs as quickly as possible. Time is obtained between two stair steps that are a known distance apart; the vertical displacement of the stairs, time, and the mass of the athlete are used to calculate power.

3. Isokinetic knee extension test: An individual sits in a chair with the knee joint line aligned with the center of rotation of a dynamometer. The thigh is strapped to a padded seat, and the ankle is strapped to a lever extending from the dynamometer. The practitioner sets a velocity, and torque is measured by the dynamometer. The torque and fixed velocity data can be used to calculate power.

4. Vertical jump(s): An athlete stands and vertically jumps as high as possible. Jump height can be determined in several ways: (1) having the athlete jump and touch a tape measure fastened to a wall and subtracting the reach height, (2) using a Vertec reach

device and the same procedures, or (3) measuring flight time with a timing mat or force plate. The mass of the subject and the vertical jump height can be used to calculate the vertical power production during the jump.

Reading these test descriptions and considering the sport of volleyball make it apparent that the vertical jump is clearly the best measure of volleyball-specific power. In other words, this test measures the construct of lower extremity vertical power production, a capability that is necessary to perform many volleyball-specific skills. It may enhance the construct validity of the test further to have the athlete perform the vertical jump on a force platform, measure the vertical force production, and compute the power from the measured ground reaction forces. Although the other tests may be used in various situations (Wingate cycle ergometer test, Margaria-Kalamen stair climb test, isokinetic knee extension test), the construct validity of the vertical jump test for a volleyball player is superior.

There are many different vertical power tests using jumps. One can use counter-movement vertical jumps, squat jumps, and one- and three-step approach jumps; each may have a different importance to the sport of volleyball and lead to a different interpretation of the effectiveness of training. Therefore, to further enhance the construct validity of the measurement, the practitioner should consider which approach is best for sport specificity.

Face Validity

Another important consideration in test selection is face validity. A test that, to a casual observer, appears to measure the desired construct has face validity (Brown et al., 2013). In the preceding example, the vertical jump test has greater face validity to measure vertical power production compared to the Wingate cycle ergometer test, Margaria-Kalamen stair climb test, and isokinetic knee extension test because performing a jump test looks the most like playing volleyball. Sometimes the relationship is less clear. Consider the following two tests:

1. A vertical jump performed in a laboratory during standing on a force platform. The force platform is used to measure the ground reaction force and compute peak power.

2. A vertical jump performed on a volleyball court: The athlete jumps and touches a tape measure fastened to the volleyball net poles. The vertical jump height and weight of the individual would be used to estimate the power.

The vertical jump performed in the laboratory has greater construct validity because the instrumentation is measuring, not estimating, power. On the other hand, the vertical jump performed on the volleyball court has greater face validity. The face validity of the test on the volleyball court might result in greater effort from the athlete because of the perception that it is measuring a construct important to volleyball. Which test should an evidence-based practitioner choose? In an ideal world, the force plate would be moved to the volleyball court and the athlete would perform the jump off the force plate but reach to the net post. This arrangement would optimize both the construct and face validity of the test. However, if this scenario is not possible, the practitioner would need to make a judgment call based on published peer-reviewed literature, his own professional experience, and knowledge of his athletes as to the most beneficial test for each particular situation.

Convergent Validity

Convergent validity assesses the relationship between two tests of the same construct (Brown et al., 2013). This concept is applicable when one is attempting to validate a new test or to validate the use of a field test in comparison to a gold standard laboratory test. In the measurement of vertical power, a vertical jump measured on a force platform is considered the gold standard for assessment. However, other field-based measurements, such as those obtained via attachment of an accelerometer near the center of mass of the individual or an estimate based on vertical jump height and body mass, can also be used to calculate power, albeit with some error (Amonette et al., 2012). Convergent validity quantifies the relationship between two assessment methods and provides an assessment of a measurement relative to an established normative measurement. Practically speaking, testing batteries should not include tests with convergent validity. If two tests are strongly correlated, they may be measuring similar constructs. The inclusion of both tests likely will not provide additional information.

Predictive Validity

The goal of many assessment tools is to predict performance, injury, or perhaps the onset of disease. Predictive validity is the ability of a test to forecast these outcomes (Brown et al., 2013). For example, we might assess whether a vertical jump test is predictive of volleyball performance. In other words, do individuals who possess greater vertical jump ability perform better in the sport of volleyball? This is often difficult to quantify—to perform such an assessment, we would need to define volleyball performance metrics (e.g., blocks, kills) that would serve as outcomes of vertical jump performance. The establishment of predictive validity may be the most important aspect of testing. If the evidence-based practitioner desires to implement an intervention in a volleyball player, the goal would almost certainly involve improving sport performance. If performance on a test such as vertical jump is predictive of volleyball performance, it would stand to reason that an intervention that improves vertical jump might also improve volleyball performance.

Discriminant Validity

Discriminant validity is the ability of a test to distinguish between two different constructs (Brown et al., 2013). Each test selected in a battery should assess a distinct construct. Vertical jump, which measures power, should not be correlated to or predictive of sit and reach scores, which assess flexibility—these are different constructs. If vertical jump were correlated to or predictive of sit and reach, it would possess poor discriminant validity, which would lead us to question its construct validity; for example, perhaps vertical jump measures a generic construct of "fitness" and not lower body power.

Reliability of Measurement

First, reliability of a test does not mean it is a valid test. A test can be reliable but invalid; however, a valid test must be reliable. It is important to clearly establish a test's validity first and, particularly in exercise science, to define its relationship to and ability to predict athletic performance. This ensures that improvement on the test metric has a practical, meaningful impact on the function or health of an individual. However, it is equally important to ensure that the test is reliable (figure 10.3). Reliability is the consistency

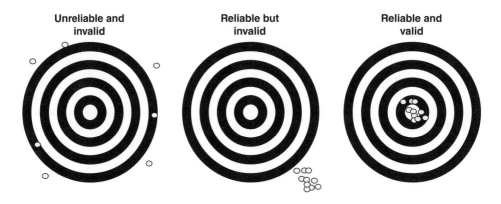

Figure 10.3 Pictorial representation of validity and reliability. Validity, ability of the test to measure what it is supposed to measure, is represented here by hitting the target. Reliability is the ability of the test to produce consistent results. A test can be *(a)* neither reliable nor valid, *(b)* reliable but not valid, or *(c)* both reliable and valid.

or reproducibility of a measurement (Brown et al., 2013). Tests that are unreliable may lead to erroneous conclusions about the effectiveness of an intervention, tool, supplement, and so on. Consider again the example of a vertical jump. Suppose you perform a vertical jump test on an individual on 3 consecutive days, with the following scores:

Monday: 18 in. (45.7 cm)

Tuesday: 21 in. (53.3 cm)

Wednesday: 24 in. (61.0 cm)

After reading the literature, you implement a new training program or technique. The athlete performs the program for 12 weeks, and you subsequently assess vertical jump. The posttraining vertical jump is 26 in. (66.0 cm). Did the new training intervention improve vertical jump height? When you tested vertical jump previously, the athlete had improved by 3 in. (7.6 cm) each day on which you tested with no intervention. Could she have tested at 27 in. (68.6 cm) on Thursday before the implementation of the new training program? Many confounding variables could have led to the variability in testing (e.g. fatigue, time of testing, hydration, injury, learning effects), but ultimately the unreliability of the test makes it impossible to determine if performance improved or if the improved vertical jump score is simply within the noise (measurement error) of the test.

Reliability is affected by the fidelity of the measurement tools, the skills of the test operator, and the consistency or variability of the subject. Test–retest reliability is quantified via performance of the same test on two or more consecutive occasions with insufficient time or opportunity for the construct being measured to have changed. Many tests of human performance confer a learning effect; if a novel test is administered repeatedly, scores will improve even though there has been no change in the underlying performance ability (Atkinson & Nevill, 1998). In such cases, it is important to administer the test several times to eliminate the influence of learning effects and to establish a true baseline. Test–retest reliability is generally higher in individuals who have routinely performed the test. For example, a 1RM back squat may improve in a novice lifter over two or three sessions if tested in consecutive weeks in a novice (Schneider et al., 2003). However, the score baselines after the third or fourth test and then becomes a stable measurement. If one is testing an experienced lifter, there may be no change over consecutive testing sessions because the test is not novel to the lifter.

Interrater reliability is the quantification of a test score between two different test operators. Interrater reliability is especially important when one is performing a subjective assessment, less important when one is performing an objective assessment. If an athlete performs a vertical jump test on a force platform with the same procedures and the same instrument and in the same conditions, two different operators should obtain near-identical results. However, if the same operators are simply watching a vertical jump and scoring the quality of the jump on a scale of 0 to 10, there may be significant variability. With the use of subjective assessments, it is important to have the same operator conduct the test in the pre- and posttest intervals; this is especially important for within-individual comparisons. A given operator may report different scores on two different occasions, but the score will probably be more consistent than with two different operators. Intrarater reliability generally improves with experience.

Designing a Testing Battery

Deciding on a testing battery is important, warranting serious thought and investigation. Ideally, the testing battery is not in response to a single need but rather a long-term testing plan. If an evidence-based practitioner defines a reliable and valid set of tests that assess an array of health and human performance constructs, longitudinal performance assessment becomes possible. Therefore, after a new intervention, supplement, or device has been incorporated into an exercise plan, a historical comparison can be performed to determine if the novel intervention meaningfully improves performance. The testing plan should be comprehensive but lean. In other words, tests should be selected that assess multiple constructs when possible, and every effort should be made to eliminate redundancy.

The constructs assessed in a comprehensive testing battery should be determined by a needs analysis, a concept developed in the early 1980s by Fleck and Kraemer and updated over the years (Fleck & Kraemer, 2014). The following factors should be considered in performance of a needs analysis (Baechle, Earle, & Wathen, 2008; Magee, Quillen, Amonette, & Spiering, 2012):

1. The metabolic systems necessary for the activity

2. Common movement patterns

3. The kinetics of the activity

4. The physical qualities (i.e., strength, power, speed, agility, and so on) necessary for performance of the activity

5. The specific flexibility and endurance necessary for the activity

Tests that closely match the needs of the individual and the sport or activity should be selected. Another important consideration, discussed previously in the chapter, is the predictive validity of a test. A test might predict performance, as in the example of a volleyball player, but it might also predict health outcomes. Resting blood pressure and body mass index are predictive of cardiovascular risk. Thus, if an evidence-based practitioner commonly works with individuals who desire to improve general health, tests of these constructs should be included in the battery because inclusion of interventions that improve these constructs is indicative of a positive impact on health. The five commonly cited components of health-related physical fitness are body composition, flexibility, muscular strength, muscular endurance, and cardiorespiratory endurance. Although often classified as a skill-related component, a strong argument could be made for power as a health-related component of fitness. The evidence-based exercise practitioner should include an assessment of each of these constructs in testing clients. The specific tests used should be determined by a thorough review of the literature; tests that are highly reliable and predictive of positive health outcomes should be selected.

Defining Intervals for Testing

Tests should be performed at regular intervals to assess changes. Frequent assessment leads to more data points for evaluation. On the other hand, regular testing reduces time available for training and may lead to inaccurate conclusions and "micromanaged" training. Testing frequency is best determined by the expected time needed to elicit a measurable change in the outcome of interest; this is determined by both physiology and the precision of the testing instrumentation.

For example, bone mineral density is an important training outcome, especially in people who are aging and those with low bone mass. The rate of detectable change with bone mineral density is highly dependent on the instrumentation and testing method. Biochemical markers of bone turnover can be detected 24 h after a training session. Peripheral quantitative computed tomography (pQCT) may detect measurable changes in bone mineral density in 8 weeks. But a detectable change in bone mineral density using dual-energy X-ray absorptiometry (DXA), the clinical gold standard measurement, may take up to 4 months. Thus, if we used DXA to assess bone mineral density by performing scans every 2 months, we might find that an intervention is not improving bone mineral density simply because the instrument is not sensitive enough to detect the change. If the same patient or client were undergoing pQCT testing, the instrument might be able to detect small changes in a short period of time due to its greater sensitivity compared to DXA. Practitioners using DXA as an outcome measure for bone mineral density should increase the testing interval to match the sensitivity of the instrument; that is, they should not reach conclusions on the efficacy of the treatment until adequate time has been allotted for measurable change to occur.

Another important and practical testing consideration is the logistical constraints of a schedule. On a training calendar, there may be logical times to implement testing. Testing is typically performed before the beginning of a new training plan or when there has been a substantial change in the plan. Physiological assessments may be required by a regulatory agency, such as an insurance company, at certain times in a treatment plan. Each testing construct and instrument may have a different time interval for detectable change. The evidence-based practitioner should rely on research evidence, when available, to inform testing intervals. However, when other constraints require different testing intervals, exercise practitioners should always interpret the results in the context of physiological response times and instrument precision. Failure to do so may result in invalid conclusions on the effectiveness of an intervention.

DECIDING TO CONTINUE OR DISCONTINUE AN INTERVENTION

Testing may be used to establish a baseline, determine training loads or intensities, or determine the effectiveness of a treatment. Exercise practitioners often work with large groups of individuals (e.g., sport teams). In such cases, the practitioner is often observing mean changes in groups. Although the mean is of some use, the most important training outcome is the individual change. The primary purpose of implementing regular testing is to detect changes within the individual. Consider the following case example.

An exercise practitioner is employed in a clinical exercise physiology group that works with individuals who are older. Ten people have been regular members of the training facility for 2 years. As part of the testing plan, the exercise practitioner measures hip bone mineral density every 4 months to assess improvements resulting from training. The training facility just received a new vibration platform, and the exercise practitioner plans to incorporate this tool into training after the next testing interval. On the basis of a published training protocol, the exercise practitioner implements a progressive vibration training regimen into the overall training plan of the clients (Verschueren et al., 2011). All participants follow the program consistently for 16 weeks, at which point bone mineral density is once again measured. The exercise practitioner is delighted that his clients, as a group, experienced a 2.1% increase in bone mineral density of the hip in 16 weeks (figure 10.4). This is in contrast to the 2.0% increase observed over the past 2 years. However, upon further inspection of the individual data, he notices that

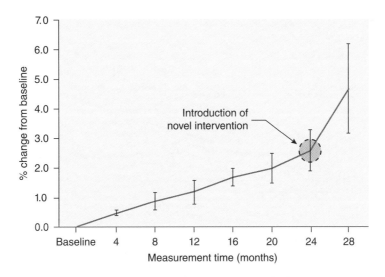

Figure 10.4 Sample longitudinal data analysis for a group of clients undergoing exercise intervention to improve bone mineral density. The point at which the novel intervention is implemented is represented on the chart.

one person actually decreased bone mineral density. After reviewing the training records, the exercise practitioner determines that there was no difference in this person's exercise plan. A conversation with the client reveals no notable changes in diet, sleep, physical activity patterns outside of training, or any other lifestyle habits. The client did, however, report feeling particularly exhausted after vibration training sessions.

Should the exercise practitioner continue to implement the vibration protocol in all of the clients? The research supports the protocol, and the group mean increased—does this not justify continued use of the protocol? Should the protocol be discontinued for all clients because of the poor results in one individual? These are important and valid questions. It is imperative not to jump to conclusions based on one data point and eliminate a potentially beneficial intervention because of one outcome. A possible solution in this scenario is to continue the intervention for all the clients but more closely monitor the individual with a negative response. If this person continues to report fatigue, then discontinue vibration training for that client but continue it in the others. If the client tolerates the training for 4 months but produces similar poor results for bone mineral density, discontinue vibration training for that client but continue it in the others. It is possible that this client is a "nonwhite swan," or that she has met her genetic potential and is now in a maintenance situation. You should not persist simply because the research supports this intervention, which is evidently ineffective in this individual (figure 10.5).

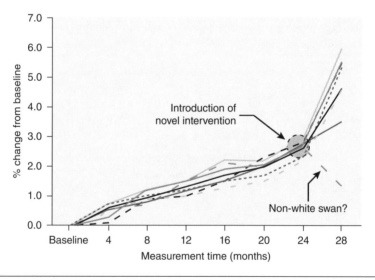

Figure 10.5 A "spaghetti graph" of individual and group longitudinal data will help to confirm the effectiveness of an intervention and identify nonresponders to the novel treatment.

COLLABORATION IN EVIDENCE-BASED PRACTICE

When one is evaluating the effectiveness of new interventions within individuals, the definition and precise collection of outcome variables are imperative. Poorly collected or nonstandardized data collection can lead to erroneous results and potential misclassification of the usefulness of an intervention. An exercise practitioner may not have the specific skill set needed or access to the equipment required to collect or interpret the ideal dependent variables. In this case it may be beneficial to seek and establish formal collaborations with scientists. As chapter 16 discusses in detail, collaboration between scientists and practitioners is a key to the progression of the evidence-based philosophy within the exercise discipline. Through collaboration, scientists gain access to real-world protocols and practical problems and also study participants who may be difficult to find in the typical laboratory setting. Exercise practitioners are afforded the opportunity to work with scientists who can assess the effectiveness of new interventions without bias. Collaboration is a positive proposition for both groups. Practitioners may seek collaborations and testing support by contacting faculty members at nearby universities. State, regional, national, or international meetings hosted by agencies such as the National Strength and Conditioning Association, American College of Sports Medicine, the UK Strength and Conditioning Association, or the Australian Strength and Conditioning Association, among others, are avenues for networking with scientists and practitioners with similar interests and developing partnerships.

CONCLUSION

The purpose of this chapter is not to teach practitioners to be scientists or to conduct formal experiments with their clients, athletes, or patients. However, approaching exercise prescription and testing in a systematic manner can be very beneficial for identifying responders and nonresponders to an intervention. One of the key tenets of research is generalizability, or the ability to extrapolate the result obtained in a sample to others in the same population. The larger the sample size of an experiment, the greater the likelihood that its findings are generalizable to the entire population. Many times the literature supporting novel interventions is scarce; thus, it is prudent to implement a testing plan within your population to confirm the evidence. Outcome measures should be carefully evaluated for reliability and validity. Research evidence should be the strongest influence on test selection and should drive testing interval duration. Failure to carefully consider each of these variables can lead to invalid conclusions and to the inclusion of ineffective interventions or the elimination of effective ones. Individuals respond differently to stimuli. Prudent practitioners should be on the alert for "nonwhite swans" and be prepared to adjust interventions to meet the needs of each individual.

Chapter 11

REEVALUATING THE EVIDENCE

Learning Objectives

1. Understand why it is necessary to reevaluate evidence.
2. Learn about prompts to reevaluate evidence.
3. Discuss practical ways to reevaluate evidence.

One of the central themes of this book is that knowledge is dynamic. An evidence-based exercise practitioner will never attain complete or perfect knowledge in any given area; however, to remain still and stagnant consigns practitioners to their current state of practice with no mechanism for advancement. Evidence-based practice (EBP) is a paradigm by which (1) questions are developed, (2) evidence is gathered, (3) evidence is evaluated and graded, (4) evidence is applied in answer to a practical question, and (5) evidence is confirmed through evaluation in the individual. Thus, after completion of the first five steps of EBP, the sixth and final step is intuitive: reevaluate the evidence (figure 11.1). A program that was state of the art 1 year ago, replete with the best available evidence, may now be improved upon by newly available information. This is the crux of the evidence-based paradigm: Never engrave anything in stone; always continue to seek new evidence that will affirm, overturn, or fine-tune current practice.

PROMPTS TO REEVALUATE

How does an evidence-based exercise practitioner know when to reevaluate evidence for a particular practice? As exercise professionals, our primary responsibility is to develop effective exercise programs for our clients, patients, and athletes; there is not much time to scour the literature on a daily basis just to see if a new study has been published that might affect our exercise programming. However, several practical situations should prompt a reinvestigation of the evidence.

New Research Evidence

The most obvious indication to reevaluate the evidence is the appearance of new research information. Although this is seemingly apparent, it is extremely easy to become set in one's ways and to not seek—or to even ignore—new evidence. Or even worse, to ignore

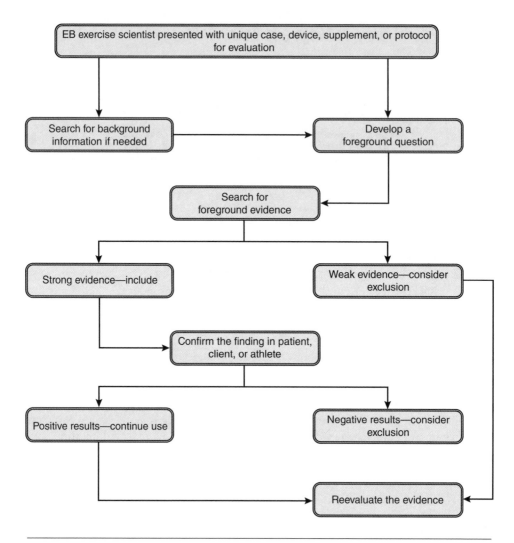

Figure 11.1 Flow chart describing the six-step evidence-based practice paradigm for exercise scientists and practitioners.

the reinterpretation of old evidence. Paradoxically, those who practice the evidence-based process can also be vulnerable to this pitfall; having completed the first five steps, it is tempting to assume that the process is final and that nothing remains to be done. But in reality, EBP is a continual process in optimizing care or exercise prescription.

Although much of the discussion in this chapter is directed toward exercise practitioners, research investigators must also be vigilant in searching and reading literature for new ideas affecting their own work. Newly published studies can prompt a host of new questions, leading exercise scientists to new hypotheses, theories, and methodologies that impel practice forward, sometimes in new directions. For example, the study of growth hormone responses to resistance exercise was well established in the late 1990s, with many mechanistic and descriptive studies demonstrating the fundamental concepts of hormonal responses to various stimuli. A study published in 2000 first explored the relationship between bioactive growth hormone, resistance exercise, and neural feedback (McCall, Grindeland, Roy, & Edgerton, 2000). Another study, published 1 year later, established that large bioactive growth hormone aggregates are present in much higher concentrations than the growth hormone monomer and that unique fractions respond to resistance exercise (Hymer et al., 2001). Exercise scientists who studied endocrinology gained new perspective from these data, and experimental designs were then (and are now) forced to account for these findings when evaluating resistance exercise. Bioactive growth hormone had been discovered decades previously but ignored in favor of a simpler assay that did not assess these larger isoforms (Ellis, Vodian, & Grindeland, 1978). However, these new studies forced attentive exercise scientists to reexamine the newly discovered complexity of this hormone. Stunningly, it was not a single hormone that was of importance but a superfamily of growth hormones spliced in aggregate forms. Thus, a new world was opened to investigators: the study of the growth hormone superfamily (Nindl, Kraemer, Marx, Tuckow, & Hymer, 2003; Kraemer et al., 2010).

New Industry Trends

While public practice and consensus is itself subject to bias to the extent that it is not even ranked within the levels of evidence, it is nevertheless prudent and useful to reflect on new trends and practices in the field by reevaluating the evidence on which we base our current practice. In some instances, observation of other professionals, or even the practices or questions of clients, patients, and athletes, may prompt new searches. Since these individuals are often searching for methods to gain a competitive advantage or to accelerate their health and fitness goals, they may search in areas that the evidence-based practitioner does not. In some instances, particularly in exercise and fitness, training concepts can attain near-mythical status, along with blind adherence, despite plentiful evidence against them. In personal communication with an author of this text, a resistance training "guru" proclaimed, "I believe in what I do and no study or studies will ever change my mind." Evidence-based practice is beyond the scope of this individual's capacity to learn or adapt; money, fear, educational background, and other factors may influence the decision-making process used to arrive at this conclusion. The key when evaluating industry trends is to place the ultimate weight of the validity of the product, supplement, or protocol on the evidence. That being said, always evaluate new trends, as some are valid and may in fact improve practice.

Passage of Time

Because new research and other types of evidence are constantly being generated, the simple passage of time should be a prompt to reexamine the evidence for a given practice. A good example is the evidence for using chains with barbell training to enhance strength–power adaptations. Searching "lifting chains" on PubMed at the time of this writing netted 62 articles, of which 5 were actually pertinent to the topic; searching "barbell chains" yielded 9 articles, 5 of which were relevant (2 of these were redundant with the "lifting chains" search). Because 7 of the 8 articles were published in *Journal of Strength and Conditioning Research (JSCR)*, a follow-up search was conducted in that journal's search engine; this yielded 173 articles, 14 of which were on topic. All 7 of the

JSCR articles from the PubMed search were also returned with the *JSCR* search, so the final result was a total of 14 *JSCR* articles plus 1 additional paper. Looking at the years of publication reveals an interesting temporal trend: The first paper was published in 2002 and the next in 2006, followed by 2 papers in 2008, 5 in 2009, 4 in 2010, and 2 in 2011 (figure 11.2).

An exercise practitioner in 2002 who stayed abreast of the cutting-edge trends would have been aware of the emerging practice of employing chains in conjunction with resistance exercise (e.g., bench press). However, she would have had only one peer-reviewed publication on which to make an evidence-based decision regarding the efficacy of chain training. To make things even more difficult, the 2002 study was an acute response study evaluating the electromyographic signals (EMG) and ground reaction forces resulting from performing traditional squats and squats plus chains. That is, although it was an appropriate (and common) type of initial evaluation, the study was not a longitudinal comparison of strength adaptations to the two different types of resistance. A follow-up search even 5 years later would have yielded only one additional study; however, from 2008 to 2010, 11 papers were published on this topic. From the practitioner's perspective, the available evidence was changing very rapidly over that 3-year period. Although the research community can be slow to investigate novel supplements or training methodologies, at a minimum, it takes several years to conduct good research studies, write them in manuscript form, submit them to peer review, and have them published. Given enough time, scientists will perform research on emergent concepts; in the meantime, strength and conditioning professionals should carefully monitor the effectiveness of the novel method in limited use with their own athletes while awaiting formal, rigorous research to evaluate its efficacy.

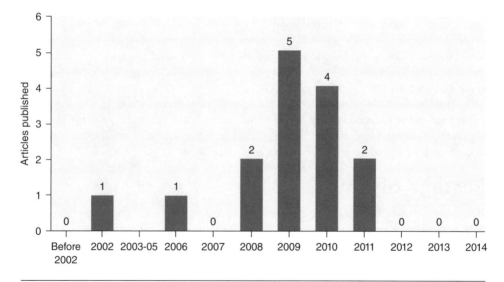

Figure 11.2 Number of relevant articles published using the terms "lifting chains" or "barbell chains." The chart indicates how the time of search can significantly affect the information obtained.

New Athlete, Patient, or Situation

Issues arising from working with a new client, patient, or athlete—or perhaps difficulties in inducing positive adaptations—may initiate a follow-up search. As discussed in chapter 6, initial questions often arise when a client, patient, or athlete presents with a novel condition or circumstance. Likewise, a follow-up question may be warranted when a new client is not responding to a treatment or intervention. There may be evidence on potential contraindications or altered protocols that are more effective in eliciting a positive response from these individuals.

Limitations in the Original Evidence

Sometimes an initial evidence search uncovers information that is limited or one-sided. Many questions with a substantial body of literature have both supporting and opposing evidence; the evidence-based practitioner must decide based on the strength of the information to incorporate or exclude. A one-sided search may indicate a very strong and effective intervention or indicate that the research regarding the intervention is in its infancy. Either way, when information is limited (e.g., evidence on chain training in 2005 as previously discussed) or one-sided, it is sensible to periodically reevaluate the evidence.

Evidence should also be reevaluated when the original work in a research area is conducted in a population or context different from the one in which a practitioner intends to apply it. A good example is the original elastic-based resistance exercise device (the interim Resistance Exercise Device, iRED) deployed on the International Space Station (ISS); due to short timelines, physiological outcome testing in humans utilizing iRED was not complete before its deployment, but it was built based on research from other similar devices (even though that research shows that elastic resistance does not produce force profiles similar to those with free weights, the gold standard of resistance exercise) (Amonette et al., 2004). Later, research conducted specifically on iRED confirmed its inferiority to free weights (Schneider et al., 2003) and led to the development of a second-generation device (the Advanced Resistive Exercise Device, ARED) that elicits training adaptations similar to those with free weights; ARED is thus the current resistance exercise hardware on the ISS (Loehr et al., 2011).

Long-Term Effects and Safety

Another reason to revisit the evidence is to gain an understanding of the long-term effects of a particular supplement or practice. A good example is the supplemental use of creatine monohydrate; creatine appeared in the 1990s as an ergogenic aid that increased muscle strength and power. Similar to the earlier example of barbell resistance training with the addition of chains, beginning in 1995 the scientific literature exploded with studies evaluating creatine in every possible dosage, population, and sport and with a wide variety of outcome measures. Generally, these investigations found that creatine supplementation resulted in improvements in muscle strength, power, and time to fatigue during short, intense exercise. Simultaneously, anecdotal reports began surfacing of creatine causing dehydration, muscle cramps, and liver and kidney dysfunction. However, not until 2000 was a review published that thoroughly examined the potential negative outcomes and long-term effects of creatine supplementation (Poortmans & Francaux, 2000). An evidence-based exercise practitioner who began suggesting creatine as a performance enhancer to his athletes in 1995 would have been well served to reevaluate the evidence over the next several years—not because of emergent work suggesting that creatine was an ineffective supplement, but to understand the long-term effects and safety of creatine usage. Over the next 10 years, a number of published studies largely concluded that long-term creatine usage is safe, with no documented side effects attributable to supplementation (Hile et al., 2006; Jager, Purpura, Shao, Inoue, & Kreider, 2011; Polyviou et al., 2012; Rawson & Volek, 2003; Volek, 2003; Watson et al., 2006). However, in a perfect illustration of the continued need for long-term reevaluation of evidence, one scientist has raised a concern about the possible risk of esophageal or stomach cancers, which have been shown in rats and mice as the result of a creatine by-product that can form in the stomach (Archer, 2004). There are certainly no data directly linking creatine and cancer (even in animals), but based on basic science research, there is a theoretical biochemical possibility that creatine ingestion could form compounds in the stomach leading to cancer; clearly long-term follow-up on this hypothesis is merited.

Technological Advances

Another major factor that necessitates reevaluation of the evidence is the advance of technology. A sound clinical example comes from the field of osteoporosis and bone mineral density. For decades, it has been understood that women rapidly begin to lose bone mineral density after the onset of menopause. A relatively inexpensive tool was developed that provides reliable and accurate bone mineral density measurements: dual-energy X-ray absorptiometry (DXA). Dual-energy X-ray absorptiometry measures grams of mineral in the bone per square centimeter (g/cm^2). Because of the widespread use of this clinical assessment tool, normative values have been established for men and women as well as different ages and races. Although DXA is a highly useful tool in the detection and monitoring of osteoporosis, it has at least one major weakness: because the measurements are two-dimensional, DXA cannot provide any information on spatial structure or bone architecture and thus of bone strength. Osteoporosis is, of course, problematic for a practical reason: Lower bone mineral density results in lower bone strength, which is associated with an increased risk for falls and fractures. Age-related decreases in bone mineral density can be safely assumed to result in lower bone strength. But for other populations such as injured athletes, astronauts, or cosmonauts in spaceflight who experience rapid losses of bone mineral density followed by relatively fast regain, it is unclear from DXA results whether changes in bone architecture occur and, if they do, how they affect bone strength and the propensity for fractures and falls (Organov et al., 1997).

A new technology is changing this; peripheral quantitative computed tomography (pQCT) facilitates a three-dimensional assessment of bone architecture, and subsequent computer modeling produces estimates of bone strength such as bending stiffness. For astronauts who lose bone mineral density during spaceflight but return to preflight baseline 1 year after landing, DXA suggests that their bones (and their bone strength) are unchanged from their preflight state. However, pQCT scans have demonstrated that this is not always the case. In spaceflight, most bone loss is from the trabecular compartment, the inner, more architecturally diverse bone that provides much of a bone's strength. After return to Earth, bone regrowth tends to be on the outer, cortical surface of the bone, with the trabeculae remaining thinned from the time in space. Thus, despite a DXA-determined return to preflight bone mineral density, pQCT has provided a deeper, more complete understanding of the changes that occur in bone both during spaceflight and upon return to Earth. As a result of these new findings, new research is under way to better understand the mechanisms of bone loss and regrowth and what interventions (such as exercise, nutrition, and pharmaceuticals) could be used to optimize bone strength during and after spaceflight. Conversely, in athletic populations, recent research employing pQCT has demonstrated that despite DXA-measured bone mineral density similar to that in active healthy controls, elite athletes possess superior bone strength and architecture (Liphardt et al., 2015). Thus, advances in technology require us to reevaluate the research evidence, as the new findings they produce can alter or expand the knowledge base in a given area.

Movement From Basic Biology to Clinical Research

An exercise practitioner should also be alert to the need to reevaluate research evidence as it matures in its scope and application. What does this mean? Often, research follows a predictable path; first, a basic science discovery (which is often secondary to the main purpose of a study if not completely unexpected or even unrelated) is made that reveals a previously unknown phenomenon or mechanism. Scientists from other areas of research notice the new finding and speculate as to what it could mean for their field. Next, they design simple, acute studies to determine whether the phenomenon will hold for their population, sport, conditions, and so on. Finally, they conduct chronic "training" studies (typically, randomized controlled trials) to evaluate the effects of the phenomenon in their environment of interest.

A good illustration of this is the timeline of research on blood flow–restricted exercise. In the 1980s, in an effort to understand physiological control mechanisms for circulation and ventilation during aerobic exercise, researchers evaluated the effects of graded supine cycling with partial blood blow occlusion of the legs. The results from this study suggested that circulation and ventilation were at least partially controlled by chemoreflexes at the local muscle level (Eiken & Bjurstedt, 1987). Eight years later, the same authors proposed a more highly developed model of blood flow–restricted exercise (Bjurstedt & Eiken, 1995). After another decade of relatively little work on the topic, several papers were published by investigators who began looking at the effects of blood flow restriction on adaptations to resistance exercise (Burgomaster et al., 2003; Moore et al., 2004; Takarada et al., 2000). This led to acute studies that (1) examined the molecular mechanisms underpinning blood flow–restricted resistance exercise-induced increases in muscle mass and strength (Fujita et al., 2007) and (2) characterized the effects of various factors (e.g., occlusion pressure, occlusion duration, and exercise intensity) that modulate the physiological adaptations (Cook, Clark, & Ploutz-Snyder, 2007).

Acute research studies often function as proofs of concept, and once an intervention or training method appears viable, the next step is chronic training studies to determine if acute observations (e.g., positive molecular events) translate to chronic training adaptations that are beneficial to athletes, patients, and clients (Cook, Brown, Deruisseau, Kanaley, & Ploutz-Snyder, 2010; Credeur, Hollis, & Welsch, 2010; Evans, Vance, &

Brown, 2010; Karabulut, Abe, Sato, & Bemben, 2010; Patterson & Ferguson, 2010). Simultaneously, studies are performed or information is inferred from the training studies concerning the safety of an intervention. This progressive march from basic biological understanding to acute phenomenon, to mechanistic elucidation, to chronic, practically applied human intervention is the "circle of science." Any given idea makes its way down this chain, with multiple loops or reinsertions into earlier steps as spinoff questions and hypotheses necessitate more basic validation. Accordingly, evidence must always be reevaluated.

In most cases, the original biological research is confirmed through a randomized controlled trial. However, there are times when the expected responses inferred from the biological evidence are unsupported by the randomized controlled trial. This may occur due to a misunderstanding of the biological evidence or ignorance of the integrated systemic physiological response. Both of these situations provide strong rationale for the lower grading of basic biological evidence in the fourth step (evaluating the evidence) of the evidence-based process. Basic biological evidence may provide initial insight for programmatic decision making, but it should be a major prompt to reevaluate the evidence. As the circle of science progresses toward human training studies, the strength of the evidence is increased and can provide a better foundation for decision making.

TECHNIQUES TO STAY CURRENT

Membership in organizations and societies that hold conferences and publish scientific journals in one's area of interest is the first step toward staying current in one's profession. However, today, with such widespread research in so many journals, practitioners must be observant beyond their own primary organizations and routinely perform broader searches. As an alternative to performing the same search (i.e., same words, search engines, and so on) repeatedly at a set time interval, numerous services notify exercise professionals of new research published in their area of practice. For example, the American College of Sports Medicine (ACSM) provides a weekly literature update for subscribers who indicate their areas of interest and subsequently notifies them of relevant new articles published that week. Certain journals, such as *Journal of Strength and Conditioning Research*, send subscribers e-mails with newly accepted articles published ahead of print. Feed readers or aggregators are another service that can create easy and manageable methods for uncovering new evidence. Most aggregator websites and mobile applications allow the user to create an account and then follow self-selected journals using keywords of interest. When a newly published article matches the chosen keywords, the user is alerted to the publication. Aggregators are available in most operating systems and mobile applications; thus, new research prompts can be sent directly to the evidence-based practitioner's computer, tablet, or phone. Practitioners can find many of these services by searching the Internet for "feed aggregators."

Another sound strategy for keeping abreast of the newest research evidence is to monitor the most recent publications of prominent researchers in one's area of interest. Tools that can help in this effort include Internet-based professional forums such as ResearchGate (www.researchgate.net). Joining ResearchGate involves creating a user profile highlighting one's own work (e.g., publications, projects, collaborators) but also provides the capability to "follow" others who are a part of the forum. Users receive notifications when researchers they are following publish new research, begin new projects, or seek new collaborations. This is ideal for the evidence-based practitioner, as the automatic updates are published on a regular basis.

Evidence-based practitioners may also discover new research evidence through professional relationships. Whether it is with a fellow strength and conditioning coach at your university or someone you meet at an international conference, maintaining relationships with others who conduct similar types of work (or at least have the same areas of interest) is a convenient and invaluable way to remain aware of new developments in the field. LinkedIn, a professional networking tool, facilitates the establishment and maintenance of professional relationships that can keep evidence-based practitioners informed of new research (www.linkedin.com).

Conference presentations and continuing education may also prompt reevaluation of the literature. Information disseminated at conferences can be practically based or may include "late-breaking" research from top exercise science laboratories. Although the information presented at conferences is not rigorously peer reviewed and is thus not a strong form of evidence, it may prompt one to search for new evidence to confirm the presentation material. Of course, some conference presentations are subsequently submitted to peer review and ultimately appear in the scientific literature.

CONCLUSION

The evidence-based approach in exercise science is a six-step process; the final step includes a periodic reevaluation of the evidence. This step is required to account for the dynamic nature of knowledge, which is due to regular scientific advancements and industry progression. Reevaluation is made easier by tools such as professional networking and information aggregators that prompt the exercise practitioner when new information emerges. Continuing education and conferences, as well as information and questions originating from clients, patients, and athletes, may also alert the evidence-based practitioner to new trends, encouraging a search for evidence. For every question that an exercise practitioner can pose, there was a time when no research evidence for it existed. Thus, the temporality of a search can meaningfully affect the answer to a question and is the primary motive for ongoing and continuous reevaluation of information.

PART III

Case Studies in Evidence-Based Practice

Learning to shoot a basketball, lift a weight, kick a soccer ball, pole-vault, or play the violin are multifaceted learning tasks. There are many theories in the literature describing the most appropriate ways to introduce novel skills and develop proficiency. At the core of each theory is practice. Most readers have likely heard of the "10,000-hour rule," a popular learning theory that suggests it takes 10,000 hours of practice to master a skill. However, before people can commit even 1 hour to practice, they need to have a fundamental understanding of the skill. Imagine traveling back in time to before Lionel Messi had ever seen a soccer ball. Imagine that he had never observed a soccer game, had never heard anyone talk about the sport, and knew nothing about soccer. If you were to have placed Messi and a soccer ball in front of a goal and said "Kick this," imagine the results. He might have been able to kick the ball into the goal, but it certainly would not have appeared like a kick from one of the greatest soccer players of all time. At some point, a coach demonstrated the skill, and Messi watched. After he watched the coach a few times, he began to imitate the skill and to practice. Then, after many, many hours of watching, practice, and correction, he became a master of the skill.

In part III of this book, we demonstrate the skill of evidence-based practice. Using real-world examples and questions pertaining to exercise program design, exercise for special populations, nutrition and supplementation, and exercise hardware evaluation, the first four steps of the evidence-based practice paradigm are shown, resulting in answers to important questions based on the current best evidence. For practical reasons, we do not demonstrate the fifth or sixth steps—confirming the evidence in the athlete, client, or patient and reevaluating the evidence. In the spirit of evidence-based practice, the answers to these questions are not meant to be definitive, final answers, but instead, to provide illustrations of how the methodology should be implemented in realistic case examples. It is always possible that new evidence may emerge, disrupting what is currently believed regarding a topic. Also, the case examples in this section are not intended to be a comprehensive review of the literature. The authors have purposefully limited the number of selected articles for the sake of brevity and have chosen to include only those articles that best answer the proposed questions. After observing the evidence-based practice methodology in action, the reader should be ready to move forward and begin practicing the skill.

Chapter 12

EXERCISE PRESCRIPTION

Learning Objectives

1. Review the need for using the best evidence to support the development of exercise prescriptions.
2. Provide five practical examples of questions that may arise in exercise prescription.
3. Understand how to develop an evidence-based practice question and find evidence.
4. Objectively assess the evidence, providing strength of evidence for five case examples.

The prescription of exercise can be a difficult and arduous task for the evidenced-based practitioner. Programming theories, techniques, and trends are abundant within the industry. There is also conflicting information concerning the number of sets and repetitions, exercise modes, and intensities that are necessary to bring about specific adaptations. The answers to many of the questions surrounding exercise prescription are even further clouded by conflicting information from different certification agencies and purported industry experts. Individuals are often influenced by people who they personally respect but who espouse misinformation or misconceptions. The decision about which model, mode, or theory to follow is not trivial; it is one that deserves much critical thought. Whether to prescribe one, two, three, four, or ten sets, whether to include or exclude strength training in a competitive runner, how long rest periods between sets should be, and any number of other daily decisions ultimately affect the long-term progress of our clients, patients, and athletes. In this chapter, we use the evidence-based approach to address five critical questions related to exercise prescription.

CASE STUDY 1: Strength Training and Cycling

Zeke is a 19-year-old category 1 road cyclist who is hoping to obtain a professional contract. He is searching for any training intervention to propel him to his goal of full-time professional cycling. Currently, he rides his road bike either outside or on an indoor trainer 6 days per week. During high-volume training weeks, Zeke rides 20 to 25 h per week. Volume fluctuates considerably during the season and during different phases of training. During his 8 years of competitive cycling, he has built a strong aerobic conditioning base, and his strength as a cyclist is his endurance. In the previous Under 23 amateur national championship road race, he missed the win by 5 s. At the end of the race, he was in a three-man group with two other cyclists, but in the last 400 m, despite good positioning, he was unable to maneuver around the other two racers. He also noticed that he struggled in the hill portion of the race and tended to lose time to his competitors.

Zeke visited a local training facility and talked to a personal trainer with expertise in strength and conditioning for athletes. The personal trainer suggested that he might want to consider replacing a portion of his endurance training with strength training. Zeke is reluctant since his training time is very valuable. Along with spending 20 to 25 h per week on the bike, he also has a part-time job, working 20 h per week. Some of his cyclist friends lift weights and are considerably larger. Zeke is afraid that if he lifts weights his body mass will increase and that this might hinder his cycling performance.

Background

Road cycling is a predominantly aerobic metabolic sport. Lower body muscle strength is important to the extent that the legs and hips are needed to repeatedly generate sub-maximal force and maintain power for long periods of time. At the professional level, the most famous race is the Tour de France. The distances in this race range from short time trials to 200+ km stages. National-level Under 23 races are often 150 to 170 km. It is not uncommon for a professional cycling event to last 5 h or more and to involve significant changes in elevation, depending on the location of the event; these changes in elevation may require bursts of high-intensity power production (Foster, Hoyos, Earnest, & Lucia, 2005). Because race durations are so long, a substantial amount of training must be devoted to developing the necessary specific endurance and aerobic capacity. To compete at the highest level, high-mileage training is a prerequisite, but does adding resistance training to cycling training improve or hinder performance in elite cyclists? The question we can pose for this evidence-based inquiry is as follows:

> **In competitive road cyclists, does adding resistance exercise to a specific cycling training program improve cycling performance as measured by time trial performance or other key cycling parameters?**

Search Strategy

The publicly available search engine Google Scholar was initially used to search and obtain research evidence. Google Scholar was used in this particular case study because some relevant peer-reviewed practice-based journals may not be indexed in PubMed. The search terms "cycling performance and strength training," and "cycling performance and resistance training," were used. With Google Scholar, the original search revealed 292 articles. The search was then narrowed through the addition of "road" to the search terms. The result was eight articles, which included one systematic review of resistance training in highly trained cyclists (Yamamoto et al., 2010). Using Yamamoto and colleagues' search terms, PubMed was searched for articles published since 2008. Because the systematic review was published in 2010, the 2 years before publication were searched to account for time from the original search to the publication of the article. This search resulted in one additional article published the same year as the systematic review, but not cited in the analysis that was relevant to the evidence-based question (Ronnestad, Hansen, & Raastad, 2010). The systematic review compared five studies; all five studies reviewed were obtained and analyzed to evaluate the quality of the individual research articles. One was excluded because the participants in the study were both cycling- and running-trained athletes (Hickson, Dvorak, Gorostiaga, Kurowski, & Foster, 1988).

Discussion of Results

Overall, five articles were reviewed as evidence for or against the inclusion of strength training in the road cycling training program. Three articles demonstrated generally positive results, whereas two articles found no significant differences between control and experimental groups. Bastiaans and colleagues (2001) studied the effects of replacing a portion of endurance training time with resistance training in competitive cyclists. Using 16 competitive male cyclists, they assigned an equal number of participants to a control group that continued their typical cycling volume and an experimental group that replaced a portion of their cycle-specific training with resistance training. The resistance training included four sets of 30 repetitions at a high intentional velocity. The exercises included were squats, step-ups, and leg press. As part of the resistance training workout, the athletes also completed two sets of abdominal exercises and intermittent 10-min bouts of cycling at 75% of maximum heart rate. The cycling volume for the experimental group was reduced such that the total training time was equivalent to that of the control group. The primary cycling-specific test was a simulated time trial and a brief power test for which the average power output during a 30-s sprint was calculated. Cycling-specific tests were performed at baseline, week 4, and week 9 of the study. The authors found that average power during the simulated time trial was improved in both the experimental and control groups. Power output during the 30-s sprint test decreased in the control group, but was maintained at pretraining levels in the experimental group.

Similarly to Bastiaans and colleagues (2001), Paton and Hopkins demonstrated improvements in time trial performance by replacing a portion of cycling time with resistance exercise (Paton & Hopkins, 2005). Using a matched-pair, nonrandomized design, 20 male competitive cyclists were allocated to either an experimental or control group. The experimental group performed 12 high-effort plyometric training sessions in which they completed single-leg jumps, single-leg depth jumps, and 5- to 30-s sprints on a cycle ergometer. Mean power output in 4 km and 1 km time trial performances as well as peak power output was measured before and after the 5-week study. No differences were observed in any of the measured variables for the control group that performed their routine cycling exercise. However, the experimental group experienced substantive improvements in all measured variables: 1 km time trial (8.7 ± 2.5%), 4 km time trial (8.4 ± 4.1%), and peak power (6.8 ± 3.6%).

The final study showing positive results measured the effects of strength training during the competitive season on cycling performance. Ronnestad, Hansen, and Raastad (2010) investigated the effects of pre- and in-season strength training programs in 12 national-level cyclists. The cyclists' sex was not specified. With participants equally divided into experimental and control groups, the control group performed endurance training alone while the experimental group performed a 25-week intervention that consisted of periodized resistance training in addition to their typical cycling volume matched to the control group. Half squats, unilateral leg press, unilateral hip flexion, and plantar flexion were performed at loads ranging from 10RM (repetitions maximum) to 4RM across the 25 weeks. The sessions were performed twice during the precompetition period and once per week during the competition period. A 40-min maximal time trial was used as the primary cycling-specific outcome. Both endurance training alone and endurance training combined with strength training improved power output during the 40-min time trial; however, the increase resulting from the combined program was greater than with endurance training alone. Secondarily, the combined resistance and endurance training group also increased the cross-sectional area of their knee extensors and flexors.

In contrast to these positive findings, two studies reported no significant differences with the addition of a strength training program. Bishop and colleagues (1999) investigated the effects of resistance training on cycling performance using 21 women. Although the authors did not specify whether the participants competed in the sport of cycling, it was noted that they had trained on average for 2.5 years. Training twice per week for 12 weeks, the experimental group completed five sets of two to eight repetitions of back squats. A 1-h time trial was used as a cycling-specific outcome; secondary measurements of leg strength, lactate threshold, and peak $\dot{V}O_2$ were obtained. There was a profound increase in muscle strength in the resistance training group (35.9%), but neither group significantly improved cycling performance, lactate threshold, or peak $\dot{V}O_2$. Similarly, Jackson and colleagues (2007) used 23 club-level cyclists (5 women, 18 men) to study the effects of resistance training on performance. They randomly assigned the participants to either a control group, a heavy resistance group, or a light resistance group. All three groups performed the same cycle exercise, but the resistance training groups performed squats, leg curls, leg press, and step-ups three times per week. The heavy resistance group performed 4 × 4 at 85% of 1RM, whereas the light resistance group performed 2 × 20 at 50% of 1RM. The load was adjusted throughout the study so that an intensity of 4RM and 20RM was maintained for the high and low resistance groups, respectively. Subjects completed a $\dot{V}O_2$ peak test in which time to exhaustion and lactate threshold were assessed as the cycling-specific measurements. The heavy resistance training group demonstrated the most significant increase in strength; there were no differences between groups in improvement in $\dot{V}O_2$ peak, lactate threshold, or time to exhaustion.

Conclusion and Strength of Evidence

There is evidence to support replacing a portion of a cyclist's strength training program with a high-repetition lower extremity training program to improve power endurance in short bursts (30 s) of all-out cycling. Higher repetitions and explosive plyometrics may also improve 1 km and 4 km time trial performance. Additionally, periodized resistance training may improve 40-min time trial performance. In contrast to these findings, Bishop and colleagues (1999) found no improvement in time trial performance from heavy resistance training even though there was a significant increase in strength and muscle cross-sectional area. This lack of improvement could be due to the increase in body weight; in other words, carrying more weight, even though it was muscle mass, negated the positive benefit of an increase in strength. It is interesting to note that there was no decrease in performance and that the subjects used in this study were all women; although there are no obvious reasons to believe women would respond differently than men, to apply these findings to Zeke the results must be generalized across the sexes. Neither high- nor low-intensity resistance training improved peak $\dot{V}O_2$ or lactate threshold. The sample sizes used in these studies were relatively small, but all used only competitive cyclists although perhaps not of near-elite caliber (like Zeke). It is also important that most of these studies did not include resistance training in addition to cycling but replaced a portion of cycling training with resistance exercise. Based on this analysis, we can conclude the following:

There is level B evidence to support the use of resistance training in addition to endurance training to improve short sprint power output and time trial performance in competitive cyclists.

Program Recommendations

Based on the evidence, it is recommended that Zeke incorporate a periodized resistance training program into his endurance training program. The cycling volumes currently used in his training program are 15% to 20% higher than those used in the studies appraised. Therefore, it is recommended that Zeke replace a portion of his endurance training. Since the body of evidence is limited, he should closely monitor his time trial performance to ensure that there are no negative effects from strength training. He may also want to monitor his long-distance performance, as no studies investigated the effects of strength training on extended distance performance.

CASE STUDY 2: Static Stretching and Soccer

Rosa is a new strength and conditioning coach for a soccer team competing in a professional league. The head soccer coach for the team has won two league championships and has worked as an assistant for the national team. Before the first practice, the coach met with Rosa to discuss the timing of the team's warm-up sessions. She had been informed that she would have 15 min before practice to "stretch the team out." The coach then gave her the stretching protocol from the previous year; the program involved 12 static stretches that were held for 30 to 45 s each and repeated twice. Just recently, Rosa attended a performance conference where a notable strength coach presented a preworkout preparation program that involved active warm-up and dynamic stretching. In the presentation he suggested that there was no evidence to support the use of static stretching to improve performance and that it may in fact hinder explosive power performance. Rosa was excited about implementing a modified version of the program, but when she shared the information with her coach, the coach responded, "We used the static stretching program and won two league championships; it is also implemented by the national team." Because she sincerely believes the new program may be more effective, Rosa plans to search for evidence supporting and opposing the two stretching techniques. The practical question that Rosa needs to address when approaching her head coach is this: Does static stretching reduce injury risks, and is there any impact on performance of either static or dynamic stretching?

Background

Static stretching is a technique that is commonly used before exercise or competition. Static stretching, as the name suggests, involves elongating a muscle and holding the tissue near the limit of its range of motion. Although recommendations vary, a static stretch is typically held for 10 to 60 s (Bandy & Irion, 1997); a common recommendation is 20 to 30 s. There is strong evidence that routine static stretching may improve the mobility of a joint and chronically increases the range of motion by remodeling the connective tissue that surrounds the muscle, fasciculi, and fibers. Stretching is not a new form of exercise as it relates to athletics or performance; it is believed that ancient armies used stretching as a form of exercise to ensure the physical fitness of their soldiers. Stretching is also commonly used in physical and occupational therapy, chiropractic rehabilitation, and physical medicine. Before athletic competitions, teams commonly gather to complete a series of 6 to 12 static stretches that focus on elongating muscles related to movement performance. Muscle strains occur when tissue is stretched beyond its normal limits of mobility, tearing the tissue. Therefore, it has long been hypothesized that stretching the muscle before movement may increase the range of motion and reduce the chance of injury by increasing the distance a muscle can be elongated before injury. Although this is theoretically sound, it is unclear what evidence supports this paradigm and, as Rosa learned at the conference she attended, some have proposed that stretching can reduce muscle power. Therefore, the following questions are appropriate:

- **In competitive athletes, does dynamic stretching reduce the risk of noncontact injuries during competitions?**
- **In college-aged athletes, does static or dynamic stretching acutely alter sprint performance?**

Search Strategy

Evidence for the initial question regarding the effectiveness of static stretching to prevent muscle injuries was obtained using PubMed. An original search using the search phrase "static stretching and injury risk" resulted in a total of 76 papers. The search was then narrowed by limiting it to "systematic review" study types, which resulted in 21 articles. Of the 21 studies uncovered with this search, two were relevant to the question (Rogan, Wust, Schwitter, & Schmidtbleicher, 2013; Small, McNaughton, & Matthews, 2008). The two articles were cross-listed in PubMed for evidence related to the topic, and another three systematic review articles were obtained for consideration (Shrier, 1999; Thacker, Gilchrist, Stroup, & Kimsey, 2004; Weldon & Hill, 2003).

A secondary search was executed for articles related to static versus dynamic or movement-based stretching and performance. An initial search using the phrase "static versus dynamic stretching and performance" was conducted in PubMed, resulting in a total of five articles. Of the five articles, only two used performance-based outcome measures, so the other three were eliminated from review (Kinser et al., 2008; Winchester, Nelson, Landin, Young, & Schexnayder, 2008). Further inspection of the article by Kinser and colleagues (2010) showed that static stretching was performed with whole-body vibration, so it was eliminated from the analysis, leaving only one relevant article (Kistler, Walsh, Horn, & Cox, 2010). Cross-referencing resulted in >100 articles, six of which used sprinting as a primary outcome measure. These seven articles were analyzed and used as primary evidence for the final question (Beckett, Schneiker, Wallman, Dawson, & Guelfi, 2009; Favero, Midgley, & Bentley, 2009; Paradisis et al., 2014; Sayers, Farley, Fuller, Jubenville, & Caputo, 2008; Sim, Dawson, Guelfi, Wallman, & Young, 2009; Winchester et al., 2008).

Discussion of Results

A summary of the results from the five systematic reviews on the topic of static stretching and injury risk can be seen in table 12.1. The systematic reviews were conducted from 1999 to 2013 and included randomized controlled trials, nonrandomized controlled

Table 12.1 Summary of Systematic Review Articles Related to Static Stretching and Injury Risk

Study	Types of articles analyzed	Articles analyzed	Summary of findings
Shrier, 1999	Prospective cohort; cross-sectional	4	Basic science and epidemiology literature does not support static stretching to prevent acute injury.
Weldon & Hill, 2003	Randomized and nonrandomized controlled trials	7	Conflicting results—three RCTs concluded that static stretching did not reduce injury risk; one RCT and three nonrandomized controlled trials concluded that there was a small reduction in injury risk.
Thacker et al., 2004	Randomized trials; prospective and retrospective cohorts	6	Evidence does not support including or discontinuing static stretching during precompetition activities.
Small et al., 2008	Randomized controlled trials (RCTs)	7	Static stretching before competitions does not reduce overall injury risk; a small effect suggests that it may reduce musculotendinous injuries.
Rogan et al., 2013	Randomized trials; retrospective cohorts	4	Findings were inconclusive due to varied study design; no recommendations for or against static stretching.

trials, prospective and retrospective cohort designs, and cross-sectional assessments. In general, the authors cite inconclusive evidence for or against the use of static stretching before competition or exercise as a preventive measure for injury. One article strongly suggests that static stretching does not reduce injury risk, while one article concludes that there is a small effect for reduction of musculotendinous injuries (muscle strains) with stretching before exercise. Although each of the systematic reviews was well-executed, there is still limited data supporting or refuting the use of static stretching. The number of articles reviewed ranged from four to seven, a relatively small number of cohort studies to develop consensus on a topic.

In contrast to the paucity of data for or against static stretching to prevent musculoskeletal injuries during exercise or competition, strong evidence suggests that it may impair sprint performance. Six of the seven studies investigated found evidence of a reduction in sprint speed resulting from static stretching before performance. Although the populations in most of the studies were not directly tied to the question, one randomized controlled trial used 20 female soccer players. Sayers and colleagues (2008) randomly assigned the 20 female professional soccer players to either a stretch or a no-stretch condition before they performed 30 m sprint tests. During the test, acceleration was computed at the 10 m and 20 m intervals. Each group performed a light mobility warm-up that included skips, shuffle, and light sprinting exercises. The no-stretch group then performed the sprints with no additional stretching intervention. However, the stretch group performed three sets of static stretching exercises for the hamstring, calf, and quadriceps muscles. The stretches were each held at 85% of maximum for 30 s each. A 20-s recovery was provided between each stretch. The authors found that overall sprint time, maximal velocity, and acceleration were significantly reduced in the group that performed stretching before the sprint runs.

Although using track athletes, Winchester and colleagues (2008) demonstrated similar detrimental results on short sprint performance. The interesting aspect of this study in relation to the evidence-based question is that dynamic and static stretching were directly compared. Within a repeated-measures randomized-balanced design, in one session the athletes performed a dynamic warm-up before sprinting, and in another they performed static stretching just before 40 m sprints. The dynamic warm-up consisted of slow jogging, squats, leg swings, and various sprint-specific warm-up drills. The static stretching routine consisted of four partner-assisted stretches for the lower extremity. Each stretch was held for 30 s and repeated three times. Following the static stretching routine, overall sprint velocity was reduced, as was velocity during the second 20 m compared to the dynamic warm-up condition. It should be noted that the overall effect sizes for the 20 to 40 m interval ($d = 0.11$) and the overall 40 m sprint ($d = 0.23$) were small. Table 12.2 presents an overall comparison of studies and outcomes.

Conclusion and Strength of Evidence

The majority of research evidence suggests that static stretching before exercise or competition does not significantly reduce injury risks, but the number of studies directed toward this topic is limited. Although there are several high-quality systematic reviews, it is difficult to develop a consensus based on the limited articles analyzed. However, the cohort studies refuting the historical claim that static stretching prevents injury are high quality and strongly support the elimination of this technique as a necessary component of athletic preparation.

If a need exists for immediate power, then static stretching right before the event may be detrimental, as more time for elastic tissue recovery is needed. Power events immediately preceded by static stretching (i.e., within 6-10 min) are negatively influenced (Torres et al., 2008). Strong evidence suggests that static stretching before exercise may impede sprint performance. Moreover, direct evidence supports a detrimental effect on performance in professional female soccer players. Therefore, we can draw these conclusions:

- **Level C evidence refutes the use of static stretching before exercise as an intervention to reduce injuries.**
- **Level B evidence supports a detrimental effect of static exercise on sprint performance in college-aged soccer players.**

Program Recommendations

The evidence supports Rosa's recommendation to include dynamic stretching before exercise and competition in place of static stretching. Rosa should approach the head coach with evidence refuting the use of static stretching to prevent injuries. She then

Table 12.2 Comparison of Basic Study Parameters and Overall Findings for Evidence Supporting and Opposing Static Stretching Before Activity

Study	Population	Outcome measure	Overall findings
Sayers et al., 2008	20 elite female soccer players	30 m sprint performance	Reduction in acceleration, maximum velocity, and overall sprint time after static stretching.
Winchester et al., 2008	11 male, 11 female National Collegiate Athletic Association track athletes	40 m sprint performance	Static stretching reduced overall speed, time in the last 20 m, and overall time compared to dynamic stretching.
Favero et al., 2009	10 trained men	40 m sprint performance	Static stretching did not affect overall sprint time, but seemed to have a greater effect on athletes with greater baseline flexibility.
Sim et al., 2009	13 male athletes	6 × 20 m sprint performance	Static stretching before or after dynamic activities tended to reduce sprint performance.
Beckett et al., 2009	12 male athletes	Repeated sprint; agility test	Static stretching between sets reduced straight-line sprint performance but did not affect change-of-direction performance.
Kistler et al., 2010	18 college track athletes	100 m sprint performance	Static stretching resulted in a small reduction in performance during the 20 to 40 m interval of the sprint.
Paradisis et al., 2014	47 adolescent boys and girls	20 m sprint performance; vertical jump	Static stretching significantly reduced 20 m sprint time and vertical jump performance; dynamic stretching did not hinder sprint performance.

can use the detrimental effects of static stretching to support the dynamic routine. Since static stretching is effective in improving flexibility, it should still be included at the end of the workout or competition. If the coach is in support of this plan, Rosa should work with the team athletic trainers to monitor the musculoskeletal injury patterns from previous seasons in which a static stretching routine was implemented. If injury increases with use of the new technique, she should discontinue it. She should also continue to routinely search the literature, as new evidence related to her questions may arise.

CASE STUDY 3: ACL Injury Prevention in Female Athletes

Will is a university strength and conditioning coach. Overall, Will is accountable for managing the strength and conditioning programs of 150 athletes. The teams have been successful, winning a combined 17 conference championships over the past 5 years. Will has a great working relationship with the entire sports medicine staff, the coaches, and the team. He is always searching for ways to gain an edge, collaborating with the athletic trainers, coaches, and athletes.

During the past 5 years, the women's basketball team has sustained seven noncontact anterior cruciate ligament (ACL) injuries during games and practices. Will plans to approach the head women's basketball coach with an intervention he has been discussing with the athletic training staff. His goal is to implement the intervention strategy during the next year to reduce injury risk. Although the head women's basketball coach is typically supportive, it may be necessary to reduce practice time to implement the new intervention due to contact time regulations. Thus, it will be important for Will to address the coach with evidence supporting the intervention. How strong is the evidence supporting ACL prevention programs in female basketball players?

Background

Jumping and landing are fundamental human movements associated with performance in most sports and are common in basketball; they are also occupationally hazardous movements, particularly for female athletes. Although relatively simple, jumping and landing from a jump involve precise coordination of multiple muscle groups. Poor hamstring muscle strength (Hewett et al., 2010) or uncoordinated activation of quadriceps and hamstring muscles may increase risk for lower extremity injuries (Hewett, Zazulak, Myer, & Ford, 2005). In fact, poor coordination of quadriceps and hamstring muscles upon landing from a jump has been implicated as a key factor (Hewett, Myer, & Ford, 2006) contributing to the four to six times greater incidence of ACL injuries in female athletes. In a closed chain (i.e., when the foot is in contact with the ground), the hamstring muscle group acts as an agonist to the ACL, preventing anterior translation of the tibia on the femur. Conversely, in a closed chain the muscle action of the quadriceps opposes the stabilizing action of the ACL, pulling the tibia anteriorly on the femur.

Both the intensity (Malinzak, Colby, Kirkendall, Yu, & Garrett, 2001) and timing (Wojtys, Huston, Taylor, & Bastian, 1996) of quadriceps and hamstring activity during landing may be potential factors in ACL injury (Gokeler et al., 2010). Landing with a knee extension moment results in a greater ground reaction force and quadriceps muscle activity, whereas landing with a knee flexion moment results in a quicker and stronger contraction of the hamstring muscles (Chappell et al., 2005; Yu, Lin, & Garrett, 2006). Thus, landing with a knee flexion moment may be protective of the ACL. Evidence suggests that women tend to land with less knee and hip flexion (Beutler, de la Motte, Marshall, Padua, & Boden, 2009) and have a tendency toward increased quadriceps activity before and upon landing (Ford, Myer, Schmitt, Uhl, & Hewett, 2011). Neuromuscular interventions designed to reduce risk factors associated with knee injuries have been developed and evaluated recently; the question is whether they are effective at reducing the incidence of knee injuries in female athletes. Will can formulate the question this way:

In female athletes, are jump and landing programs effective in reducing noncontact knee injuries compared to routine strength and conditioning programs alone over the course of a competitive season?

Search Strategy

The PubMed Clinical Queries database was searched for systematic reviews using the search terms "knee injuries" and "female athletes" and "neuromuscular interventions." The Clinical Studies category (Therapy) and scope (Broad) were selected to narrow the results. This particular search strategy was used because of the abundance of information relating to ACL injuries in female athletes. The search tools were used to narrow the topic, resulting in a leaner, more directed search. The search resulted in a total of 11 papers. Two were eliminated because they focused on all-inclusive injury prevention programs as opposed to knee injury alone. The outcome was five systematic reviews (Hewett, Myer, & Ford, 2005; Hubscher et al., 2010; Kelly, 2008; Sugimoto, Myer, McKeon, & Hewett, 2012; Ter Stege, Dallinga, Benjaminse, & Lemmink, 2014), one meta-analysis (Hewett, Ford, & Myer, 2006), one narrative review (Bien, 2011), and one randomized controlled trial (Pfile et al., 2013). Because some of the studies were included in multiple systematic reviews, only three of the systematic reviews were evaluated (Hubscher et al., 2010; Sugimoto et al., 2012; Ter Stege et al., 2014).

Discussion of Results

A comparison of the three systematic reviews is provided in table 12.3. Hubscher and colleagues (2010) reviewed neuromuscular interventions that included mechanical training in jumping and landing, strength, and balance training. Their search resulted in 32 appropriate studies, but only seven were considered of sufficient quality for inclusion. These seven studies varied significantly as to the intervention, evaluation time, and study methodologies (Hubscher et al., 2010). The ethnically diverse study included participants from Norway, Canada, Finland, and the United States and used school and club sport athletes from basketball, volleyball, soccer, team handball, hockey, and floorball. Statistical analysis and pooled results suggested that a combined neuromuscular intervention using balance and mechanical training may reduce overall acute knee injuries (Relative Risk [RR] = 0.46) and ankle injuries (RR = 0.50). In other words, the risk for an acute knee injury according to the studies evaluated was 54% less in those who received a neuromuscular treatment than in those who did not.

Ter Stege and colleagues (2014) used a similar search strategy but included only studies in which athletes competed in team ball sports including soccer, basketball, and volleyball. Their review focused on "prevention programs," which could include neuromuscular interventions, feedback, or both. In contrast to the previous review, studies included in this analysis used both males and females. The authors concluded that neuromuscular intervention programs including plyometric, agility, and feedback training may be effective in improving knee injury outcomes (Ter Stege et al., 2014). However, the literature did not strongly support these conclusions due to the wide variety of methodologies in the studies analyzed.

Possibly the strongest and most relevant systematic review was conducted by Sugimoto and colleagues (2012). They performed a number-needed-to-treat (NNT) analysis to determine how many individuals it would be necessary to treat in order to prevent one ACL injury. A total of 12 studies met the inclusion criteria and thus were included in the analysis. The studies analyzed to determine the NNT were predominantly prospective and retrospective cohort designs. Pooled analysis resulted in 10,019 female control and 8064 female participants who received a neuromuscular intervention. Athletes from a variety of sports were used, including basketball, soccer, handball, and volleyball; they ranged in age from 14 to 22 years. From these data, it was determined that the NNT to prevent one ACL injury is 120. Therefore, it is estimated that for

Table 12.3   Comparison of Systematic Review Studies Assessing Prevention of ACL Injuries in Female Athletes

Study	Studies included	Participants	Intervention	Outcome
Hubscher et al., 2010	7 randomized controlled trials	7500+	Multi-interventional programs including plyometric, "core" stability, strength training, stretching, and balance training	Athletes receiving intervention were 54% less likely to sustain knee injury.
Sugimoto et al., 2012	12 controlled trials	18,000+	Multi-interventional programs	Neuromuscular training may reduce ACL injuries; treatment may prevent one knee injury in every ~120 athletes.
Ter Stege et al., 2014	35 controlled trials	Unspecified	Multi-interventional programs including jumping, landing, cutting, and feedback	Programs using plyometric and agility drills, with addition of a biofeedback component with jumping or landing, tend to be most effective.

approximately every 120 athletes who receive the neuromuscular intervention, one ACL injury will be prevented.

The NNT analysis included one study of particular interest (Gilchrist et al., 2008). Although this project used female soccer players, it provides strong evidence supporting the use of an intervention during warm-up to prevent knee injuries in female athletes. With 61 collegiate soccer teams in a randomized controlled cluster design, 35 teams served as controls and 26 received the intervention. A total of 583 controls and 852 intervention subjects participated; all were Division I athletes. Control teams performed their customary warm-up, but the intervention group used a 20-min warm-up that included a general warm-up, stretching, strengthening exercises, plyometrics, and agility training. The exercises were carefully monitored and the athletes were coached to ensure they were using appropriate mechanics associated with reduction of knee injuries. In the control group, there were 58 total knee injuries; of these, 18 were ACL injuries and 10 of these were noncontact injuries. The intervention group sustained a total of 40 knee injuries, including 7 ACL injuries, 2 of which were noncontact in nature.

This suggests that the simple 20-min program used as a component of the warm-up may significantly reduce knee injuries in female athletes.

Conclusion and Strength of Evidence

During the past 5 years, several well-conducted systematic reviews have evaluated the effectiveness of neuromuscular interventions to prevent knee injuries. The overall effectiveness of the systematic reviews in providing recommendations is limited by the heterogeneity of study designs and variation in reporting methodologies. Most of the research associated with knee injuries uses a cohort design, although there are a few randomized controlled trials. The research strongly supports the use of a combined neuromuscular training program that includes mechanical training in jumping and landing and leads us to the following conclusions:

> **Level B evidence supports the use of a jumping and landing program to reduce noncontact knee injuries versus routine strength and conditioning programs alone over the course of a competitive season in female athletes.**

Program Recommendations

There is strong evidence to support Will's plan of implementing a combined neuromuscular intervention. Will should approach the coaches for each sport with the evidence and propose the use of a modified version of the Gilchrest protocol (table 12.4) for basketball as a starting point for the program. The intervention would require 20 min and could be performed as a part of the warm-up (which as of now is 10 min) two or three times per week. If some coaches are not supportive of implementing the program, as a part of his normal strength and conditioning time Will may consider screening athletes and identifying those with mechanics that may increase risk for knee injuries. He may consider contacting a biomechanist or applied sport scientist from the university's exercise science program for possible assistance in data collection and interpretation. Will could then approach the coach about dedicated time for the athletes identified as at higher levels of risk.

CASE STUDY 4: Single Versus Multiple Sets for Strength Improvement

Mei is a collegiate kinesiology student who is recreationally active. She enjoys running, swimming, and playing basketball, volleyball, and tennis. After completing a sophomore-level resistance training class, she has become interested in beginning a long-term strength training program to increase her muscle size and strength. In the class she learned the basics of program design, but she has been looking on the Internet for more information. While searching the Internet she stumbled upon www.bodybuilding.com, where she read a number of articles on high-intensity training (HIT). Proponents of this technique perform a single maximal set to failure and claim that the results are similar to or greater than those with performance of multiple sets. Mei is interested in finding more information on training volume and approaches Billy, a graduate student in the kinesiology department, for information. Mei outlines her goals of increasing muscle size and strength long-term and asks Billy his opinion on single versus multiple sets of resistance exercise to improve strength and hypertrophy. What advice should Billy provide for Mei, and what is the supporting evidence?

Table 12.4　Protocol to Prevent ACL Injuries in Basketball Players

Exercise	Distance, duration
WARM-UP AND STRETCHING	
Full-court jog forward and backward for 5 min at a 4 to 6 rating of perceived exertion (1-10 scale). Light stretching completed in athletes with functional limitations; all others progress directly to the strength component. Static stretching completed by all athletes at the end of the workout in accordance with recommendations of the latest research literature.	
STRENGTH	
Walking lunges forward and backward	1-2 × full court
Walking single-leg Romanian deadlift	1-2 × full court
Single-leg heel raises	1-2 × 20-30 reps
PLYOMETRIC	
Lateral line hops	1-2 × 10-20 reps
Front/back line hops	1-2 × 10-20 reps
Backboard taps	1-2 × 10 reps
Lunge jumps	1-2 × 10-20 reps
AGILITY	
Forward–backward shuttle run	5 × half court
Diagonal runs	5 × full court
Bounding	5 × full court

Focused attention on reduction of landing forces and prevention of knee valgus movements and hip internal rotation during plyometric and agility exercises.

Based on Gilchrist et al. 2008

Background

Single set to failure versus multiple sets is a commonly debated topic in the strength and conditioning, personal training, and bodybuilding industries (Byrd et al., 1999; Carpinelli & Otto, 1998). The debate originated in the 1970s with Arthur Jones, the founder of Nautilus, Inc., a fitness device manufacturer. It was made popular by Mike Mentzer, a 1970s bodybuilder who developed the Heavy-Duty Training System (Mentzer & Little, 2002). The single-set system uses controlled multiple- and single-joint exercises, but, after a warm-up, a single set to momentary muscular failure is performed for each exercise. Proponents of the single-set nonperiodized systems assert their effectiveness on the basis of contrived explanations and misinterpretations of the literature regarding basic biological science. They suggest that a single set of an exercise performed to failure exhaustively recruits the entire motor unit pool; therefore, any additional sets lead to unnecessary stimulation and potentially unnecessary muscular damage.

Multiple-set periodized systems, which are more traditionally used, originated from the work of DeLorme and Watkins (1948). Although many periodized resistance training models have been proposed, most are based on Hans Selye's model of stress and adaptation (Selye and Fortier, 1950). Selye's model proposes that when an organism is stressed beyond its customary levels, there is an immediate suppression of physiological function. With rest and recovery, an adaptive supercompensatory response occurs to allow the organism to respond more capably to future stressors. However, if the organism is not allowed to recover, an exhaustion response occurs and there is no adaptation. Periodized models are based on this paradigm (Stone, O'Bryant, Garhammer, McMillan, & Rozenek, 1982). Individuals are subjected to increasingly intensive resistance training loads to stress the neuromuscular system and then are allowed to recover through deloading weeks of submaximal lifting intensities. Thus the individual lifts to momentary muscular failure during the high-intensity weeks but does not lift to failure during the submaximal weeks to allow for recovery. Both volume and intensity are varied in a systematic pattern over the course of a training period to allow adequate stimulus and recovery. The question proposed in this scenario uses a time period of 12 weeks to allow sufficient time for differences between the two protocols to manifest:

In a college-aged recreational athlete, is single- or multiple-set resistance exercise more effective to improve muscle strength and hypertrophy during a >12-week period?

Search Strategy

Initially the position stand from the American College of Sports Medicine (ACSM) on exercise progressions was obtained (Ratamess et al., 2009). The position stand was used because it reflects an evidence-based search performed by a large group of experts in the field of strength training. Although the position stand should not be used alone as the basis for a strong opinion on the topic, the substantial body of research reviewed by the expert panel provides a solid foundation on which to form an opinion. The ACSM position stand is a great starting point for a search since it exhaustively reviewed the literature up to the year 2009. From the position stand, two meta-analyses were obtained and reviewed because of their interesting approach in comparing dose responses from multiple studies (Rhea et al., 2003; Peterson, Rhea, and Alvar, 2004). PubMed was then searched using the phrase "single versus multiple sets and resistance training." It was then narrowed via a search limited to human studies and to systematic reviews published since the position stand guidelines. Since 2009, two quality meta-analyses had been published on the topic. (Remember that a meta-analysis is a systematic review that compiles data and subjects from multiple studies. Analytical techniques are then used to pool data from multiple studies and compute summary statistics.) The meta-analyses were advantageous to this search since an abundance of studies have been published on this topic. These two meta-analyses compared muscle strength (Krieger, 2009) and hypertrophy outcomes (Krieger, 2010) from single versus multiple sets, providing a quantitative analysis of the different treatments.

Discussion of Results

The ACSM position stand on resistance exercise progressions is perhaps the most comprehensive consolidation of the literature available. It examines scientific studies evaluating strategies to best improve strength, power, and hypertrophy in various populations.

The acute programming variables of exercise selection, order, rest periods, intensity, frequency, and volume are all evaluated; the authors also provide consensus statements with specified levels of evidence. With respect to training volume, the position stand authors recommend, based on the literature, that one to three sets is effective for immediate strength gains in novices (level A), but higher-volume programs with systematic variation are required for optimal gains in intermediate and advanced athletes (level A).

Krieger (2009) used a meta-regression approach to determine the effects of multiple versus single sets on strength outcomes. Using studies that involved both trained and untrained subjects, he calculated effect size changes in 30 treatment groups obtained from 14 different studies. The studies included in his analysis were limited to those that compared single versus multiple sets, assessed strength with a 1RM, used healthy subjects, and were at least 4 weeks in length. When comparing single versus multiple sets, Krieger determined that there was a significant effect of two or three sets versus one set, irrespective of training status. In fact, it was estimated that the strength gain resulting from two or three sets was 46% greater than that with one set. No additional benefit on strength from greater than three sets was observed, possibly due to the limited number of studies evaluating these protocols.

Comparing hypertrophy gains from single versus multiple sets, Krieger (2010) used a similar meta-analytic approach. Eight studies comprising 15 treatments and 55 effect sizes were compared. Almost identical study inclusion characteristics were used, except the outcome measures in this study required the measurement of hypertrophy using commonly employed techniques. There was a 40% greater gain in hypertrophy with two or three sets compared to one set, irrespective of training status. Due to a lower number of studies, fewer subjects sampled, and various outcome measures for hypertrophy, the second meta-analysis had lower statistical power.

In addition, two meta-analyses comparing and providing a dose–response relationship between volume (i.e., number of sets) and strength were published by Peterson and colleagues (2003) and Peterson, Rhea, and Alvar (2004). Using a large body of literature that included more than 370 effect sizes for pooled analysis, the studies determined optimal volume and intensity for untrained and trained individuals. These data support the use of loads of approximately 80% to 85% of 1RM in trained individuals with eight sets per muscle group for optimal strength gains. For untrained individuals, loads of 60% 1RM, with four sets per exercise, may be sufficient in the early phases of training.

Conclusion and Strength of Evidence

Strength and hypertrophy can be achieved from programs using single or multiple sets, but a large body of evidence strongly supports the use of multiple sets for significantly greater improvement in both outcomes. This is supported by position stand recommendations and meta-analyses of data. Although there is some evidence supporting single-set systems, the overwhelming outcome of randomized controlled trials supports multiple-set systems. Based on the evidence, we can reach this conclusion:

> **Level A evidence supports the use of multiple versus single sets to improve muscular strength and hypertrophy in college-aged recreational athletes.**

Program Recommendations

On the basis of the evidence, Billy should recommend that Mei perform approximately three sets per exercise, as the outcome from multiple sets is clearly better. However, if

Mei's time is limited, she may use single sets for the first month of the training program, as novices may see gains from such programs before progressing to higher-volume multiple-set programs. Billy should also encourage the kinesiology student to search for information on PubMed or Google Scholar rather than websites that espouse non–peer-reviewed evidence.

CASE STUDY 5: Models of Periodization

Luke is a fitness manager and trainer for a local city department. In his job, he operates the city gym, which is open to all employees, including the police and fire department. He also manages the small gyms that are housed in each of the local fire stations. He was recently tasked with developing the strength and conditioning program for the SWAT team and the city firefighters. After much research, he developed a program based on a traditional periodization model. The program is arranged in approximately 12-week mesocycles with three blocks of training within each mesocycle. During each phase, he works primarily on one physical capability, e.g., strength, power, or endurance. Due to the erratic schedules for the SWAT team and police department, Luke is frustrated that these employees are missing many workouts. He's also noticed that when the tactical athletes arrive at the facilities for their workout, it may be after a 3-day off-period or after a 24-h shift. There is no way to predict how tired the participants will be and to program accordingly since the schedule changes so frequently. Luke is searching for a solution to the problem and a way to introduce more flexibility into the program to fit the unique needs of his athletes. Recently, he read a short article online about a "daily undulating periodization model." It appears that the planned nonlinear periodization model may provide greater flexibility while maintaining the positive benefits of systematic variation. However, he does not want to implement daily undulating periodization unless he is sure that it will not compromise the positive results from the program. Should Luke implement a daily undulating model of periodization?

Background

The traditional model of periodization was developed in the late 1950s by a Russian sport scientist named Matveyev (Stone, O'Bryant, & Garhammer, 1981; Stone et al., 1982). The original model, which is still used predominantly today, was based on systematic variation in volume, intensity, and skill training. In the early off-season, training volume was high; intensity and skill training (practice) were low. As the athlete progressed toward the competitive season, volume gradually decreased, and intensity gradually increased along with skill training. As the most important competition approached, there was a decrease in intensity (taper) as skill-related training was at its peak for the season. There is much research on the effectiveness of this model. It works well for athletes who must peak for a single event, such as weightlifters, track athletes, and athletes in other Olympic sports. It is also modified and used routinely for strength athletes.

Daily undulating periodization, now called nonlinear periodization, is a newer model of training that has emerged to account for limitations of Matveyev's model for non-traditional athletes. In a daily undulating model, an athlete may have wide variations in the intensity and volume of training within a week. For example, athletes may focus on strength by lifting heavier weights and performing lower reps on Monday, then on Friday work on endurance by lifting light weights with higher repetitions. This wide range of volumes and intensities is a marked contrast to the traditional model, in which an entire 8- to 16-week period may be spent working on a single physical quality such as strength before the set–rep scheme is dramatically changed to work on another physical quality. Luke needs to determine the answer to this question:

In adults, is a daily undulating periodization model as effective as a traditional periodization model to improve strength as determined by 1RM?

Search Strategy

The Google Scholar database was searched for articles related to the evidence-based question. Google Scholar was used as the original search engine because it was possible that some newer research articles related to the topic had been published in peer-reviewed journals not indexed in PubMed. PubMed was used secondarily to determine if any additional articles were available that did not appear in Google Scholar using the search terms. The original search terms were "undulating periodization" and "strength." This resulted in a total of 222 articles. The search was narrowed to include articles with a direct comparison to traditional periodization by adding the term "traditional periodization." The results pointed to 16 articles. Of these 16 articles, three were narrative reviews and two were dissertations, so they were eliminated from comparison. Three articles directly compared strength outcomes in undulating and traditional training programs (Apel, Lacey, & Kell, 2011; Bartolomei, Hoffman, Merni, & Stout, 2014; Franchini et al., 2015). These articles were then cross-searched in PubMed, from which three additional articles were obtained (Buford, Rossi, Smith, & Warren, 2007; Peterson, Dodd, Alvar, Rhea, & Favre, 2008; Rhea, Ball, Phillips, & Burkett, 2002). After review, it was determined that the article by Bartolomei and colleagues was not directly relevant to the search; thus five articles were reviewed for analysis.

Discussion of Results

Rhea and colleagues compared a 12-week traditional periodization (TP) program to a daily undulating periodization (DUP) model in 20 college-aged men (Rhea et al., 2002). The subjects had a minimum of 2 years of prior strength training experience. The subjects were randomly assigned to one of two groups: traditional periodization or DUP. Each group performed the program 3 days per week, and subjects were not allowed to perform any other type of exercise for the duration of the study. The general training protocol for the two groups is provided in table 12.5. The authors found that both the TP and DUP groups improved strength compared to baseline, but the differences were significantly greater in the DUP group (Rhea et al., 2002). Using the same protocols (table 12.5) but adding a weekly undulating periodization (WUP) group, Buford and colleagues (2007) compared a total of 30 college-aged individuals (Buford et al., 2007). They found significant improvements in strength within all groups, but in contrast to Rhea and colleagues they observed no differences between the groups in overall strength gains.

Apel, Lacey, and Kell (2011) completed a similar investigation, comparing a traditional periodization model to a weekly undulating model. Forty-two recreationally active men were divided into control (CON), traditional periodization, and daily undulating periodization groups. The subjects all had greater than 6 months experience with resistance exercise before completing the 12-week study using free weights and machines. Although there were no significant differences between training groups, there was a

Table 12.5 Traditional and Daily Undulating Periodization Protocol Used by Rhea and Colleagues

Week	Day	Traditional periodization (TP)	Daily undulating periodization (DUP)
Weeks 1 to 4	Day 1	3 × 8RM	3 × 8RM
	Day 2	3 × 8RM	3 × 6RM
	Day 3	3 × 8RM	3 × 3RM
Weeks 5 to 8	Day 1	3 × 6RM	3 × 8RM
	Day 2	3 × 6RM	3 × 6RM
	Day 3	3 × 6RM	3 × 3RM
Weeks 9 to 12	Day 1	3 × 3RM	3 × 8RM
	Day 2	3 × 3RM	3 × 6RM
	Day 3	3 × 3RM	3 × 3RM

RM = repetition maximum.
Based on Rhea et al. 2002.

general trend toward greater strength improvements for the traditional periodization group. Another study that was appraised compared traditional periodization to daily undulating periodization in 13 judo athletes (Franchini et al., 2015). The 8-week study showed significant improvements regardless of the periodization model and no differences between groups in strength outcomes.

The most population-specific project was completed by Peterson and coworkers (2008). They compared strength and performance outcomes in 14 firefighters (Peterson et al., 2008). Performing 1RM tests and a firefighter-specific testing battery (i.e., the "Grinder"), subjects were divided into traditional and daily undulating periodization groups before completing the 9-week, 27-session protocol. Both groups experienced significant improvements in strength, with a trend toward greater advances with undulating periodization. Traditional and undulating periodization resulted in greater firefighter-specific fitness test improvement, but the overall magnitude of changes was greater in the undulating periodization group. An overall comparison of the five studies is shown in table 12.6.

Conclusion and Strength of Evidence

There is strong evidence supporting the use of daily undulating, weekly undulating, and traditional periodization to improve strength outcomes in adults. The volume of data supporting the traditional model is greater since this model has been investigated for many years. In any case, the data seem to support improvements in strength regardless of the systematic variation implied. Some studies show greater improvements using the undulating or nonlinear model; others show traditional periodization to be more effective for improving strength. The single study investigating functional outcomes

Table 12.6 Appraised Studies Comparing Traditional to Undulating Periodization

Study	Subjects	Time	Comparison	Primary outcome measure	Results
Rhea et al., 2002	20 college-aged men	12 weeks	Traditional (TP) versus daily undulating (DUP)	Bench press and leg press 1RM	Bench press: TP = Δ14.4 ± 10.4%; DUP = Δ28.8 ± 19.9%; Leg press: TP = Δ25.7 ± 19.0%; DUP Δ55.8 ± 22.8%
Buford et al., 2007	20 men and 10 women, college aged	12 weeks	TP versus weekly undulating (WUP) and DUP	Bench press and leg press 1RM	Bench press: TP = Δ24.2%; WUP = Δ17.5%; DUP = Δ17.5% Leg press: TP = Δ85.3%; WUP = Δ99.7%; DUP = Δ79.0%
Peterson et al., 2008	14 firefighters	9 weeks	TP versus DUP	Bench press and back squat 1RM, "Grinder"	Back squat: TP = Δ16.8%; DUP = Δ20.7% Bench press: TP = Δ8.1%; DUP = Δ16.6% "Grinder": TP = Δ28.0%; DUP = Δ40.7%
Apel et al., 2011	42 college-aged men	12 weeks	Control (CON) versus TP versus WUP	Squat, bench press, lat pulldown	Back squat: CON = Δ3.2%; TP = Δ53.6%; WUP = Δ33.7% Bench press: CON = Δ45.1%; TP = Δ23.9%; WUP = Δ18.6% Lat pulldown: CON = Δ3.8%; TP = Δ22.2%; WUP = Δ19.3%
Franchini et al., 2015	13 male judo athletes	8 weeks	TP versus DUP	Back squat, bench press, lat pulldown	Back squat: TP = Δ5.2%; DUP = Δ8.2% Bench press: TP = Δ12.5%; DUP = Δ9.5% Lat pulldown: TP = Δ12.2%; DUP = Δ12.5%

in firefighters demonstrated a significantly greater improvement using an undulating model of periodization. We can conclude as follows:

Level B evidence suggests that undulating periodization is as effective as traditional periodization for improving strength outcomes over a short period of time (8-15 weeks).

Program Recommendations

Based on the research evidence, Luke is likely to observe positive training outcomes regardless of the selected model—traditional or undulating (nonlinear). One study investigating firefighter performance supports an undulating model, but given the small sample size this is not enough to warrant elimination of the traditional model. Luke may consider implementing the daily undulating periodization model as it affords greater flexibility. It is unclear how long-term outcomes in strength are affected by the greater level of change, but in the short term, the evidence supports the undulating model as similar in effectiveness to the traditional periodization model. Because there is an increased interest in training for the tactical athlete, Luke should routinely search for literature, as larger studies that better answer his question may be published.

CONCLUSION

Determining the number of sets and reps, velocities, rest periods, and other acute program-related variables is a daily task for many exercise practitioners. Simply memorizing information from academic programs or certification material studied many years ago is not enough to allow practitioners to stay current with the growing body of new programming theories. Many theories emerge that have research support or are backed by research several years after they have appeared in practice. Others are refuted by research demonstrating that they are ineffective or potentially dangerous. In order to stay up to date with the growing industry, exercise practitioners must routinely assess the research to find evidence for or against new training philosophies and techniques. In this chapter we provided five examples of how the process of evidence-based practice might be implemented in practice. The reviews are not comprehensive, and the evidence-based practitioner should reassess the literature as new evidence for these questions becomes available.

Chapter 13

EXERCISE FOR SPECIAL POPULATIONS

Learning Objectives

1. Discuss the importance of carefully evaluating information when prescribing exercise for persons with disease or disabilities.
2. Provide three practical examples of questions that may arise in the prescription of exercise for special populations.
3. Understand how to develop an evidence-based practice question and find evidence.
4. Objectively assess the evidence, providing strength of evidence for three case examples.

Decisions regarding the prescription of exercise can be challenging. Along with considering which treatment is most likely to result in a positive training outcome, the exercise practitioner must also ponder the potential short- and long-term negative responses to the treatment. Exercise has inherent danger; the American College of Sports Medicine (ACSM) suggests that, during a maximal exercise test, there is always a remote possibility of significant cardiovascular events, such as a heart attack or heart rhythm disturbance (Thompson, 2010). Specifically, it is estimated that one death and five hospitalizations will occur per 10,000 exercise tests in people referred for clinical exercise testing (Thompson, 2010). It is also possible that people may trip, fall, sustain an acute or overuse injury, or have osteoarthritic changes in the joints. Most individuals hedge their bet on the positive outcomes of exercise; the probability of long-term improvements in health far outweighs the possibility of negative health events. Thus, you lace up your running shoes and "roll the dice" expecting an improvement in health.

In the preceding paragraph, we may have slightly overdramatized for effect, but there are some conditions that may increase the risk of negative outcomes. Suppose an individual has previously had a cerebrovascular event or a myocardial infarction or is hypertensive. What if someone has suffered a traumatic brain injury or spinal cord injury resulting in a functional limitation or mobility impairment? Should they still engage in exercise, and if so, what kind, how much, how intense, and how often? Chronic health conditions and long-term disabilities can create difficulties in the performance and prescription of exercise. However, when people with disease or disabilities avoid physical activity, they are prone to the same morbid conditions that affect anyone who is physically inactive or who does not participate in regular exercise. A careful inspection of the research evidence suggests that exercise can be profoundly effective in improving health and reducing morbidity and mortality in many populations with special needs. Thus it is important to find safe and effective ways to improve health in these individuals through exercise, even though the task can be challenging. Moreover, the answer provided by the evidence for what is "safe and effective" is not always the industry norm.

This chapter highlights three special populations commonly encountered by exercise practitioners: people who are elderly, people with cancer, and individuals recovering from traumatic brain injuries. These are three of the numerous special cases practitioners may work with. As in the previous chapter, the case study examples are not meant to deliver a comprehensive evaluation of the literature or an all-inclusive systematic review. Instead, they are examples of how questions arise in practice and of the methodology that exercise practitioners can use to find, evaluate, and incorporate evidence into practice.

CASE STUDY 1: Resistance Exercise and Functional Outcomes in Those Who Are Elderly

Jenny is an exercise practitioner employed as a personal trainer at a reputable fitness facility. She works with clients of both sexes and of all ages and needs. She is approached by Hernan, who is a 70-year-old member of the fitness facility. Hernan has been physically active and has exercised for most of his life. He retired at the age of 65 from a career as an oil and gas executive. Early in his career, his job was physically demanding, but then he was promoted to a management position in which he spent most of the day behind a computer. As he aged, he experienced a noticeable decrease in muscle mass and strength. After retirement, he began a resistance training program. He has performed an eight-machine circuit-type workout for the past 5 years. His set and rep scheme is 2 × 15, and he performs each lift slowly "to avoid injury." Despite his efforts, Hernan has noticed a decline in functional performance over the past 5 years. It is now harder to rise from a chair, walk up stairs, and get in and out of his vehicle. Hernan's friends are all experiencing similar declines, and when he broaches the subject they say, "It's just part of getting old." Hernan has several grandkids, enjoys playing golf, and wants to remain physically active. He is willing to do whatever it takes to improve his function. He has approached Jenny for advice about how he might alter his program to do that.

Background

Since the 1950s, there has been a steady increase worldwide in the number of individuals who are elderly (over the age of 65 years) (Department of Health and Human Services, 2005). The United Nations Department of Economic and Social Affairs Population Division (2001) estimates that the elderly population in developed countries will increase by 70% and that in underdeveloped nations it may triple by 2050. This dramatic shift in the world's population is the result of remarkable improvements in the quality and availability of the health care system. The unprecedented shift in the population is of great importance due to the rising costs of health care, reduced income, increasing need for health care, and increasing rate of disease and disability associated with aging. It is reported that 33% of individuals over the age of 65 derive 90% of their total income from Social Security in the United States (National Center for Health Statistics, 2006).

All but 1% of individuals who are elderly have some form of health insurance, but in 2003 they still paid on average 12.5% of their total income to cover health care expenses (National Center for Health Statistics, 2006). The increased need for health care is the result of an increased disability and disease rate. Fifty-four percent of U.S. individuals who are elderly report at least one disability, and 37% report more than one disability (Department of Health and Human Services, 2005). The number of people reporting disabilities increases each decade of life after the fifth decade. Almost all individuals who are elderly report having been diagnosed with one of the following conditions: hypertension, arthritis, heart disease, cancer, diabetes, or sinusitis. Most report having more than one of these conditions (Department of Health and Human Services, 2005). Increased disease and disability rates result in decreased independence, decreased physical work capacity, and consequently the need to hire home health care or reside in an assisted living facility or nursing home.

Aging is normally associated with a number of declines in physiological function such as a decrease in balance (Nakano, Otonari, Takara, Carmo, & Tanaka, 2014), reduction in bone mineral density (Harada, 2014), and dysregulation of hormone activity (Maggio et al., 2014). Aging is also associated with a normal loss of skeletal muscle mass and a resultant reduction in strength and power (Gauchard, Tessier, Jeandel, & Perrin, 2003). The loss of skeletal muscle mass and associated decline in strength, power, and function is a normal phenomenon, termed sarcopenia (Marcell, 2003; Correa-de-Araujo & Hadley, 2014). Increased disease and disability rates in those who are elderly results in a reduction in the ability to complete normal activities of daily living and, consequently, in a decreased hypertrophic or eutrophic stimulus on skeletal muscle. Resistance training is effective to preserve muscle and improve strength in a variety of populations. The underlying question is what intensity is best for improving strength in individuals who are elderly and whether resistance exercise is safe in this population. The following questions can be asked:

- **In individuals who are elderly, is heavy or lighter resistance more effective to improve functional capabilities as measured using functional outcome measures such as the sit-to-stand and stair climb tests after >4-6 weeks of training?**
- **What intensity of resistance exercise is safe in individuals who are elderly?**

Search Strategy

PubMed was searched using the following terms: "elderly" and "strength training" and "functional outcomes" and "intensity." The search resulted in 53 articles. Forty-five of the articles were eliminated because they focused on a subpopulation such as patients with peripheral artery disease, cancer, multiple sclerosis, or Parkinson's disease. The remaining eight articles included two systematic reviews (Liu & Latham, 2009; Steib, Schoene, & Pfeifer, 2010), four randomized controlled trials (Engels, Drouin, Zhu, & Kazmierski, 1998; Meuleman, Brechue, Kubilis, & Lowenthal, 2000; Orr et al., 2006; Seynnes et al., 2004), one controlled trial (Newton et al., 2002), and a narrative review (Reid & Fielding, 2012). A search of cross-listed articles returned an additional systematic review of randomized controlled trials assessing the effectiveness of high-intensity resistance exercise in persons who are elderly (Raymond, Bramley-Tzerefos, Jeffs, Winter, & Holland, 2013). Because the randomized controlled trials were included in the results from the meta-analyses and systematic reviews, we do not consider these studies further. We note, though, that one can find an abundance of well-designed and well-executed studies supporting the question that are outside of the limited scope of this chapter (Mero et al., 2013; Holviala et al., 2012; Sallinen et al., 2007; Walker, Santolamazza, Kraemer, & Häkkinen, 2014). A second search to address the safety of exercise intensity was deemed unnecessary as evidence was uncovered in the primary search that provided an answer to that question.

Discussion of Results

All the analyses and studies examined strongly support the use of resistance exercise for improving strength and functional outcomes in adults who are elderly. Perhaps the most comprehensive of these was a Cochrane review performed by Liu and Latham (2009). Their analysis included 121 studies comprising a total of 6700 participants (Liu & Latham, 2009). Eighty-three of the 121 studies included high-intensity resistance exercise as a primary intervention; most were performed two or three times per week. Most studies were longer than 12 weeks. Combined analyses across the 121 studies indicated that progressive resistance exercise had a large effect on 6-min walk distance, gait speed, and timed up-and-go. Exercise intensity significantly affected strength outcomes. Studies using heavier resistance showed greater improvements in strength compared to studies using lighter resistance. Strength training was more effective in individuals reporting a good health status versus poor and was more advantageous in those with no functional limitations.

Raymond and colleagues (2013) reviewed 21 randomized controlled trials assessing the effectiveness of high-intensity progressive resistance exercise on strength and functional outcomes. They clearly demonstrate that higher-intensity strength training resulted in greater strength outcomes compared to lower intensities (Raymond et al., 2013). Higher intensities (i.e., heavier resistances) were also associated with greater functional improvements. In support of these findings, Steib and colleagues (2010) evaluated 29 randomized controlled trials using different "doses" of resistance exercise to determine independent effects of volume and intensity on strength and functional outcomes. Along with strength training protocols, the paper also considered three predominantly power training protocols in which the individuals lifted at a high intentional velocity. The results suggest that resistance exercise performed at 60% to 80% 1-repetition maximum (1RM) is most effective at improving strength and that training two times per week is required for optimal gains (Steib et al., 2010). Not surprisingly, power training results in greater improvements in power compared to heavy, lower-velocity resistance training.

The concept of higher-velocity resistance was assessed in two separate review studies by Reid and Fielding (2012). The authors describe studies using higher- versus lower-velocity resistance exercise to improve power and functional outcomes. They highlight several studies supporting the superiority of power (versus heavy strength or aerobic capacity) to predict functional performance in those who are elderly (Foldvari et al., 2000), a concept that was introduced in 1992 (Bassey et al., 1992). Moreover, the review highlights evidence suggesting that high-velocity power training improves physical function to a greater extent than traditional heavy resistance exercise.

Perhaps the best-controlled example of the effect of high-velocity resistance exercise is a study of 20 inactive males (60-76 years) (Bottaro, Machado, Nogueira, Scales, & Veloso, 2007). Subjects were randomly assigned to either a traditional resistance training group (TRT) or power training group (PT). Both groups performed four familiarization training sessions over 2 weeks before strength and power testing. A total of seven upper and lower body exercises with equal volume (3 × 8-10) and relative intensity (60% of 1RM) were performed. Technogym exercise machines, a form of dynamic constant external resistance, were used in this study. The only difference between groups was the exercise velocity. Exercise velocity is sometimes expressed as a ratio of the prescribed speed during the eccentric (lowering) and concentric (raising) portion of the lift. The PT group performed exercise at a 2- to 3-s eccentric and explosive concentric velocity while the TRT group performed strength training exercise at a 2- to 3-s eccentric and 2- to 3-s concentric controlled velocity. Absolute strength was assessed via 1RM on the leg press and bench press exercises using Technogym equipment. Average power was determined on the leg press and bench press exercises based on two sets of four-repetition explosive concentric lifts at 60% of 1RM, and functional power was assessed using timed 8-ft (2.4 m) up-and-go and 30-s sit-to-stand tests. Significant increases from baseline strength and power were realized in both groups compared to control. However, the PT group experienced significant increases in power compared to TRT. Analysis of the functional power tests showed that PT and TRT resulted in significant decreases in 8-ft up-and-go test time and an increase in number of sit-to-stand repetitions performed in 30 s. Similar to the leg press power tests, the PT group showed a significantly greater increase than TRT in both variables.

An exemplary study providing evidence supporting the effects of a combined strength and high-velocity power protocol was completed by Newton and colleagues (2002). They used a nonlinear periodization model in young (*n* = 8; 30 ± 5 years) and older men (*n* = 10; 61 ± 4 years). Three days per week, subjects completed both machine-based and free weight exercises. One day per week was devoted to hypertrophy (8-10RM), strength (3-5RM), and power (six to eight reps), respectively. Power training was completed with a submaximal load at a high intentional velocity. Both young and older men improved isometric strength and explosive power. In fact, the older men improved power by 25% during the 10-week study.

Support for the safety of heavy or higher-velocity (lighter) resistance training is found primarily in the lack of reported injuries. Although the reporting methodology for injuries in studies involving those who are elderly is inconsistent, in the 121 articles evaluated by Liu and Latham (2009), no serious injuries from exercise were reported. However, the authors note that many of the studies reported muscle soreness and joint pain, which were monitored but did not result in participants dropping out. Therefore, soreness and pain, which may accompany exercise at any age, did not rise to the level of severity forcing the participants to discontinue exercise. In practice, periodizing the protocol and minimizing the number of sets carried to muscle failure may reduce the negative effects.

Conclusion and Strength of Evidence

The evidence strongly supports the use of heavy, progressive resistance exercise to improve strength in persons who are elderly. Interestingly, lighter but high-velocity resistance exercise appears to be more effective in improving functional capabilities in these individuals. Although there is some debate because of lack of uniformity in how adverse events are reported in the literature, when carefully monitored by qualified exercise professionals (i.e., degreed and certified through top-tier agencies) and progressed correctly, heavy or high-velocity resistance exercise (or both) is likely safe and effective. After establishment of a strong conditioning base, a mixed model of training may also be effective to elicit positive benefits in strength and power. There is no evidence suggesting otherwise. Based on the research, we can state the following conclusions:

- **There is level A evidence to support the use of heavy versus light (low velocity) resistance exercise to improve strength in individuals who are elderly.**
- **There is level B evidence to support the use of high- versus lower-velocity resistance exercise to improve functional capabilities in individuals who are elderly.**
- **There is level D evidence supporting the safety of resistance exercise in individuals who are elderly.**

Program Recommendations

On the basis of the evidence, Jenny should strongly suggest that Hernan increase his exercise intensity. Hernan should advance his resistance training program, progressively increasing the intensity up to 80% or more of 1RM. Once strength gains are sufficient, he should lower the intensity but lift at a high intentional velocity for a period of time. This is likely best to mitigate the functional declines noticed over the past 5 years. Hernan may also consider employing a mixed model of strength and power training, as evidence supports such protocols. In general, the program should include a variety of lower and upper body exercises, with two or three sets per exercise performed at least two times per week. Jenny should closely monitor Hernan's program and ensure that he is progressing safely and logically. She should also communicate with Hernan and make sure that he reports any adverse effects from the exercise and make adjustments as needed.

CASE STUDY 2: Exercise and Cancer Cachexia

Aviva is a 38-year-old woman who is premenopausal and was recently diagnosed with stage III invasive breast cancer. Her doctor has developed a treatment plan that will include chemotherapy before and after surgery to remove the tumors; there is no time-table on the treatment plan. Aviva has always been physically active and has worked out at the YMCA 3 days per week for the past 10 years. She has read about the negative side effects of cancer treatment on the Internet and really wants to alter her exercise program to ensure that she maintains her strength, endurance, and quality of life as she fights the disease. She approaches Louise, a personal trainer at the YMCA, for advice. Should Louise recommend exercise as a countermeasure to prevent losses in strength and endurance during and after the treatment?

Background

Cancer is a disease characterized by unabated proliferation of cells. There are hundreds of forms of cancer; in many forms the tumor may disrupt organ function, leading to death. In 2011, cancer was the second leading cause of death in the United States, quickly gaining ground against the leading cause, heart disease. Treatment for cancer often involves surgical removal of cancerous tumors, chemotherapy medications, or localized high-dose radiation. The degree of success of such treatments is highly dependent on the maturity of the tumor and the type of cancer. In any case, the treatment can have devastating effects on other organ systems, resulting in a myriad of disabilities. Cancer cachexia is a side effect linked with cancer and its treatments. Cachexia is characterized by the loss of skeletal muscle mass, resulting in poorer functional performance and reduced quality of life; depending on the severity of loss, it may also result in death (Tisdale, 2002). Exercise, nutrition, and drug therapy have all been proposed as potential countermeasures to cancer cachexia. In this particular case, Louise needs to answer the following question:

In women undergoing treatment for breast cancer, is exercise an effective countermeasure to prevent strength and endurance loss during and after treatment?

Search Strategy

PubMed was searched, using filters for Clinical Trials, Systematic Reviews, and Humans, for information supporting the evidence-based question. PubMed was used as the primary search tool because it was suspected that most articles addressing exercise and cancer are found in medically based journals indexed in this database. PubMed allowed for narrowing of the topical search for a more focused discovery of articles. The following search phrases and terms were used: "exercise and breast cancer" and "strength" and "endurance". The result was nine articles related to the terms. Reading the papers made it evident that four of the studies were irrelevant to the proposed question. One paper was eliminated because its clearest outcome measure was the economic impact of exercise therapy. Two others were eliminated because they assessed the effectiveness of exercise to improve hormone release but did not include functional outcomes of exercise. The five studies used to make programming decisions included four randomized controlled trials (Cantarero-Villanueva et al., 2013; Cormie et al., 2013; Herrero et al., 2006; Schmidt, Weisser, Jonat, Baumann, & Mundhenke, 2012) and one case series (De Backer et al., 2007). Through a search of cross-listed articles, one narrative review specifically addressing exercise in patients with breast cancer was obtained and reviewed (Eyigor & Kanyilmaz, 2014).

Discussion of Results

Exercise intensity is an important concept that has been studied in relation to exercise in women recovering from breast cancer. Two randomized controlled trials addressed the outcomes from low-load resistance exercise. Schmidt and colleagues (2012) studied a 6-month low-load resistance exercise protocol in patients rehabilitating from breast cancer. The patients, who were all postoperative and had completed chemotherapy and radiation therapy, were randomized into either a strength training group or a control group. The control group completed gymnastics exercise, but its specifics were not explained. Intervention participants completed one set of 20 repetitions at approximately 50% of their 1RM. The exercises included a variety of upper and lower body machines. A submaximal cycle test was used to measure aerobic performance, and a quality of life questionnaire was administered. Both "gymnastics" and low-load strength training improved submaximal aerobic endurance in breast cancer survivors (Schmidt et al., 2012).

Low- versus high-load resistance training was compared to no exercise by Cormie and colleagues (2013). These investigators implemented a 3-month randomized controlled trial during the cancer treatment phase in 62 women diagnosed with breast cancer. Both strength training groups completed 60-min sessions twice per week in a supervised setting. After warming up, patients completed six of 10 exercises per session: chest press, seated row, lat pulldown, shoulder press, lateral raise, biceps curl, triceps extension, wrist curl, leg press, and squat. The only difference between groups was that the high-load group used a resistance of 75% to 85% of 1RM within a 6- to 10-repetition range whereas the low-load group used 55% to 65% of 1RM for 15 to 20 repetitions. Physical function was assessed using 1RM testing on major muscle groups and an endurance test consisting of a 70% load lifted for repetitions to failure. Additionally, quality of life was measured using a questionnaire, and severity of symptoms resulting from treatment were tracked. Strength and endurance improved in both groups compared to control, but there were no between-group differences (Cormie et al., 2013). Heavy or light weight training did not exacerbate the severity of symptoms, was safe, and improved health-related quality of life.

A combined aerobic and resistance training intervention was assessed by Herrero and colleagues (2006). Sixteen volunteers who were breast cancer survivors were randomly assigned to either an exercise group or nonexercising control. The exercise group trained for 8 weeks, three times per week, in 90-min sessions. The resistance exercise component consisted of 11 upper and lower body exercises. The intensity varied throughout the program from 8- to 15RM (~55-80% 1RM) and was progressively increased as the patients increased strength. The progressive steady-state cycling exercises consisted of 30 min of continuous pedaling, and the intensity was increased throughout the training period such that it peaked at 80% of maximum heart rate in the final week. Strength, aerobic fitness, and quality of life all improved in the exercise group compared to the control (Herrero et al., 2006). It is noted, however, that there was large variation in improvements within both groups, indicating that there were responders and nonresponders to the protocol.

The case series investigation used high-intensity resistance exercise with interval training. Of 57 patients (both males and females), 34 were patients with breast cancer, and all were in the posttreatment phase for cancer. The participants completed an 18-week study that included six lower body exercises performed at 65% to 80% 1RM and interval training one or two times per week (De Backer et al., 2007). The interval protocol was completed in two 8-min bouts; the first was performed before strength training and the second afterward. To alter intensity, the protocol called for performing the first 30 s of each minute at 65% peak workload followed by 30 s at 30% peak workload. A variety of outcome measures were assessed, including strength, body composition, cardiopulmonary function, and quality of life. Eleven patients did not complete the study. Six dropped out because of cancer reoccurrence and five for personal reasons. Men and women who completed the study saw significant improvements in aerobic capacity, maximal workload, strength, and quality of life. Strength improved to the greatest extent; on average, lunge strength improved by 105% and pullover strength by 93%.

Water aerobics may also be an effective treatment option for patients undergoing treatment for breast cancer. Cantarero-Villanueva and colleagues (2013) used a three times per week 8-week aquatic program in women who were breast cancer survivors. Comparisons were made to a randomized nonexercising group. Aquatic exercise consisted of common strength and endurance exercises performed in the low-impact pool environment. Lower body muscle strength was assessed using a sit-to-stand test and trunk endurance using a curl-up test. Aquatic exercise significantly improved trunk endurance and leg strength, but there were no significant improvements in the nonexercising control group.

The findings of these studies are supported by Eyigor and Kanyilmaz (2014). They highlight the evidence supporting the efficacy of resistance exercise in patients with breast cancer. However, they also suggest that other more novel interventions, such as yoga, Pilates, and tai chi, have been effective in improving outcomes in breast cancer survivors. Each of these exercise systems has been shown to improve strength and endurance. Table 13.1 presents a comparison of studies.

Table 13.1   Appraised Evidence Supporting or Refuting the Use of Resistance Exercise to Improve Functional Outcomes in Women Diagnosed With Breast Cancer

Study	Population	Design	Time	Intervention	Groups	Outcome
Cormie et al., 2013	62 women diagnosed with breast cancer	Randomized controlled trial	3 months	Resistance exercise	Low load (55-65%) High load (75-85%) Usual treatment	QOL increased in both exercise groups; strength increased in both groups; no increased symptoms.
Cantarero-Villanueva et al., 2013	68 women breast cancer survivors	Randomized controlled trial	2 months	Aquatic exercise	60 min of endurance exercise in the pool No exercise	Exercise group reduced fatigue, increased leg strength and trunk endurance.
Schmidt et al., 2002	30 women diagnosed with breast cancer	Randomized controlled trial	6 months	Resistance exercise	Low load (50%) Gymnastics	Both groups improved endurance, subjective effort, and QOL.
De Backer et al., 2007	57 individuals diagnosed with various forms of cancer	Case series	18 weeks	Resistance exercise	60%-80% 1RM and interval training on cycle	Large increase in strength, moderate increase in endurance and physical functioning.
Herrero et al., 2006	8 breast cancer survivors	Randomized controlled trial	2 months	Combined resistance and aerobic training	Strength training and endurance exercise No exercise	Exercise improved aerobic capacity, leg press strength, sit-to-stand, and QOL.

QOL = quality of life

Conclusion and Strength of Evidence

There is evidence supporting a variety of exercise types and modalities in patients recovering from breast cancer in the postoperative phase, and some research supports the safe and effective use of gentle strength training throughout the treatment phase. Strength, interval, and aquatics training have all been shown to be effective in improving physical function. Additionally, there is evidence supporting yoga, tai chi, and Pilates. There are no studies specifically comparing the outcomes from the different training interventions and few quality randomized controlled trials. However, it appears from those studies that have been published that exercise is safe and effective at maintaining or improving physical function during and after treatment. Patients should closely follow their surgeon's postoperative advice and not begin exercise training until they are cleared by their physician.

> **Level C evidence supports exercise as an effective countermeasure to prevent strength and endurance loss, compared to medical treatment alone, during and after treatment for breast cancer.**

Program Recommendations

Louise should encourage Aviva to begin an exercise routine to be performed throughout the treatment phase. Before developing the program, it may be advantageous to determine Aviva's exercise interests. Because there is evidence supporting a wide variety of modes and systems, Louise can be creative in her program design. If Aviva would like to engage in strength training, the intensity of exercise may not be as important as adherence in the early phases of training. Evidence supports both heavy and light programs. Louise may also want to implement a testing regimen before training to ensure that gains are being made throughout recovery. As every patient is different and every treatment plan is unique, Louise should be prepared to be flexible and change the program as needed to adapt to the treatment plan. Aviva should be encouraged to continue the structured exercise plan after treatment, as the evidence supports continued improvements in physical function in cancer survivors.

CASE STUDY 3: Exercise and Traumatic Brain Injury

Camila is employed as an exercise specialist in a pay-for-service transitional therapy center. The clinic partners with a major neurorehabilitation hospital that refers patients for continued fitness training after physical and occupational therapy are completed. Several neurologists from the area also refer patients to the center, and it has gained a reputation as a top fitness center for patients with disabilities. Camila is approached by a potential client, Anthony, who is a veteran, having served 15 years in the armed services. Anthony was deployed three times during his career. Toward the end of his final deployment, he was approached by a vehicle for a routine inspection. While he was walking toward the car, it exploded, knocking Anthony to the ground and rendering him unconscious. He was evacuated from the field, and although the true extent of the injury was not clearly defined, he was diagnosed with a moderate to severe traumatic brain injury. Two years after the injury, he still has occasional headaches and is visiting a neuropsychologist for posttraumatic stress disorder. Along with some cognitive impairments, Anthony has also noticed some physical declines. He has always been an avid runner and has engaged in resistance exercise for the past 20 years. Since the injury, his exercise frequency has declined, and he has seen a noticeable drop in his aerobic fitness. Although his motor impairments are not severe, he does have some disturbances in balance and loses concentration at times. He has approached Camila and asked her to develop a program to improve his fitness and reduce fatigue.

Background

Traumatic brain injuries (TBI) are at the forefront of popular media due to the increased public awareness of the harmful chronic effects of sport-related concussions. Public awareness has also increased recently with the emergence of information concerning blast-related TBI. Because of the changing landscape of modern warfare, more individuals are sustaining TBIs from the energy associated with detonation of an improvised explosive device (IED). In years past, many of the people injured by an IED did not survive. However, improvements in battlefield emergency medicine have significantly improved the survival rate after injury. The number of blast-related TBIs has increased to such an extent that this injury has been called "the signature wound of the war" in Iraq and Afghanistan (Snell & Halter, 2010). In these wars, the injury rate was calculated at approximately 35 per 10,000 soldier-years (Kozminski, 2010). Posttraumatic stress disorder is common following a blast TBI, along with other neuropsychological and physical impairments (Caldroney & Radike, 2010; Helfer et al., 2011; Sayer et al., 2008; Scherer et al., 2011; Theeler, Flynn, & Erickson, 2010).

Traumatic brain injury may be caused by an explosion, blunt head trauma, a gunshot, a wound, or some other mechanism. The effects of TBI vary widely. Disabilities resulting from a TBI may include memory problems, attention deficit, depression, time–space disorientation, aggressiveness, motor impairments, and fatigue (LaChapelle & Finlayson, 1998; Leon-Carrion, 2002). A reduced aerobic capacity and lower anaerobic threshold have also been reported in patients recovering from a TBI, which may, in part, contribute to fatigue (Amonette & Mossberg, 2013; Mossberg, Ayala, Baker, Heard, & Masel, 2007). With the numerous comorbidities, Camila needs to know whether patients with TBI benefit from exercise similarly to apparently healthy individuals. The question can be posed in this way:

In patients recovering from a TBI, can conditioning exercise improve aerobic capacity compared to sedentary nonexercising conditions?

Search Strategy

PubMed was searched using the following terms and phrases: "aerobic capacity" and "exercise" and "traumatic brain injury." The search resulted in 15 papers. Five were directly relevant to the study question. After careful inspection, it was determined that one paper used patients with an acquired brain injury (ABI), which may result from a cerebrovascular accident, brain tumors, hypoxia, or any number of conditions resulting in damage to the brain tissue. Although some of the clinical manifestations are similar, there are enough dissimilarities between a TBI and ABI that this paper was eliminated. The four remaining papers included two case studies (Mossberg, Orlander, & Norcross, 2008; Scherer, 2007) and two case series studies (Bhambhani, Rowland, & Farag, 2005; Jankowski & Sullivan, 1990). After a careful search of the cross-referenced articles in PubMed, a Cochrane review paper was also acquired that related to fitness training and TBI (Hassett, Moseley, Tate, & Harmer, 2008).

Discussion of Results

Two of the papers acquired were case series investigating the effects of circuit training on physical work capacities in patients with a TBI. Janakowski and Sullivan studied 14 sedentary adults with a TBI who underwent a 16-week circuit program. Measuring peak $\dot{V}O_2$, muscle endurance, and submaximal oxidative capacity showed that the program was effective at improving endurance and aerobic capacity but failed to reduce the oxygen cost of walking (Jankowski & Sullivan, 1990). Using a similar circuit design, Bhambhani and colleagues studied 14 patients with a TBI. The patients were monitored to ensure that heart rate remained above 60% of reserve during the exercise sessions. The 18-week program was successful at improving peak aerobic capacity and workload, but only after 12 weeks (Bhambhani et al., 2005). Despite the rigorous training program, there were no reductions in body fat or increases in lean body mass observed.

Both case study investigations focused on a body weight support treadmill training program. Using the device, the investigators utilized a program based on the ACSM recommendations for exercise. Mossberg and colleagues (2008) found that body weight support treadmill training improved peak aerobic capacity in two patients recovering from a TBI. It also improved 6-min walk time in patients recovering from a blast-related TBI (Mossberg, Orlander, and Norcross, 2008). A second case study using body weight support treadmill training in a patient recovering from a TBI demonstrated improvements in gait walking capabilities (Scherer, 2007). The intervention was implemented at the midpoint of physical therapy in a patient who had experienced a blast-related TBI. It included six sessions with progressive increases in time, speed, and level of body weight support; these were performed concurrently with the standard-of-care physical therapy. Six-minute walk distance, maximum distance walked, and gait quality were all improved after implementation of the body weight support intervention.

The Cochrane review included six studies; one study combined TBI and ABI patients (Hassett et al., 2008). In the search, any study with TBI and an exercise intervention was included. Although the results do not provide any more information than already obtained from the appraised studies, the authors do conclude that none of the studies reported adverse events related to exercise. In fact, four of the six studies did not report a single dropout. A comparison of the studies can be seen in table 13.2.

Conclusion and Strength of Evidence

There are few quality studies supporting exercise as an intervention to improve aerobic fitness or reduce fatigue in patients recovering from a TBI. The few data addressing such hypotheses show moderate improvements in aerobic capacity. One potential factor that could blunt the fitness adaptation resulting from exercise is that patients with a TBI have a high incidence of growth hormone deficiencies, which is related to a reduced aerobic capacity (Mossberg, Masel, Gilkison, & Urban, 2008). Patients with minimal adaption from exercise may need to be screened for growth hormone deficiency and perhaps receive replacement therapy to improve outcomes (Bhagia et al., 2010; High et al., 2010).

In patients with a TBI, there is level C evidence to support conditioning exercise to improve aerobic capacity compared to sedentary nonexercising conditions.

Table 13.2 Summary of Critical Appraisal of Studies Related to Exercise and Traumatic Brain Injury

Study	Design	Participants	Duration	Program characteristics	Outcome measures	Results
Janakowski et al., 1990	Case series	14 sedentary patients with a TBI	16 weeks	16 weeks, 3 times per week, supervised circuit program	Muscle endurance, submax and peak $\dot{V}O_2$	Increase in peak $\dot{V}O_2$, increase in muscle endurance, no change in submax $\dot{V}O_2$
Bhambhani et al., 2005	Case series	14 patients with a TBI	18 weeks	18 weeks, 3 times per week, circuit training	Body composition, peak exercise responses	Increase in peak aerobic capacity, no change in body composition
Scherer, 2007	Case study	1 patient with a blast-related TBI	18 weeks	Body weight support treadmill training	6-min walk distance, total distance covered in treatment session	Increase in both 6-min walk distance and distance walked in session.
Mossberg, Orlander, and Norcross et al., 2008	Case study	2 patients with a TBI	12 weeks	Body weight support treadmill training	Peak $\dot{V}O_2$ and workload	Improvement in work capacity, improved peak work

Program Recommendations

There are only a few quality studies evaluating exercise to improve aerobic capacity, but there is no reason to believe it would be detrimental. The data available suggest that adaptations occur but that they may take longer than for individuals without a TBI. Anthony is somewhat dissimilar to the patient populations tested in the literature since he has minimal motor impairments; therefore, he may adapt differently than the participants tested in the literature. Before initiating Anthony's program, Camila should implement a comprehensive testing plan to assess aerobic fitness, strength, and body composition. Every 6 to 8 weeks, testing should be repeated. Based on the literature, it may require at least 12 weeks to observe improvements in aerobic capacity. Because Anthony is an experienced exerciser, he should be educated to help him understand that his impairments may increase the time needed for adaptation compared to what he was accustomed to before the injury. He will need to be patient, but adaptations will likely occur with consistency in training. The program may include conditioning and strength exercise. Since Anthony has balance impairments, Camila should ensure that the mode of exercise is safe. If she prescribes treadmill exercise, Anthony may need constant monitoring to ensure that he maintains concentration and does not fall. If Anthony has not experienced significant improvements in 12 to 18 weeks, he should consider discussing endocrine screening with his physician to ensure that there are no underlying deficiencies blunting exercise responses. Camila should routinely reassess the literature; novel investigations may provide more evidence to support new programs and protocols in patients with a TBI, since this is an emerging and growing research field.

CONCLUSION

Exercise for patients with disease, disabilities, and injuries is a growing area of opportunity for exercise practitioners. This is evidenced by population data suggesting a large shift in the number of individuals who are older and an increase in individuals living with one or more disabilities. Postrehabilitation care for people completing physical and occupational therapy may grow in the coming years, and changes in the health care system may create greater opportunity for exercise practitioners. The ACSM has offered certification courses for the Clinical Exercise Specialist and the Registered Clinical Exercise Physiologist for many years; it also offers specialty certifications particular to cancer, disabilities, and public health. The National Strength and Conditioning Association in the United States has launched an Exercise for Special Populations certification. Obtaining academic training and certification in these areas provides a solid background for exercise practitioners but cannot fully prepare them for the breadth of scenarios that may arise when working with individuals who are disabled. Even with the finest training, situations will arise in which an exercise practitioner is asked to evaluate a novel patient condition. Additionally, the literature supporting safe and effective use of exercise in many patient populations is just beginning to emerge, and many of the protocols that are most effective in these populations are not well known and may not currently be the standard of care. When preparing to face these challenges, the exercise practitioner must be confident in finding, evaluating, and implementing evidence in practice in order to increase the probability of a safe and successful outcome in patients.

Chapter 14

NUTRITION AND SUPPLEMENTATION

Learning Objectives

1. Discuss the nutritional supplement industry and why evidence-based practice is particularly important in this context.
2. Provide four practical examples of a nutritional supplement question that may arise in practice.
3. Understand how to develop an evidence-based practice question and find evidence.
4. Objectively assess the evidence, providing strength of evidence for four case examples.

The areas of nutrition and supplementation are key to optimal health and athletic performance but, unfortunately, terribly misunderstood. Dietary recommendations are often based on fad books or theories promoted by models with dramatic results using a prescription provided in a book or on a website. The supplement industry is poorly regulated, with new products appearing at a staggering rate. Additionally, oversight of the ingredients actually packaged in the supplement container is minimal. Conversely, many supplement manufacturers conduct research, fund independent laboratories, and are legitimately concerned with the quality and safety of their products. They also diligently search for better and more effective solutions through perpetual in-house research and collaborations with independent labs. Unfortunately, some do not take this rigorous approach, which leads to some products on the shelves that are effective and safe and others that are ineffective and potentially dangerous. The problem is that the general public does not know which is which.

This chapter demonstrates the methods that should be employed to find evidence-based answers related to nutrition and supplements. As in all the case study chapters, the information here is not intended to provide definitive answers to evidence-based questions, as the entire point of this book is that scientific knowledge is dynamic.

CASE STUDY 1: Creatine Monohydrate and Cycling

Cornelius is a 41-year-old, 75 kg Category 2 road cyclist who competes at the state and regional level. His specialty is sprinting, as he has always been one of the fastest guys among his training and racing buddies; as a road racer, his races are typically 70 to 90 miles (113-145 km) long. Because Cornelius is a sprinter, he tends to avoid races with longer climbs. In performance tests, Cornelius could previously generate 1500 W for 5 s (20 W/kg) in a sprint, but he has noticed that this number has dropped a bit. As a husband and father, Cornelius is training as much as his time will allow. In addition to his training on the bike, he participates in a strength training program during the off-season.

Flatter races that end in sprints are Cornelius's forte. Thus, in addition to brief, very high power output for the sprint, he must have reasonably high aerobic capacity and endurance in order to stay with the pack over the course of a 3- to 4-h race. Accordingly, Cornelius spends 10 to 15 h per week training, combining shorter 1- to 2-h rides on weekdays with long, 3- to 4-h rides on the weekends; this provides him with the strong aerobic base necessary for him to make it to the final sprint. He also performs sprint workouts on the bike, with some of his sprint work conducted while he is fresh to max-imize power output and some performed at the end of long, hard rides to develop the neuromuscular drive necessary for producing large power outputs in a fatigued state.

Cornelius has heard some lifters discussing the use of creatine monohydrate to improve muscle strength and power. He consumes a healthy diet and has never used supplements. He wonders if creatine monohydrate might be able to improve his per-formance on the bike because, from what he has heard, it is mostly used by athletes in strength–power sports like American football.

Background

Road cycling is predominantly an endurance sport, as the ability to complete races rang-ing from 1 to 7 h (at the professional level) requires a large aerobic capacity. However, all road races are not created equal. Time trials are ridden at threshold power for ~1 h (high aerobic capacity and ventilatory threshold); mountainous races are hours long and necessitate high sustained power-to-weight ratio (W/kg); hilly races require a measure of all these traits plus acceptable anaerobic capacity (to power up short 1- to 5-min climbs). Flatter races, with terrain unable to separate the peloton, often end in a bunch sprint. Maximal cycling power is known to decline ~7.5% per decade in the general male population (Martin, Farrar, Wagner, & Spirduso, 2000); although the decline is probably much less (slower) in trained cyclists, it is likely present.

Creatine monohydrate is the supplemental form of creatine, a naturally occurring nitrogenous organic acid in humans involved in chemical reactions that rapidly resyn-thesize adenosine triphosphate (ATP) in high energy demand tissues such as skeletal and cardiac muscle (Brosnan & Brosnan, 2007). Historically, creatine supplementation often begins with a loading dose (4×5 g daily for 5 days) followed by a maintenance dose of ~5 g/day (Preen, Dawson, Goodman, Beilby, & Ching, 2003). Creatine is typically coingested with carbohydrate as this has been shown to better increase total muscle creatine (Preen et al., 2003), likely due to insulin action (Green, Hultman, Macdonald, Sewell, & Greenhaff, 1996); when it is coupled with exercise, uptake is even greater (Brosnan & Brosnan, 2007; Harris, Soderlund, & Hultman, 1992). The increase in total creatine varies across individuals, as those with higher initial levels of muscle phospho-creatine see the smallest increase in total muscle creatine with supplementation; muscle total ATP is not increased (Harris et al., 1992). Creatine supplementation has several physiological effects: (1) an increase in lean body mass (Branch, 2003); (2) an increase in phosphocreatine levels that facilitates the performance of higher-intensity exercise (or a higher volume of high-intensity exercise), both of which lead to enhanced training adaptations (Branch, 2003); and (3) an increase in muscle fiber diameter with resistance training that is greater than the increase elicited by resistance training alone (Volek et al., 1999). The evidence-based question derived for this case study is as follows:

In competitive road cyclists, does creatine monohydrate supplementation in addition to specific cycle training improve sprinting performance as measured by 1- to 10-s peak power output?

Search Strategy

Typing "creatine" into PubMed revealed a long list of MeSH terms from which to choose. Scanning this list, "creatine supplementation sprint" was selected as it is most specific to the search interests; this returns 78 articles. Refining the search to "creatine supplementation sprint cycling" reduced the number of articles to 14. A quick scan of the article titles and abstracts revealed that 11 studies either used noncyclists as subjects or did not have maximal sprint performance as an outcome variable. Of the three remaining articles, two were randomized controlled trials and one was a review paper; a fourth relevant article was found on the reference list of one of the original three papers (Finn et al., 2001; Juhn & Tarnopolsky, 1998; Vandebuerie, Vanden Eynde, Vandenberghe, & Hespel, 1998; Van Schuylenbergh, Van Leemputte, & Hespel, 2003).

Discussion of Results

A summary of the four selected articles is provided in table 14.1. The first study (Finn et al., 2001) examined the effects of creatine supplementation (4×5 g/day, 5 days, with carbohydrate) on 4×20-s sprint performance in well-trained triathletes. It should be noted that although triathletes are typically not excellent sprinters (this is evidenced by the peak power values reported in this study), they are well-trained endurance athletes. The evidence-based question pertains to a single sprint, not repeat sprint performance. Fortunately for the inquiry, Finn and colleagues reported and compared the values of each of the four sprints; the peak values of only the first of the four 20-s sprints are examined here. Creatine supplementation increased relative 1-s peak power (W/kg) and showed a tendency to increase absolute 1-s peak power ($P = 0.07$); absolute 5-s peak power also tended to increase with creatine supplementation ($P = 0.08$). Because the authors were more interested in performance over the full 20-s sprint and the subsequent three 20-s sprints, they concluded that creatine supplementation had no effect on repeated sprint performance (Finn et al., 2001). However, for our question, this study suggests that creatine supplementation might benefit single sprint performance in trained cyclists.

Van Schuylenbergh and colleagues (2003) evaluated the effects of a creatine–pyruvate supplement (2×3.5 g/day for 1 week) on maximal sprint performance after a maximal 1-h time trial in trained male cyclists. The supplement was 60% creatine and 40% pyruvate; the authors cite commercial advertisements stating that this blend increases creatine bioavailability. This is not the sort of reference that one typically finds in a peer-reviewed scientific article; but because the researchers were evaluating a commercial product, it was reasonable to relate this. After a familiarization session, subjects were randomized

Table 14.1   Summary of Critically Appraised Studies Related to Creatine and Sprinting Performance in Cyclists

Study	Population	Outcome measure	Findings
Finn et al., 2001	16 triathletes	1-s and 5-s peak power	Creatine improved relative 1-s peak power; absolute 1-s and 5-s peak power tended to increase ($P = 0.07$ and 0.08, respectively).
Van Schuylen-bergh et al., 2003	14 cyclists and triath-letes	Peak power and mean 10-s power	No changes in sprint performance for creatine or control group.
Vandebuerie et al., 1998	12 trained cyclists	Peak power and mean 10-s power	Creatine increased peak and mean power by 8% to 9% compared to control ($P < 0.05$).
Juhn & Tarnopolsky, 1998	Review	Review	No findings directly pertinent to the evidence-based question.

to either a supplemental or placebo group and completed a pretreatment 1-h time trial at maximal effort. Following a 10-min active recovery, subjects performed five 10-s maximal sprints, each separated by 2 min. Subjects then consumed the supplement or placebo for 1 week and returned to the laboratory to repeat the time trial and five sprints. Results showed no differences between the supplemented group and placebo group either before or after treatment for time trial power, total work, or any sprint power parameter. The obvious problem with this study was the low dose of creatine; subjects were ingesting only 4.2 g creatine per day, which, as noted in the discussion section, is likely insufficient to increase muscle creatine content. The authors conclude that this is the probable explanation for their negative findings (Van Schuylenbergh et al., 2003).

The same group of investigators previously used a crossover study design to evaluate maximal sprint performance in well-trained cyclists ($\dot{V}O_2max$: 69 mL · kg^{-1} · min^{-1}) before and after 5 days of creatine monohydrate supplementation (25 g/day) (Vandebuerie et al., 1998). Subjects completed three test conditions: (1) creatine loaded (25 g/day × 5 days), (2) creatine loaded plus acute intake during performance testing (5 g/h), or (3) placebo. During each of the conditions, which were separated by 5-week washout periods, subjects performed an exhausting 2.5-h endurance bout followed immediately by five 10-s maximal sprints, each separated by 2 min. Only creatine supplementation (not creatine plus acute intake) improved peak and mean sprint power for all five sprints (8-9% better than placebo); endurance performance was unaffected by treatment (Van-debuerie et al., 1998).

Juhn and Tarnopolsky's 1998 review was published before or simultaneously with these randomized controlled trials (Juhn & Tarnopolsky, 1998). Although thorough and well written, it does not include more recent work and primarily provides a higher-level view of creatine's effects on performance; for example, the studies reviewed employed many different subject populations, from untrained individuals to resistance-trained college football players. Thus, while providing helpful background information, a large (early) bibliography, and a general overview of findings, it does not substantially contribute to answering our evidence-based question.

Conclusion and Strength of Evidence

Although an abundance of research supports creatine's effectiveness in a variety of athletic populations, there is only a small body of limited evidence to support its effectiveness to improve sprint performance in endurance-trained athletes. Moreover, individual responses to creatine vary widely. However, there appears to be very little potential downside to creatine supplementation in cyclists (Dalbo, Roberts, Stout, & Kerksick, 2008; Kim, Kim, Carpentier, & Poortmans, 2011). Although anecdotal reports and an older position stand warned against creatine supplementation in athletes performing strenuous exercise in hot environments (Terjung et al., 2000), subsequent evidence demonstrated that creatine supplementation does not impair thermoregulation or performance during exercise in the heat (Volek et al., 2001).

There is level B evidence to support the use of creatine supplementation to improve sprint performance in competitive road cyclists.

Program Recommendations

Evidence for the effectiveness of creatine to improve cycling sprint performance is limited, but because there is apparently little risk of impaired thermoregulation or performance, Cornelius should begin supplementing with creatine monohydrate (4×5 g/day, 5 days, followed by a maintenance dose of 5 g/day). He should do this during the off-season or before a relatively unimportant race and observe its effects, taking special care to perform and document the results of standardized sprinting tests. Optimally, he should perform tests both fresh and after hard training rides that mimic a typical road race. Cornelius should continue his cycle endurance training. He should evaluate his on-bike sprint training and consider altering variables such as the volume and frequency with which he performs these particular workouts. Cornelius could also evaluate other ergogenic aids such as caffeine, which has a rich body of evidence supporting performance enhancement in endurance exercise, if not specifically in sprint performance. Further, he should evaluate the scientific evidence to determine the effectiveness of plyometric–power training to improve sprint performance in competitive cyclists and consider making these strategies part of his training. Lastly, although an experienced sprinter, Cornelius should consider strategic and tactical aspects that may yield performance benefits; these have been documented in the scientific literature (Menaspa, Abbiss, & Martin, 2013).

CASE STUDY 2: Caffeine, Coffee, and Performance Enhancement

Sameer is a 24-year-old Category 4 road cyclist who competes at races of 40 to 60 miles (64-97 km). He has been racing at the state and regional level for the last 2 years. Not excelling at any one aspect of the sport, Sameer is an all-rounder who does reasonably well at all facets of cycling: climbing, time trialing, sprinting, and short, hard efforts. He is also a second-year graduate student with a research assistantship and drinks a good bit of coffee to maintain his alertness during long hours of classes, studying, and working in the laboratory.

Sameer is well aware that caffeine is a potent ergogenic aid for endurance exercise, but he wonders if he needs to ingest extra caffeine at races because of how much he already consumes (360-480 mg/day). Moreover, with his understanding of physiology, he has begun to wonder if his chronic coffee consumption might attenuate or negate the performance-enhancing effects of caffeine. Sameer really wants to improve and advance in cycling and would be willing to give up drinking coffee if it would facilitate an ergogenic effect of caffeine supplementation on race days. Should Sameer consume extra caffeine before races or consider limiting consumption on a daily basis to enhance the ergogenic effect on race day?

Background

The basic physiology of cycling is discussed in Case Study 1 in chapter 12. Briefly, road cycling necessitates a large aerobic capacity to sustain a high level of energy production over races that last 2 to 5 h at the amateur level. More so than $\dot{V}O_2max$, a high lactate threshold has been shown to be the primary determinant of prolonged cycling performance (Coyle, Coggan, Hopper, & Walters, 1988). Coyle and coworkers showed that even in cyclists with very high aerobic capacities, 5 or more years of cycling experience was associated with a higher lactate threshold (% $\dot{V}O_2max$), lower levels of glycogenolysis, higher percentage of type I fibers, and greater muscle capillary density (Coyle et al., 1988). This is pertinent to Sameer, as he has only 2 years of cycling experience; he will benefit greatly from additional years of high-volume (and high-intensity) training.

Caffeine is probably the most widely researched ergogenic aid for endurance sports, with the scientific literature clearly demonstrating its potent performance-enhancing effects (Sokmen et al., 2008). Previously, it was believed that caffeine exerted its ergogenic effects via increased lipolysis and concomitant glycogen sparing; currently it is understood that caffeine acts as an adenosine receptor antagonist, which results in the following effects: (1) decreased pain and perceived exertion (Doherty & Smith, 2005) and (2) improved motor recruitment and excitation–contraction coupling (Mohr, Nielsen, & Bangsbo, 2011; Tallis, James, Cox, & Duncan, 2012; Tarnopolsky, 2008; Tarnopolsky & Cupido, 2000). In this case, the two-part evidence-based practice (EBP) question is as follows:

> **In competitive adult endurance athletes, (1) does the caffeine in coffee provide a performance-enhancing effect compared to non-caffeine use and (2) does chronic coffee intake impair the ergogenic effects of acute caffeine ingestion on endurance performance parameters?**

Search Strategy

Typing "caffeine" into PubMed revealed a long list of MeSH terms from which to choose. Scanning this list, we select "caffeine endurance," as it is most pertinent to our search interests; this returns 309 articles. Adding "AND coffee" to the search string ("caffeine endurance AND coffee") reduces the article count to 21. Scanning the article titles and abstracts reveals that 17 papers are not appropriate for answering our evidence-based question, most because they do not differentiate between caffeine and coffee intake; one review paper was excluded as it was relatively old (1998) and provided no information beyond that in the four selected articles. Of the four remaining articles, three are randomized controlled trials and one is a review paper (Graham, Hibbert, & Sathasivam, 1998; Hodgson, Randell, & Jeukendrup, 2013; McLellan & Bell, 2004; Tunnicliffe, Erdman, Reimer, Lun, & Shearer, 2008).

Discussion of Results

The first randomized controlled trial studied nine male endurance runners (69.1 mL · kg^{-1} · min^{-1}) and one female (52.5 mL · kg^{-1} · min^{-1}) endurance runner with a very straightforward crossover research design that addresses the first part of our evidence-based question (Graham et al., 1998). Subjects completed five sessions in which they ran at 85% $\dot{V}O_2$max to voluntary exhaustion after consuming one of five different experimental beverages: (1) decaffeinated coffee, (2) decaffeinated coffee + caffeine, (3) regular coffee, (4) caffeine in water, and (5) placebo in water. The randomized beverage was ingested 1 h before each treadmill run; for the three caffeinated beverages, caffeine dose was constant across drinks at 4.45 mg/kg. This design facilitated elucidation of the distinct effects of caffeine alone, coffee alone (decaffeinated), and caffeine in coffee (both regular and added to decaffeinated coffee). The results were clear-cut; despite similar plasma caffeine concentrations in all three caffeinated conditions, there was only one between-group difference: The caffeine-in-water group ran significantly longer than all four other groups. This study clearly demonstrates that caffeine is ergogenic only when consumed without coffee and that some ingredient in coffee may interfere with the otherwise performance-enhancing effects of caffeine.

McLellan and colleagues studied the ergogenic effects of caffeine in subjects who were habitual users of caffeine, primarily from coffee (9 males: 52 mL · kg^{-1} · min^{-1} $\dot{V}O_2$max; 4 females: 40 mL · kg^{-1} · min^{-1}; all: 660 ± 446 mg caffeine/day) (McLellan & Bell, 2004). After abstaining from coffee and caffeine for 12 h, subjects completed six testing sessions during which they ingested 150 to 250 mL of either decaffeinated coffee, regular coffee, or water, followed 30 min later by capsules containing either a caffeine supplement or a placebo. After a 60-min rest (90 min after the initial drink), subjects cycled to exhaustion at 80% $\dot{V}O_2$max. The six conditions were (1) decaffeinated coffee + placebo capsules, (2) decaffeinated coffee + caffeine capsules (5 mg/kg), (3) coffee (1.1 mg/kg) + caffeine capsules (5 mg/kg), (4) coffee (1.1 mg/kg) + caffeine capsules (3 mg/kg), (5) coffee (1.1 mg/kg) + caffeine capsules (7 mg/kg), and (6) brown-colored water + caffeine capsules (5 mg/kg). This permitted the investigators to study the effects of prior coffee consumption on the ergogenic effects of various doses of supplemental caffeine. This study is ideal to address the second part of our evidence-based question. In all five conditions in which subjects received caffeine (via either coffee, caffeine capsules, or both), time to exhaustion was improved by 24% (27.0 ± 8.4 min vs. 21.7 ± 8.1 min) compared to control (decaffeinated coffee + placebo capsules); caffeine dosage had no effect on time to exhaustion, nor did the initial beverage (coffee, decaffeinated coffee, or water) (McLellan & Bell, 2004). Given the large amounts of coffee or caffeine that these subjects habitually consumed, coupled with the research design that incorporated a typical serving of coffee (~ 1 cup) before capsule supplementation (a design with excellent external validity to a typical racing cyclist), the results of this study are quite compelling.

With relatively minor changes to the study design, Hodgson and colleagues attempted to replicate the findings of Graham and colleagues, stating from the outset their hypothesis that there would be no difference between caffeine supplementation in water and via normal coffee (Hodgson et al., 2013). They studied eight trained cyclists or triathletes ($\dot{V}O_2$max = 58.3 mL · kg^{-1} · min^{-1}) in four conditions: (1) decaffeinated coffee, (2) regular coffee, (3) caffeine in water, and (4) placebo in water. The randomized beverage was ingested 1 h before each exercise test; for the two caffeinated beverages, caffeine dose was constant at 5.0 mg/kg. The exercise test consisted of a 30-min steady-state bout at ~55% $\dot{V}O_2$max followed immediately by an energy-based time trial (650 ± 37 kJ) that would take 45 min if performed at 70% of the maximum wattage attained in the $\dot{V}O_2$max test; subjects were instructed to give maximum effort and complete the time trial as quickly as possible. Power outputs for both coffee and caffeine groups (291 ± 7 and 294 ± 6 W, respectively) were higher ($P < 0.05$) than for decaffeinated and placebo (276 ± 7, 277 ± 4 W, respectively) and not different from each other. This pattern was also seen in time trial time to completion, which was significantly reduced (4-5%) for both the coffee and caffeine groups compared to the decaffeinated and placebo groups. In attempting to reconcile the conflict between their findings and those of Graham and coworkers, Hodgson and colleagues suggest that perhaps the discrepancy is due to their different outcome measures; they note that exercise to exhaustion tests have been shown to have a moderate degree of variability, while the time trial they employed is a better approximation of real-life racing situations and has much lower test–retest variability (Hodgson et al., 2013).

Tunnicliffe and colleagues reviewed coffee and caffeine use in athletes and endeavored to differentiate their individual and combined effects on performance (Tunnicliffe et al., 2008). They evaluated the two randomized controlled trials already discussed (Graham et al., 1998; McClellan & Bell, 2004), as well as other studies less relevant to the evidence-based question. Their conclusions were to question the ergogenic effects of coffee compared to caffeine since coffee contains may other compounds that could interact with or confound the ergogenic effects of caffeine (Tunnicliffe et al., 2008).

Conclusion and Strength of Evidence

This case study has conflicting evidence. McLellan and colleagues answered the second part of the question posed (what are the effects of chronic coffee consumption on the ergogenic effects of caffeine?) by demonstrating that chronic coffee drinkers with concomitant high caffeine intakes did in fact realize an endurance performance benefit from supplemental caffeine capsules even after consuming one cup of coffee (McLellan & Bell, 2004). The other part of the question (does coffee provide an ergogenic benefit?) was supported by two randomized controlled trials demonstrating that either coffee itself (containing an ergogenic dose of caffeine: ~5 mg/kg) or coffee plus supplemental caffeine did enhance endurance performance (Hodgson et al., 2013; McLellan & Bell, 2004). But another randomized controlled trial showed that only supplemental caffeine in water improved performance; indeed, in that study, neither coffee nor decaffeinated coffee with supplemental caffeine was any different from decaffeinated coffee or water placebo (Graham et al., 1998). Although the review paper concluded that coffee was an inferior or ineffective ergogenic aid in comparison to caffeine (Tunnicliffe et al., 2008), it seemed to gloss over the positive findings of McLellan and Bell and obviously could not comment on the results of Hodgson and coworkers.

Thus, the preponderance of the evidence supports the use of supplemental caffeine either with or without coffee consumption to improve endurance performance; integrating the findings of both Graham and colleagues and McLellan and Bell, it appears parsimonious to limit coffee intake to one cup (if coffee is desired) before races and to achieve the remainder of a total caffeine dose of ~5 mg/kg via supplemental caffeine in capsule form. Chronic coffee use does not appear to impair the acute ergogenic actions of caffeine. It should be noted that the evidence obtained was from relatively small randomized controlled trials; thus the evidence-based practitioner should monitor the literature for new evidence that may address the question.

> **In competitive adult endurance athletes, level B evidence supports the proposition that caffeine in coffee provides a performance-enhancing effect compared to non-caffeine use, and it appears that chronic coffee intake does not impair the ergogenic effects of acute caffeine ingestion on endurance performance parameters.**

Program Recommendations

For Sameer, the most important question was answered: Will his chronic coffee drinking impair the ergogenic effects of caffeine? This does not appear to be a problem—he can continue drinking coffee. However, it might be best to limit his coffee intake to one cup before races, in combination with a supplemental caffeine dose of ~5 mg/kg. Sameer could also conduct his own *n*-of-1 study; on three different days he should evaluate his average power output over a 20- to 40-min time trial on either a lightly trafficked outdoor course or his trainer under three conditions: (1) no coffee or other caffeine within 8 to 12 h (control), (2) a cup of coffee + supplemental caffeine to equal a total of 5 mg/kg of caffeine 1 h before training, and (3) an equivalent dose of supplemental caffeine in water 1 h before training. For all of these conditions, he should abstain from coffee and other caffeine for 8 to 12 h before the training session. He should take care to ensure that his nutrition, hydration, and overall energy levels are similar between the three trials and that environmental conditions are also similar; he should randomize the order of the trials and separate them by at least 2 or 3 days. In this way, he can determine his individual response to these various conditions.

CASE STUDY 3: Protein Intake for Endurance Athletes

Isabella is a strength and conditioning coach at a small college. She, along with several assistants, is responsible for team sports (e.g., basketball and baseball), individual sports (e.g., tennis and golf), and endurance sports (e.g., cross country, cycling). The college does not employ a registered dietician, so she is also responsible for providing nutrition recommendations for the athletes. Isabella was recently approached by several endurance athletes who reported that they have not been eating well and want a nutrition plan to optimize their athletic performance. Isabella is quickly able to formulate recommendations for total energy and carbohydrate intake, but she is not sure how much protein an endurance athlete should consume. What recommendation should Isabella make for protein consumption in her endurance athletes?

Background

Dietary protein, which supplies the body with essential amino acids (amino acids that cannot be synthesized in vivo and must be provided exogenously), is one of the three macronutrients along with carbohydrate and fat. The most recent update from the U.S. Institute of Medicine of the National Academies, published in 2005, sets the Recommended Dietary Allowance (RDA) for protein at 0.8 g/kg per day for the general population; 20 studies using nitrogen balance as a primary outcome measure and "designed to estimate [protein] requirement by feeding a number of individuals several different intake levels" (p. 1252) were selected in addition to 39 others that employed designs less suited to estimating protein requirements (but that also measured nitrogen balance) (Institute of Medicine, 2005). In addition to this lengthy U.S. government report, a 2003 meta-analysis that aimed to establish the protein requirements of healthy adults resulted in the selection of a nearly identical set of studies, with the final recommendation of a daily protein intake of 0.83 g/kg (Rand, Pellett, & Young, 2003). Based on this collection of nitrogen balance studies, there appears to be little disagreement that a daily protein intake of 0.8 g/kg is adequate to maintain nitrogen balance in the general adult population. Outside the United States, the European Food Safety Authority published a report in 2012 recommending a daily protein intake of 0.83 g/kg per day (EFSA Panel, 2012).

For athletes, the same guidelines may not apply. Amino acid oxidation supplies 1% to 6% of the energy during aerobic exercise (the remainder and vast majority coming from carbohydrate and fat oxidation) (Tarnopolsky, 2004) and exercise itself increases the turnover of muscle protein (Carraro, Hartl, et al., 1990; Carraro, Stuart, Hartl, Rosenblatt, & Wolfe, 1990), suggesting the possibility that protein intake above that of the general population may be needed to promote postexercise recovery.

Another parameter for recommended protein intake is the Acceptable Macronutrient Distribution Range (AMDR), which is defined by the U.S. Institute of Medicine as "a range of intakes for a particular energy source that is associated with a reduced risk of chronic diseases while providing adequate intakes of essential nutrients" (Institute of Medicine, 2005, p. 14). The AMDR for protein in adults is 10% to 35% of total energy intake. This is obviously an enormous range that "permits" a protein intake of 250 to 875 kcal (63-219 g) for a standard 2500 kcal diet. If the reference person is 75 kg, this would result in a per kilogram protein intake range of 0.84 to 2.92. However, as Wolfe and Miller (2008) argue, the AMDR actually implies uncertainty regarding the optimal protein intake and reflects the fact that protein intake in excess of the RDA is not deleterious in healthy adults; indeed, protein intakes above the RDA have been shown to be beneficial for a wide variety of outcomes including muscle mass and strength (Wolfe, 2006), bone health (Dawson-Hughes, 2003), and cardiovascular function (Stamler et al., 1996).

Although it is typically "prescribed" on a per day basis, research in the area of dietary protein has shown that, more important than total daily intake, maximal stimulation of synthetic pathways is achieved with a per meal protein serving of 25 to 30 g (Symons, Sheffield-Moore, Wolfe, & Paddon-Jones, 2009). Taking this further, Mamerow and colleagues demonstrated that over a 24-h period, three meals containing ~30 g protein each elicited a 25% greater muscle protein synthetic response than the same amount of protein consumed in a "skewed" fashion (i.e., 10, 15, and 65 g at breakfast, lunch, and dinner, respectively) (Mamerow et al., 2014). Others have argued that although synthesis is maximized with 25 to 30 g protein servings, muscle protein breakdown is progressively inhibited with increasing doses of protein intake, resulting in greater net anabolism (Deutz & Wolfe, 2013). Given this, Isabella should ask the question:

Is protein intake above the recommendation for the general population necessary for young competitive male endurance athletes to maximize adaptations to high-intensity endurance training?

Search Strategy

Typing "protein intake" into PubMed does not result in a MeSH term that is pertinent to our EBP question. Thus, we add "AND" to the search term; this displays additional MeSH subheadings. Still, none of the MeSH subheadings appear relevant to dietary protein intake in endurance athletes, so our search will need to use keywords assigned by authors. Next we add "endurance athletes" to the search string ("protein intake AND endurance athletes"); this yields 165 articles. Skimming the article titles quickly yields eight review articles published in the last 10 years that are directly pertinent to our evidence-based question; of these, four were authored by Phillips, so only his most recent review will be evaluated (Phillips, 2012). One of the reviews does not discuss protein requirements and thus is excluded. The other four review papers, including a joint position stand, are included (Kreider & Campbell, 2009; Phillips, 2012; Rodriguez, Di Marco, & Langley, 2009; Tarnopolsky, 2004).

Discussion of Results

Tarnopolsky's review (2004) provides a thorough discussion of skeletal muscle protein metabolism and explains amino acid oxidation during endurance exercise. The author also details the two methodologies most widely employed to study the question of protein adequacy. The first, nitrogen balance, measures all nitrogen that enters the body and is excreted from the body. Negative nitrogen balance demonstrates that dietary protein intake is too low, while positive nitrogen balance suggests that it is adequate (although some would argue that it is too high). The second method is the use of amino acid tracers, a technique pioneered by Robert Wolfe that allows investigators to quantify the synthesis and oxidation of amino acids (Wolfe and Chinkes, 2005). With this methodology, large increases in amino acid oxidation or a plateau in protein synthesis or both are used as indices of "protein excess" (Tarnopolsky, 2004). Tarnopolsky describes three studies that measured nitrogen balance in high-level endurance athletes. The findings resulted in the following daily protein recommendations: (1) 1.6 g/kg in six male endurance athletes ($\dot{V}O_2$max = 76.2 mL $\cdot$ kg^{-1} $\cdot$ min^{-1}, ~12 h/week training volume) (Tarnopolsky, MacDougall, & Atkinson, 1988), (2) 1.49 g/kg in five well-trained runners (Friedman & Lemon, 1989), and (3) 1.5 to 1.8 g/kg in cyclists ($\dot{V}O_2$max = 65.1 mL $\cdot$ kg^{-1} $\cdot$ min^{-1}) (Brouns et al., 1989a, 1989b).

Kreider and Campbell provide a thorough overview of protein, from its molecular structure to protein quality, digestion and absorption, types of protein, and its relationship to exercise (Kreider & Campbell, 2009). Specific to protein and endurance exercise, they review several of the same studies as Tarnopolsky (Friedman & Lemon, 1989; Tarnopolsky et al., 1988) and conclude with the recommendation that aerobic athletes consume 1.4 to 2.0 g protein/kg per day (Kreider & Campbell, 2009); it is not clear why they recommend this slightly higher range.

Phillips's review moves past nitrogen balance studies intended to evaluate protein intake adequacy and looks at potential advantages to be gained by athletes consuming protein in excess of the RDA of 0.8 g $\cdot$ kg^{-1} $\cdot$ day^{-1} (Phillips, 2012). Phillips echoes the importance of a meal-based approach to protein ingestion (Symons et al., 2009; Paddon-Jones and Rasmussen, 2009; Mamerow et al., 2014) and cites work demonstrating that muscle protein synthesis is maximally stimulated with 10 g of essential amino acids (Cuthbertson et al., 2005); this equates to ~25 g of high-quality protein given that most are ~40% essential amino acids by content (Phillips, 2012). Further, he notes that muscle protein synthesis returns to postabsorptive values ~4 h after a meal; thus, consuming four or five meals per day with 25 g protein per meal would equal 125 g protein, 1.67 g/kg per day for a 75 kg athlete. Phillips closes by stating that "emerging dietary guidelines for protein are in the range of 1.2-1.6 g/kg/d" (p. S164).

The joint position stand of the American Dietetic Association, Dietitians of Canada, and the American College of Sports Medicine evaluates in detail all aspects of nutrition and athletic performance, from hydration to energy requirements; body composition; micronutrient intake; nutrition before, during, and after training; and ergogenic aids. A small section of the paper is devoted to protein intake in endurance athletes, with a recommendation of daily intakes of 1.2 to 1.4 g/kg (Rodriguez et al., 2009).

Conclusion and Strength of Evidence

Based on the review papers discussed, it appears that both the intake and the timing of the protein are important for endurance athletes. In fact, endurance athletes appear to require a protein intake greater than the RDA in order to avoid negative nitrogen balance. This is evident from the pooled results of several high-quality randomized controlled trials. We can state the following conclusion:

Level A evidence supports daily protein intake greater than the RDA (≥1.2 g/kg) for competitive endurance athletes who engage in high-intensity and high-volume training.

Program Recommendations

Isabella should recommend that her endurance athletes consume at least 1.2 g protein/kg per day and that, to the extent possible, they consume it in servings of at least 25 to 30 g in order to maximize the muscle protein synthetic response at each meal. For athletes consuming four or five meals per day (which may be optimal and can facilitate appropriate pre- and postworkout nutrition), these recommendations will be mutually reinforcing. She may want to collect a 7-day dietary recall from the athletes to ensure they are meeting the requirements.

CASE STUDY 4: HMB and Strength–Power Athletes

Isaac is a 19-year-old (sophomore) National Collegiate Athletic Association (NCAA) Division 1 American football player; his position is outside linebacker. Currently, Isaac is second team, and he thinks that if he can improve his strength and explosive power it would help him to become a starter. Some of the upperclassmen on the team are taking β-hydroxy-β-methylbutyrate (HMB), and Isaac wonders if doing this might be able to help him train harder, recover quicker, and ultimately become stronger and more powerful.

Background

American football is a sport that requires high levels of strength and power. Although the average duration of a play is short—on the order of several seconds—placing high demands on the phosphagen immediate energy system, the repetitive and prolonged nature of the sport also taxes the glycolytic anaerobic energy system and even the aerobic system (primarily to metabolize excess lactate and facilitate thermoregulation) (Pincivero & Bompa, 1997).

The essential amino acid leucine is highly anabolic and potently stimulates muscle protein synthesis even in the absence of other anabolic agents (e.g., exercise) (Bolster, Vary, Kimball, & Jefferson, 2004). In addition to increasing muscle protein synthesis, a leucine-enriched essential amino acid serving decreases muscle protein breakdown, unlike an equivalent essential amino acid drink with half the leucine content (Glynn et al., 2010). β-hydroxy-β-methylbutyrate is a leucine metabolite and as such could contribute to a reduction in muscle protein breakdown, potentially increasing skeletal muscle protein accretion, strength, and power. Although poorly studied, the proposed mechanisms for HMB's actions include (1) upregulation of insulin-like growth factor 1 (IGF-1) gene expression, (2) stimulation of the mammalian target of rapamycin (mTOR) pathway and muscle protein synthesis, and (3) attenuation of muscle protein breakdown by inhibition of the ubiquitin–proteasome system (Zanchi et al., 2011). To determine whether Isaac should consume HMB, we can pose the following question:

In young male strength and power athletes, does HMB, in conjunction with appropriate resistance, plyometric, and sport-specific training, improve muscle strength and power compared to a control nonsupplemented condition?

Search Strategy

Searching "HMB supplementation" in PubMed yields 119 articles; activating the Age filter (Adult: 19-44 years) reduces this number to 36. Adding "AND strength" ("HMB supplementation AND strength") to the search string reduces the search return to 16 articles. Articles were obtained only if they investigated HMB alone as a supplement. Scanning the article titles and abstracts reveals that eight articles (two reviews and six randomized controlled trials) are pertinent to the evidence-based question; the other eight articles are discarded because they evaluated untrained subjects or a proprietary blend of HMB that included additional ingredients. In addition to the results of this search, one of the co-authors of this text, a noted sport and exercise scientist with a wealth of experience in nutrition research, pointed out several relevant studies that were not discovered in the search described—one a randomized controlled trial and the other a position stand of an international professional society. This illustrates the fact that searches and evaluations of the literature are not foolproof and require real "digging" (and are often improved with a team approach)!

Discussion of Results

Kreider and colleagues studied 40 young males with an average of 5.5 years of resistance training experience who trained 6.9 h/week (Kreider, Ferreira, Wilson, & Almada, 1999). Subjects were matched for body mass, fat-free mass, years of training, hours per week of training, and training type and volume and were randomly assigned to one of three groups: (1) 0 HMB (control), (2) 3 g/day HMB, or (3) 6 g/day HMB. All subjects consumed HMB or placebo according to group assignment three times per day at mealtimes for 28 days; the HMB or placebo was provided in a double-blind fashion and incorporated into a commercial supplement product that provided 27 g carbohydrate, 25 g protein, and 1 g fat at each of the three daily meals. Subjects maintained their normal diets (~1.5 g protein/kg per day before supplementation, 2.3 g/kg per day with supplementation) and training regimens throughout the study. Although all subjects increased muscle mass and leg press strength, there were no differences between groups (Kreider et al., 1999).

Panton and colleagues evaluated 39 men and 36 women aged 20 to 40 years during a 4-week training study (Panton, Rathmacher, Baier, & Nissen, 2000). Subjects were randomized to a placebo or HMB group (1 g/meal = 3 g/day); the groups were also balanced by sex and resistance training status. All subjects trained 3 days a week with a mix of free weights and Cybex machines. Three sets of 11 exercises were performed at 90% 1-repetition maximum (1RM) for three to six repetitions; all sets were performed to failure. Primary outcome measures were strength (1RM) and body composition (underwater weighing). Results showed that there was not a differential response to HMB supplementation due to sex or training status. A pooled analysis of all subjects showed that the increase in bench press 1RM for the HMB-supplemented group was significantly greater than for the placebo group (7.5 kg. vs. 5.2 kg, $P < 0.05$). Fat-free mass tended to increase more in the HMB group compared to the placebo group (1.4 kg vs. 0.9 kg, $P = 0.08$). The authors speculated that the divergence between their results and those of Kreider and colleagues might be an effect of differences in the training programs; that is, Panton and colleagues used supervised, high-intensity training while Kreider and colleagues did not supervise training or report training loads—subjects were simply matched for training volume between groups (Panton et al., 2000).

Slater and colleagues evaluated 27 male national-level water polo players and rowers over a 6-week period of resistance training; all were experienced lifters and in the precompetition phase of training (Slater et al., 2001). Subjects were randomly assigned to one of three groups: (1) standard HMB (1 g/meal = 3 g/day), (2) time-release capsule HMB (1 g/meal = 3 g/day), or (3) placebo. All groups contained a similar proportion of polo players and rowers. Subjects trained 2 or 3 days a week under the supervision of strength and conditioning coaches. Protein intake before the study was 1.7 g/kg per day; the addition of a meal replacement shake to all subjects' diets increased protein intake during the study to 2.4 g/kg per day. Outcome measures included strength (3RM for bench press, leg press, and chin-ups), body composition (dual-energy X-ray absorptiometry [DXA]), and various blood markers (e.g., creatine kinase [CK] and lactate dehydrogenase [LDH]). After baseline measures, these data were obtained again at the midpoint (3 weeks) and at completion (6 weeks) of training. Lean mass and total body mass increased significantly with training, but there were no differences between groups. A similar result was found for 3RM strength for all three exercises: Significant training-induced increases were observed, with no differences between groups. No differences between groups were observed for CK or LDH (Slater et al., 2001).

Ransone, Neighbors, Lefavi, and Chromiak (2003) studied 35 male NCAA Division I American football players; all subjects had at least 4 years of lifting experience and averaged 20 h of exercise per week. In a double-blind, randomized crossover design, subjects performed 4 weeks of training with one condition (placebo or HMB), completed a 1-week washout, and then crossed over to the second condition (placebo or HMB); supplementation was provided at each of three daily meals (1.5 g at breakfast, 750 mg each at lunch and dinner = 3 g/day). The study was conducted during the summer before the competitive season at a single university. Training was supervised by certified strength and conditioning coaches and consisted of ~4 h per day, 4 days per week; this included warm-up, stretching, and strength and endurance exercise. Each session included 10 strength exercises performed at 70% to 90% 1RM (8-12 sets per exercise, 2-10 repetitions per set); endurance drills included speed and tempo exercises. Food frequency questionnaires estimated a mean daily energy intake of 2600 kcal. Outcome measures included changes in strength (bench press, squat, and power clean; 3- to 5RM) and body composition (skinfolds). The investigators reported no significant differences between groups for any outcome measure (Ransone et al., 2003).

Hoffman and colleagues also evaluated the effects of HMB supplementation in college American football players (Hoffman, Cooper, Wendell, Im, & Kang, 2004). They studied 26 athletes before and after a 10-day preseason football training camp consisting of two daily practices that were each 2 h long. Subjects were randomly assigned to either placebo or HMB (3×1 g = 3 g/day) in a single-blind fashion. Outcome measures included anaerobic power (Wingate cycle test) and serum levels of testosterone, cortisol, myoglobin, and creatine kinase. Five-day dietary recall showed that HMB and placebo subjects ingested 3543 and 3254 kcal and 1.89 and 1.45 g protein/kg per day, respectively. There were no group differences either pre- or postsupplementation for any outcome; all Wingate variables (e.g., peak power, relative peak power, mean power, rate of fatigue, and total work) were unchanged. Cortisol was decreased after the training camp in both groups while creatine kinase was significantly increased in both groups (Hoffman et al., 2004).

O'Connor and Crowe evaluated the effects of HMB and HMB + creatine supplementation in 30 male rugby players with more than 4 years of lifting experience, 2 years of state- or national-level play, and $\dot{V}O_2$max values of 47.1 to 59.3 mL $\cdot$ kg^{-1} $\cdot$ min^{-1} (O'Connor & Crowe, 2007). Several subjects did not want to take any supplements and thus self-selected to the control group; this prevented the implementation of a double-blind, randomized protocol. The remaining subjects were randomly assigned to either HMB (3 g/day) or HMB + creatine (3 g HMB + 3 g creatine + 6 g carbohydrate/day = 12 g/day) supplementation; all subjects underwent 6 weeks of supplementation and training. Each week of training consisted of eight sessions in a 5-day period followed by a 2-day rest period. Each week's eight sessions were composed of three total body strength training sessions, one speed–power session, and four team conditioning and skill sessions. Outcome measures included muscle strength (bench press, deadlift, dumbbell shoulder press, and prone row 3RM), muscle endurance (chin-ups to exhaustion), leg power (10-s cycle ergometer test), and anthropometric measures. Results showed that bench press, deadlift, shoulder press, chin-ups, and peak power and total work (10-s cycle test) significantly increased with training; there were no posttraining differences or interactions between groups. Similarly, the sum of skinfolds, arm girth, waist girth, and hip girth all decreased significantly, with no differences between groups (O'Connor & Crowe, 2007).

Zanchi and associates published a very thorough review of HMB, covering its metabolism, effects on immune responses, kinetics, effects on muscle mass and strength, clinical and therapeutic effects, and mechanisms of action (Zanchi et al., 2011). The authors reviewed a number of the studies already discussed, but did not provide any conclusions or summary statements regarding the use of HMB for the improvement of strength and power in athletes. A second brief review by Palisin and Stacy examined

HMB and its use in athletics (Palisin & Stacy, 2005). The authors reviewed all of the articles already discussed and concluded that the scientific literature did not support HMB as an effective ergogenic aid.

Wilson and colleagues (2014) evaluated the effects of HMB-free acid (HMB-FA) over 12 weeks of supervised periodized resistance training in 20 resistance-trained young men; HMB in the free acid form provides greater bioavailability with resulting higher plasma levels of HMB. Subjects were randomized in a double-blind fashion to either HMB-FA (3 × 1 g/day) or placebo (3 × 1 g/day). On training days, supplements were consumed 30 min before training and at the lunch and dinner meals; on nontraining days, supplements were consumed at each of the three daily meals. Subjects were given individual diet plans and consumed 45%, 33%, and 22% of energy from carbohydrate, fat, and protein, respectively. Training consisted of 8 weeks of an undulating (nonlinear) periodized program (3 days/week), 2 weeks of overreaching (5 days/week), and 2 weeks of reduced training volume. Outcome measures included muscle strength (squat, bench press, and deadlift 1RM), body composition (DXA), and power (modified Wingate, vertical jump). Twelve weeks of training resulted in significant group × training interactions ($P < 0.05$) for each of these outcome measures; that is, training-induced increases in muscle strength, muscle mass, and muscle power were greater with HMB-FA supplementation (Wilson et al., 2014).

Wilson and colleagues also authored the International Society of Sports Nutrition's position stand on HMB; it is worth noting that three of the coauthors were Hoffman, Kreider, and Zanchi, who also wrote papers on HMB that we discussed earlier (Wilson et al., 2013). A thorough review of the literature, including the influence of factors such as subject training status, study duration, training intensity, and supplement timing on HMB efficacy, yielded evidence in support of HMB to improve training-induced adaptations in muscle mass, strength, and power in resistance-trained individuals (Wilson et al., 2013). This conclusion appears at odds with the results of most of the studies we found in our evidence-based search. However, if we examine table 2 in the paper by Wilson and colleagues, we see that they found or included seven articles that we did not and that most of these reported positive effects of HMB supplementation; the table also includes all of the studies with negative results that we discussed previously.

Conclusion and Strength of Evidence

Although several studies showed little or no beneficial effect of HMB supplementation, they either did not incorporate intensities above those to which subjects were already accustomed, were relatively short in duration, or did not use the newer, more bioavailable form of HMB. The preponderance of the evidence, particularly the thorough, evidence-based review and position stand of the International Society of Sports Nutrition (which included a number of studies that we did not find in our search) supports the use of HMB.

This case study serves as a good example of the importance of a team-based approach in addressing EBP questions, as several key pieces of evidence (e.g., the International Society of Sports Nutrition position stand) were not discovered via the initial literature search but were found after a more experienced team member suggested review. Also, this case study illustrates the utility of a position stand (or a well-conducted systematic review) from a reputable professional society: Such papers have not only gathered pertinent evidence; they also evaluate and synthesize the evidence and formulate a mature, comprehensive position based on it (Kraemer et al., 2009). The International Society of Sports Nutrition position stand alone and evidence contained therein would have been adequate to lead to the following conclusion:

Level B evidence supports the use of HMB as a supplement to improve strength and power in trained strength and power athletes when employed in conjunction with intense resistance, plyometric, or sport-specific training or some combination of these.

Program Recommendations

Isaac should use HMB consistent with the recommendations provided in the International Society of Sports Nutrition position stand (e.g., dosage of 38 mg/kg per day). Although evidence is still scant, he should consider using the newer free acid form of HMB in lieu of the more commonly studied calcium form. He may also consider a more thorough search of HMB in conjunction with a comprehensive amino acid blend, as there is some evidence supporting its efficacy (Kraemer et al., 2009). Also, Isaac should consider creatine monohydrate, a supplement with well-documented efficacy for improving strength and power. Before using any ergogenic aid, Isaac should check with the strength and conditioning and athletic training staff. The NCAA has strict rules on supplement consumption, and he should verify that the supplement does not contain ingredients banned by the NCAA. Isaac should also reevaluate the literature on a consistent basis, as new work is likely to continue to become available.

CONCLUSION

The nutrition, nutritional supplement, and ergogenic aid industries are three of the fastest-growing markets in exercise and sport science. Companies are constantly developing new products that may help athletes improve their performance or health. Some products contain "newly discovered" ingredients that are potentially ergogenic or may enhance recovery from exercise. Others contain different quantities, ratios, or formulas of ingredients from previously marketed products. The reputable supplement companies diligently work to ensure that their products are safe, effective, and free of banned substances, with many providing funding to independent testing programs (e.g., Informed Choice) to confirm results. However, the industry is poorly regulated, so some products are developed and marketed before their safety and effectiveness are supported. Thus, it is important for the evidence-based practitioner to be skilled in evaluation of the literature concerning supplements, ergogenic aids, and nutrition plans. While some new products may be effective and legitimately enhance the health and fitness of clients, athletes, and patients, others may be cost-ineffective or even dangerous.

Chapter 15

EXERCISE DEVICES, EQUIPMENT, AND APPAREL

Learning Objectives

1. Understand the need to use the best evidence to support the selection of exercise devices, equipment, and apparel.
2. Provide four practical examples of device and equipment questions that may arise in practice.
3. Understand how to develop an evidence-based practice question and find evidence.
4. Objectively assess the evidence, providing strength of evidence for four case examples.

Many elite and amateur athletes perform much or all of their training using traditional equipment such as barbell and dumbbell free weights for resistance exercise and standard, cushioned shoes for long-distance running. However, coaches and athletes are constantly in search of equipment and apparel that can improve training-induced physiological adaptations, prevent injury, or even introduce variation into stale, repetitive training programs. Add to the mix equipment manufacturers that are keen to generate profits, and the result is a plethora of available devices and equipment with anecdotal support for their efficacy. A continual quest for improvement is at the heart of exercise and sport, so it is wholly appropriate that athletes explore new training devices; however, it is essential that new devices be rigorously examined and compared to proven training equipment to prevent inadvertent performance regression due to training with inferior equipment.

In this chapter, four real-world case studies related to exercise devices, equipment, and apparel are introduced. Using the methodology defined in this book, the case examples are evaluated for quality of information related to their efficacy and safety. Based on this evidence, recommendations are made to include or exclude from training. As in previous case study chapters, the review of the literature is not exhaustive. Instead, several key papers were obtained using a systematic search process and are critically appraised. Because the literature was examined at a single point in time, new papers may surface supporting or refuting the outcome of these case study examples. Thus, the intention of the recommendations is not to provide the final statement on the product. The evidence should be reevaluated on a routine basis as new research emerges.

CASE STUDY 1: Chains

Rafael is a baseball player at a large university. During high school, he was a star of his team, pitching and then playing outfield. Rafael was recruited for his bat and foot speed, not as a pitcher, so now, as a freshman outfielder, he recognizes the need to increase his upper and lower body strength and power in order to improve his hitting power and defensive abilities in center field. Apart from the higher level of competition at the college level, college baseball fields are ~50 ft (15.2 m) deeper in center field than high school fields, thus requiring Rafael to cover more ground, make longer throws, and offensively to hit the ball farther to attain the extra base hits that made him successful as a high school player. In high school, Rafael played multiple sports and performed resistance exercise as part of that training, but none of his resistance exercise was baseball specific. Now that he is a single-sport athlete, he wants to maximize his training for baseball success.

A secondary concern for Rafael is shoulder pain. During high school, he endured chronic shoulder pain; concerns over his shoulder are what prompted colleges to recruit him as an outfielder and not a pitcher. Thus, the training that he does to improve strength and power needs to benefit and not exacerbate any existing shoulder issues.

Rafael has observed some of the athletes in the athletic training facility using chains for bench press, squat, and deadlifts and wonders if he might benefit from using them.

Background

Baseball is a team sport that is dominated by the need for strength, speed, and power. Although demands vary somewhat by position, offensively, all players must possess speed and power to hit and run the bases and, defensively, to cover their respective position in the field and throw quickly and accurately over moderate (infielders) or large distances (outfielders) (Coleman & Amonette, 2012). Research has shown that resistance training programs can increase throwing velocity in young baseball players (Escamilla et al., 2012; Prokopy et al., 2008) and that greater upper body strength and power are associated with greater bat speed and hitting more home runs (Miyaguchi & Demura, 2012); resistance training may also increase maximal running velocity.

Although employed in practice earlier, resistance training with the addition of chains made its initial appearance in the exercise science literature in a 2004 *Strength and Conditioning Journal* topical paper (Berning, Coker, & Adams, 2004). In the absence of peer-reviewed literature examining the effectiveness of resistance training with chains, Berning, Coker, and Adams discuss exercises that could most benefit from the addition of chains (those with ascending strength curves, e.g., squat, bench press, and deadlift); considerations when training with chains (chain gauges, chain selection, safety); and, most importantly, the scientific rationale behind the use of chains. The primary rationale for the use of chains with resistance exercise is more precise matching of the strength curve of a movement to maximally stimulate the muscle over the entire range of motion. Due to their inertia and resultant oscillations during lifting, chains also appear to increase the recruitment of stabilizer muscles. Further, during an exercise like the bench press, the shoulders are in a somewhat vulnerable, mechanically disadvantaged position at the bottom of the movement; it is possible that training with chains for this exercise will facilitate optimal adaptations in strength and power while minimizing stress on the shoulder. In Rafael's case, this is the evidence-based question:

Does resistance exercise with the addition of chains result in increases in muscle strength and power that are superior to those with traditional resistance training?

Search Strategy

Typing "resistance exercise" into PubMed does not yield an applicable MeSH subheading (e.g., "resistance exercise chains"), so we add "AND chains" to the search string ("resistance exercise AND chains"), which returns 108 articles; in this case, "chains" is searched as a keyword or a word in the title or abstract. Activating the Age filter (Adult: 19-44 years) reduces this number to 50. Scanning these abstracts reveals eight papers directly related to chains with resistance exercise; however, two of them are survey results, and five are acute kinetic or kinematic characterization studies. Scanning the reference list of the one remaining article yields two other articles not turned up in our original search (Ataee, Koozehchian, Kreider, & Zuo, 2014; Ghigiarelli et al., 2009; McCurdy, Langford, Ernest, Jenkerson, & Doscher, 2009).

Discussion of Results

The results of the three studies reviewed are summarized in table 15.1. Ghigiarelli and colleagues (2009) evaluated the effects of resistance exercise with chains in male college American football players during off-season training (Ghigiarelli et al., 2009). Subjects ($N = 36$) were from one Division 1-AA university and represented the football players with the 36 highest bench press 1RM values relative to body weight on the team determined during winter testing sessions. They were randomly assigned to 7 weeks of (1) traditional resistance training (control), (2) resistance training with chains, or (3) resistance training with elastic bands; all groups trained identically (four or five sessions/week, seven or eight exercises/session, ~85% 1 repetition maximum [1RM]) in an undulating (nonlinear) periodized design. The training program included both strength and power elements. The first two workouts of the week focused on strength development with heavy lower and upper body lifting (four to six repetitions), respectively; the second two workouts of each week were performed explosively with lighter weight (two to four repetitions) for lower body and upper body power development, respectively. Only during the fourth workout of the week (upper body power) was the assigned treatment employed (i.e., control, chains, or elastic bands) and then, during only five or six sets of explosive bench press exercise. The chains attached to the bar during bench press had a total mass of ~40 kg. Outcome measures included muscle strength (5- to 7RM bench press with predicted 1RM value) and muscle power (five-repetition bench press speed test at 50% predicted 1RM). Predicted bench press 1RM increased significantly in all subjects ($P < 0.05$), with no differences between groups. Average power (average of the peak power of all five repetitions of the bench press speed test) was unchanged with training; however, analysis of peak power of the single highest repetition of the set demonstrated a trend for improvement in both the chain and elastic band groups ($P = 0.11$) (Ghigiarelli et al., 2009).

McCurdy and colleagues (2009) examined the effects of resistance training with chain-loaded and plate-loaded exercise in male college baseball players ($N = 27$) with

Table 15.1   Summary of Critically Appraised Studies Related to Chain Training and Strength-Power in Baseball Players

Study	Population	Outcome measures	Findings
Ghigiarelli et al., 2009	36 male college football players	Upper body muscle strength and power (bench press)	Chain training and traditional resistance exercise both increased predicted 1RM and trended toward an increase in peak power ($P = 0.11$); changes with chain training were not different than those elicited with traditional resistance training.
McCurdy et al., 2009	27 male college baseball players	Upper body muscle strength (bench press) and shoulder pain	Chain training and traditional resistance exercise both increased bench press 1RM with no difference between groups. Shoulder pain tended to be less with chain training ($P = 0.07$).
Ataee et al., 2014	8 kung fu athletes and 8 wrestlers	Upper and lower body strength (bench press and squat) and power (med ball throw and vertical jump)	Changes in lower body power and upper body strength and power did not differ between chain training and traditional resistance exercise. Lower body strength was increased more with chain training ($P = 0.04$).

an average of almost 5 years of lifting experience (McCurdy et al., 2009). Subjects were counterbalanced for strength and were assigned to one of two groups: (1) a chain-loaded group that trained with an Olympic bar and attached chains only or (2) a plate-loaded group that trained with an Olympic bar and standard Olympic plates. All subjects trained 2 days per week for 9 weeks during the baseball off-season. The training program incorporated a variety of exercises including bench press; training was traditionally periodized and performed at 60% to 95% 1RM. Outcome measures, which were assessed pre- and posttraining, included bench press 1RM (measured separately in plate-loaded and chain-loaded configurations in both groups) and ratings of shoulder pain and muscle soreness (0-4 scale). Plate-loaded bench press resistance was measured at maximal elbow extension, that is, at its heaviest point. Both groups had similar, significant increases in bench press 1RM tested in the plate-loaded and chain-loaded configurations. Soreness did not differ between groups, but there was a trend ($P = 0.07$) toward less shoulder pain in the chain-loaded group compared to the plate-loaded group (McCurdy et al., 2009).

Ataee and associates (2014) studied the effects of traditional resistance exercise and resistance exercise with chains in 16 trained male athletes (8 kung fu and 8 wrestlers) (Ataee et al., 2014). Subjects were randomly assigned to either a traditional resistance exercise or a resistance exercise with chains group; both groups trained three times per week for 4 weeks. Training consisted of the squat and bench press exercises (three sets × five repetitions at 85% 1RM each session). The chain resistance group lifted with the same 85% 1RM barbell and plate load, but also with chains that added 20% 1RM to the load when the barbell was fully extended and the maximum amount of chain was lifted off of the floor. Outcome measures included upper and lower body strength (bench press and squat 1RM, respectively) and upper and lower body power (medicine ball throw and vertical jump, respectively). Changes in lower body power and upper body strength and power did not differ between the groups; however, lower body strength increased to a greater extent in the chain-trained group compared to traditional training ($P = 0.04$). The authors did note that effect sizes for upper and lower body power were moderate (0.62 and 0.64, respectively), which indicated a positive effect of chain training (and likely an underpowered study) (Ataee et al., 2014).

Conclusion and Strength of Evidence

Generally, these studies show that resistance exercise with chains does not elicit superior adaptations in muscle strength or power in young, trained athletes. However, in no case did chain training result in poorer outcomes than traditional resistance exercise. Although two of the examined studies are randomized controlled trials, the other used counterbalanced assignment, and none of them reported nutritional status, which can have a profound influence on training adaptations. One study showed a trend for less shoulder pain with chain training compared to traditional resistance exercise.

> **Level B evidence supports the use of resistance training with chains as a method to elicit positive adaptations in strength and power that are similar to those elicited by traditional resistance exercise. However, evidence does not suggest that training with chains is superior to traditional resistance exercise. Resistance training with chains may reduce shoulder pain in comparison to traditional resistance exercise.**

Program Recommendations

During the preparatory phase of his training, Rafael could incorporate chain training into his resistance exercise program and carefully monitor changes in strength, power, and shoulder pain. Depending on these objective and subjective findings, chain training may be a useful resistance exercise training variation. He may want to speak with one of the strength and conditioning coaches at his university for guidance and instruction to ensure that he is correctly performing the exercises. Rafael should not replace traditional free weight training with chain training alone, since there is an abundance of literature supporting the efficacy of free weight exercise to improve performance.

CASE STUDY 2: Vibration and Muscle Strength and Power

Charity is a university strength and conditioning coach. Recently, a professor in the school's exercise physiology program acquired several vibration plates for a planned research study. The vibration plates are housed in the athletic training facility and are available to the athletes. Charity heard about training with vibration plates at a national conference she attended recently; the presenter described a variety of upper and lower body exercises that could be performed on the plate and anecdotally claimed that such training results in noticeable performance improvements in athletes participating in a variety of sports. As an evidence-based practitioner, Charity realizes that although the vibration plate presentation was compelling, she needs to search the literature to determine what kind of scientific support exists for using vibration plate training as a modality to improve physiological and athletic performance.

Background

Vibration is thought to enhance acute neuromuscular performance via the stimulation of Ia afferent nerves, which effects a myotatic reflex contraction (Abercromby et al., 2007). Acutely, vibration improves maximal force and power output during concentric contractions; this facilitative effect is greater in elite athletes than in amateurs (Issurin & Tenenbaum, 1999; Liebermann & Issurin, 1997). However, vibration is ineffective when applied acutely to promote recovery or improve subsequent running performance after a strenuous exercise bout in highly fit runners (Edge, Mundel, Weir, & Cochrane, 2009). In competitive athletes, routine vibration training can increase strength, jump performance, and flexibility (Fagnani, Giombini, Di Cesare, Pigozzi, & Di Salvo, 2006), as well as improve proprioception and balance after anterior cruciate ligament reconstruction in comparison to a conventional rehabilitation program (Moezy, Olyaei, Hadian, Razi, & Faghihzadeh, 2008).

Charity currently employs evidence-based strength and conditioning programs with her athletes; these programs consist largely of resistance exercise—both traditional strength and explosive Olympic weightlifting exercises. Thus, to make a valid comparison in her athletic population, Charity should compare the effectiveness of vibration training to that of traditional resistance exercise. Given the limitations of vibration training (cost, availability, spatial restrictions), only if vibration training (either alone or in combination with traditional training) proves superior to traditional training would it be prudent to incorporate it on a large scale. Thus the evidence-based question is as follows.

> **In trained athletes, does vibration training, either alone or as an adjunct to resistance exercise, elicit improvements in muscle strength or power that are superior to those realized with traditional resistance training?**

Search Strategy

Typing "vibration" into PubMed yields a list of MeSH subheadings; we select "vibration training." This returns 1141 studies. Activating the Age filter (Adult: 19-44 years) and Language filter (English) reduces this number to 377. Next, we add "AND athletes" to the search string ("vibration training AND athletes"), which returns 20 articles. Scanning these abstracts reveals 3 papers directly related to the chronic effects of vibration training (Delecluse et al., 2005; Issurin, Liebermann, & Tenenbaum, 1994; Preatoni et al., 2012). Of the 17 articles not included, most are acute studies and several others examine the effects of vibration on postworkout recovery and postsurgical rehabilitation outcomes; another study was excluded because it is not clear what exercise the control group performed.

Discussion of Results

The three articles reviewed are summarized in table 15.2. Preatoni and colleagues (2012) evaluated the effects of vibration training alone and in combination with traditional resistance training in 18 national-level female athletes (12 soccer and 6 softball athletes, with equal proportions in all groups). Subjects were randomized to one of three groups: (1) whole-body vibration training group, (2) traditional strength training group, and (3) combined whole-body vibration and strength training group. All training was performed during the winter preparatory period and in combination with other sport-specific field training such as speed drills, aerobic work, and technical and tactical skill practice. The periodized resistance training program was performed 2 days per week for 8 weeks; for the first week it consisted of six sets × six repetitions of squats performed at body weight with vibration (vibration group), at 60% 1RM (strength training group), or 30% 1RM with vibration (combined whole-body vibration and strength training group). Every 2 weeks, the external load was increased 6% (for the strength training group) and 3% (for the combined whole-body vibration and strength training group); greater intensity for the vibration group was achieved by increasing vibration frequency 5 Hz every 2 weeks (frequency was also increased for the combined group). Outcome measures included isometric strength (leg press); dynamic force, velocity, and power (explosive leg press with loads of 100% to 200% body weight in 20% increments); and power and power endurance (vertical jump and continuous 15-s vertical jumps, respectively). Training increased isometric strength (main effect, $P = 0.02$) with no differences between groups. No changes were observed for any parameter of the explosive leg press test. Performance on both vertical jump tests increased with training (main effect, $P < 0.002$), but there were no differences between groups. Maximum jump height (vertical jump test) and mean jump height and power (continuous 15-s vertical jump test) were increased only in the strength-trained group. The investigators also evaluated the characteristics of the vibration device (i.e., frequency, amplitude, and acceleration) and found variations up to 20% from the selected value; this was particularly true at higher frequencies. On the basis of other published data (Blottner et al., 2006; Mulder et al., 2008, 2007; Rittweger et al., 2006), the authors conclude that vibration exercise can elicit similar or improved outcomes compared to traditional strength training only when similar external loads are used; that is, they attribute the lack of an effect in their study to the lower external loads lifted by the combined vibration + strength training group (Preatoni et al., 2012).

Delecluse and colleagues (2005) examined the additive effect of a whole-body vibration training program over 5 weeks in 20 sprinters. Male and female sprinters (mean 100 m times: female = 12.46 ± 0.59 s, male = 11.45 ± 0.42 s) were randomly assigned to either a vibration or a control group. Both groups maintained their conventional training program, which consisted of intervals (10-60 s), speed training (two or three sessions per week with efforts near race pace), speed drills (two sessions per week), plyometric drills (one session per week), and explosive resistance training (three sessions per week) at 75% to 95% 1RM (three to five sets × two to five repetitions). In addition to their typical training, the vibration group completed three sessions per week of unloaded static and dynamic leg exercise on a vibration platform. The exercises employed were high squat, deep squat, wide stance squat, single-leg squat, lunge, and heel raise. The vibration program was implemented progressively through increases in the duration of vibration time and concomitant decreases in the rest periods; vibration amplitude (displacement) and frequency were also increased over the 5-week program. The study was conducted during the precompetitive phase of training. Outcome measures included strength (isometric and isokinetic knee extensor-flexor), maximal knee extension velocity (at 1%, 20%, 40%, and 60% of maximum isometric force), vertical jump, starting parameters (start time, horizontal start velocity, and horizontal start acceleration), and maximum velocity in a 30 m sprint. There were no changes in either group after the 5-week training program, nor were there any interaction effects or differences between groups posttraining.

Table 15.2   Summary of Critically Appraised Studies Related to Vibration Training and Strength-Power in Trained Athletes

Study	Population	Outcome measures	Findings
Preatoni et al., 2012	18 national-level female soccer and softball players	Isometric strength (leg press), dynamic force, and power and power endurance (explosive leg press, vertical jump, and continuous 15-s vertical jumps)	Vibration only, traditional resistance exercise, and resistance exercise + vibration all increased strength ($P = 0.02$) with no group differences. No changes for dynamic force or leg press power. Only traditional resistance exercise increased vertical jump and continuous vertical jump endurance.
Delecluse et al., 2005	20 male and female sprinters	Strength (isometric and isokinetic knee extensor-flexor), maximal knee extension velocity (at 1%, 20%, 40%, and 60% of maximum isometric force), vertical jump, starting parameters, and maximum velocity in a 30 m sprint	Neither sprint training with explosive resistance exercise nor sprint training with explosive resistance exercise + vibration training elicited changes in any outcome measure.
Issurin et al., 1994	28 male athletes	Strength (bench pull 1RM) and flexibility (two-leg split distance and sit and reach)	Numerically greater improvements in outcome measures for resistance exercise + vibration in comparison to resistance exercise only, but no statistical contrasts to substantiate them; equivocal results.

Issurin, Liebermann, and Tenenbaum (1994) examined the effects of vibration training in 28 young male athletes who regularly participated in a wide cross section of club or varsity sports such as judo, swimming, volleyball, tennis, soccer, track and field, and cycling. Subjects were randomized to one of three groups: (1) upper body strength training with vibration and lower body flexibility training, (2) upper body strength training and lower body flexibility training with vibration, and (3) a calisthenics–basketball game control. The training program, which was conducted three times per week for 3 weeks, consisted of a ~10-min warm-up, a single upper body strength exercise (seated bench pull: six sets × six repetitions at 80% to 100% 1RM performed to failure), and ~20 min of specific static and ballistic stretching of the upper leg musculature. The program was performed by both experimental groups; vibration was added to either the upper or lower body activity according to group assignment. Outcome measures included strength (bench pull 1RM) and flexibility (two-leg split distance and sit and reach distance). Collectively, 1RM strength and both flexibility measures increased with training (main effects); there were also differences between groups for each outcome (group × time interaction effect). Unfortunately, the authors did not provide statistical contrasts between groups (e.g., strength training vs. strength training with vibration) to elucidate the between-group differences; because of the influence of the control group (which changed very little for any measure) on both main and interaction effects, it is difficult to interpret the study findings.

Luo, NcNamara, and Moran (2005) published a review evaluating the effects of vibration training on muscle strength and power, examining the effects of both chronic and acute vibration training. In 2005, only three papers had been published on the chronic adaptations to vibration training; of these, only one study employed trained athletes as subjects and has already been discussed (Issurin et al., 1994).

Conclusion and Strength of Evidence

There has been an explosion of literature regarding whole-body vibration in recent years. Some studies have been well conducted; others are poorly designed or uncontrolled, leading to erroneous or equivocal conclusions. There is still minimal understanding of the appropriate frequency, amplitude, direction, and length and mode of exercise needed for positive adaptations to vibration training. Despite the widespread use of vibration in athletic and private training settings, the evidence at the current time does not suggest that it is a strong tool to improve strength in athletically trained populations, although there is evidence for use in other populations.

> **There is level B evidence to refute the use of vibration training as a stand-alone or adjunct training method to increase muscle strength or power in athletes, as it is not demonstrably superior to traditional resistance exercise.**

Program Recommendations

Although evidence supports vibration as a tool to improve strength and power in untrained populations, the data do not support its use to improve performance in athletes versus traditional strength methods. There is some evidence to support its use as a tool to acutely potentiate a power response; thus if Charity chooses to implement vibration, based on the literature it should be used as a postexcitatory potentiation tool. In general, Charity should continue training her athletes using traditional strength and conditioning programs but continue to watch the literature for emergent studies and protocols.

CASE STUDY 3: Instability Training and Muscle Strength

Kathryn is a personal trainer at a local gym. Although she is new to the profession, she has a personal training certification from a well-renowned national society that requires a bachelor's degree in an exercise science-related field to sit for the exam. As a new trainer, Kathryn looks to the other more experienced trainers for advice and guidance. She has noticed that most spend large amounts of time with their clients performing balance and strengthening exercises using Swiss balls, wobble boards, and other instability equipment. During her years as a college athlete, Kathryn trained mostly with traditional free weight resistance exercises performed on stable surfaces (e.g., squat, deadlift, bench press, power clean), and she is generally unfamiliar with the use of instability equipment in strength training. Most of Kathryn's clients are working-age adults who simply want to be healthy and "functional" in activities of daily living; with these clients, she employs broad fitness programs that develop strength, power, flexibility, balance, and agility. To date, Kathryn has not incorporated instability training into her programs, and she wonders if her clients might do as well or better with training on unstable surfaces as they do with their current traditional training.

Background

The theory behind instability training is that training on an unstable surface should elicit greater muscle recruitment, which will lead to larger gains in muscle strength and balance. Acutely, maximum isotonic strength and muscle activation are reduced during upper body exercise on a Swiss ball (Saeterbakken & Fimland, 2013a), although other studies show no differences (Uribe et al., 2010) or an increase (Marshall & Murphy, 2006) in muscle activation with submaximal loads. For the lower body, maximum strength is similarly reduced during isometric exercise on a BOSU ball, but muscle activation is no different than on a stable surface (Saeterbakken & Fimland, 2013b). Strength training (3 days per week for 12 weeks) on a Swiss ball has been shown to elicit increases in trunk and lower body strength and endurance, flexibility, and dynamic balance in sedentary subjects (Sekendiz, Cug, & Korkusuz, 2010). Instability training may also be an effective treatment modality for individuals with neurologic disorders; a case study in a patient with multiple system atrophy (a condition involving Parkinson-like ataxia, balance, and coordination issues) reported increases in quadriceps muscle mass, strength, balance, functional performance, improved performance in activities of daily living, and decreased motor symptoms and risk of falls after 6 months of such training (Silva-Batista et al., 2014). Instability training is often associated with trunk muscle or "core" strength; however, Keogh and colleagues showed that static measures of core stability are not correlated to shoulder press performance on a Swiss ball, indicating a high degree of specificity in core strength and suggesting that static or single-joint core training may not translate to improved dynamic performance (Keogh, Aickin, & Oldham, 2010). For Kathryn, this is the relevant research question:

> **In untrained or recreationally active adults, does routine instability training elicit improvements in muscle strength or power that are equal or superior to those realized with traditional resistance training?**

Search Strategy

Typing "strength training" into PubMed yields a list of MeSH subheadings, but none are relevant to instability training. Adding "AND instability" to the search string ("strength training AND instability") returns 219 articles; activating the Age filter (Adult: 19+ years) and Language filter (English) reduces this number to 109. Many of the returned articles are orthopedic studies addressing joint instability in athletes. Adding "NOT athletes NOT joint instability" ("strength training AND instability NOT athletes NOT joint instability") yields 39 articles; 3 papers are relevant to the evidence-based question (Cowley, Swensen, & Sforzo, 2007; Kibele & Behm, 2009; Sparkes & Behm, 2010). Another search of "strength training AND Swiss ball" with the same filters activated (Age filter, Adult: 19+ years and Language filter, English) yields 16 articles, none of which are relevant. Again, most of the excluded articles are acute studies that look at electromyographic differences between stable and unstable resistance exercise. Two other articles are found using the "Related Citations in PubMed" feature, one of which is a position stand of the Canadian Society for Exercise Physiology (Behm, Drinkwater, Willardson, & Cowley, 2010; Mate-Munoz, Monroy, Jodra Jimenez, & Garnacho-Castano, 2014).

Discussion of Results

Table 15.3 summarizes the selected articles. Cowley, Swensen, and Sforzo (2007) studied the effects of upper body strength training on a stability ball in a cohort of untrained subjects. Fourteen young women were matched for 1RM chest press strength and assigned to one of two groups: a flat bench group and a stability ball group. Subjects trained seven times over the course of 3 weeks; workouts consisted of three sets × three to five repetitions of chest press at ≥85% 1RM performed on the platform corresponding to group assignment. Outcome measures included upper body strength (chest press 1RM performed on both platforms by both groups), upper body endurance (YMCA bench press test), and abdominal power (front and side abdominal power tests). Pretraining, there were no-between group differences in 1RM chest press strength for either platform. Training elicited significant increases in 1RM strength and endurance on both platforms; there were no differences in response between groups. Similarly, both groups increased performance in the front abdominal power test with no-between group differences; there was no change in side abdominal power in either group. This study showed positive effects of training for both flat bench and stability ball platforms for strength, endurance, and abdominal power in young untrained females (Cowley et al., 2007).

Kibele and Behm (2009) evaluated the effectiveness of a strength training program with lower body instability in a large group of young inexperienced lifters. Forty male and female sport science students were randomly assigned to one of two groups: traditional resistance training and resistance training on an unstable surface. The training program consisted of (1) back squats (5 × 12 repetitions at 75% 1RM for the traditional group and 50% 1RM on a wobble board, DynaDisc, or BOSU ball for the instability group); (2) vertical jump (three × six repetitions on a 30 cm box), and (3) upper body exercises (pulldowns, butterfly, and bench press: 3 × 15 repetitions at 70% 1RM). The instability group also performed four trunk stabilization exercises on a Swiss ball (supine hip extension–knee flexion combination, T bridge fall-off, prone hip and knee flexion combination with both legs, and prone hip and knee flexion combination with single leg). Subjects trained twice per week for 7 weeks. Outcome measures included lower body strength (standing isometric leg extension and squat 1RM), static and dynamic balance, abdominal muscle endurance (sit-ups in 40 s), lower body power (long jump), and functional measures (hopping test, shuttle run, and sprint). Collectively, training improved all outcome measures (main effects) except sprint times and shuttle runs. Also, the instability group improved its performance in sit-ups and right-leg hopping to a greater extent than the traditional training group (group × time interaction effects). In conclusion, this study found no differences in the muscle strength and power adaptations to traditional and lower body instability resistance training programs (Kibele & Behm, 2009).

Sparks and Behm (2010) studied the effects of resistance training with instability in untrained, recreationally active young adult men and women. Subjects were balanced for sex and randomized to one of two groups: (1) a stable group that performed whole-body resistance training primarily using machines and (2) an instability group that performed whole-body resistance training largely using stability balls, inflatable discs, and dumbbells. The unstable resistance training group employed eight upper body and six lower body exercises; the stable resistance training program incorporated eight upper body and seven lower body exercises. Subjects performed 2 × 10 repetitions of each exercise with a load that did not permit completion of an 11th repetition with proper form. Training was performed for 1 h per day, 3 days per week for 8 weeks, with sessions alternating between upper and lower body workouts. Outcome measures

Table 15.3   Summary of Critically Appraised Studies Related to Instability Training and Strength-Power in Untrained or Recreationally Active Individuals

Study	Population	Outcome measures	Findings
Cowley et al., 2007	14 untrained young women	Strength (chest press 1RM performed on stability ball and flat bench by both groups), upper body endurance (YMCA bench press test), and abdominal power	Both training on a flat bench and on a stability ball increased strength (on both platforms), upper body endurance, and abdominal power; no differences between groups.
Kibele and Behm, 2009	40 untrained young men and women	Lower body strength (standing isometric leg extension and squat 1RM), static and dynamic balance, abdominal muscle endurance (sit-ups in 40 s), lower body power (long jump), and functional measures (hopping test, shuttle run, and sprint)	Both traditional resistance and instability training increased lower body strength, static and dynamic balance, abdominal muscle endurance, lower body power, and hopping performance. Sit-ups and hopping performance were increased to a greater extent with instability training.
Sparks and Behm, 2010	18 active young men and women	Upper body strength (isometric chest press on both flat bench and stability ball, bench press 3RM), lower body strength (squat 3RM), balance (30-s wobble board test), lower body power (countermovement jump and drop jump), and a functional test (one-leg overhead medicine ball throw)	Both traditional resistance exercise and resistance exercise training with instability increased all outcome measures with no differences between groups.
Mate-Munoz et al., 2014	36 active young men	Upper and lower body strength (velocity-predicted bench press and squat 1RM), movement velocity (bench press and squat average and peak velocity), power (bench press and squat average and peak power), and lower body power (squat and countermovement jump height)	Both circuit resistance exercise and resistance exercise training with instability increased all outcome measures with no differences between groups.

included upper body strength (isometric chest press on both flat bench and stability ball, bench press 3RM), lower body strength (squat 3RM), balance (30-s wobble board test), lower body power (countermovement jump and drop jump), and a functional test (one-leg overhead medicine ball throw). All outcome variables improved for the groups collectively (main effects), with no differences between training groups. This study demonstrated that similar improvements in upper and lower body strength, lower body power, and functional performance can be realized with traditional and instability resistance training (Sparkes & Behm, 2010).

Mate-Munoz and colleagues studied the effects of a circuit training program in 36 physically active but non–resistance-trained young male sport science students (Mate-Munoz et al., 2014). Subjects were randomized to one of three groups: (1) a traditional group that trained with machines and free weights, (2) an instability group that trained with BOSU and TRX devices, and (3) a nonexercising control group. Training programs for both exercise groups consisted of two eight-exercise workouts (performed in alternating fashion) that included upper and lower body exercises; each exercise was performed for 3 × 15 repetitions. Exercise intensity was based on rating of perceived exertion and was increased by either greater load or increased instability. Rest periods between exercises (30 s to 0 s) and sets (2 min to 1 min) were progressively reduced with each week of training. Subjects trained 3 days per week for 7 weeks. Outcome variables included upper and lower body muscle strength (velocity-predicted bench press and squat 1RM), movement velocity (bench press and squat average and peak velocity), power (bench press and squat average and peak power), and lower body power (squat and countermovement jump height). Training induced significant improvements in all outcome variables for the two exercise groups, but there were no differences in posttraining values between them. In untrained subjects, circuit training with instability evoked increases in strength, power, movement velocity, and jumping ability that were similar to those after a program of traditional stable circuit training (Mate-Munoz et al., 2014).

The position stand of the Canadian Society for Exercise Physiology on the use of instability to train the core musculature was written by Cowley and Behm, who also wrote three of the articles already reviewed (Behm et al., 2010). For the general (nonathletic) population, the authors emphasize the importance of ground-based free weight movements such as the squat, deadlift, and Olympic weightlifting exercises as the basis of a resistance exercise program due to their whole-body muscle activation and similarities to activities of daily living. However, they acknowledge that instability training with reduced loads can elicit beneficial training effects and functional health benefits, although gains in maximum strength and power may be compromised (Behm et al., 2010).

Conclusion and Strength of Evidence

Several studies compare traditional free weight to instability exercises. There does not seem to be a strong difference in early phase training adaptations between the two methodologies. This is a common observation in early-phase exercise interventions; these studies do not address long-term changes in strength. Thus we can conclude the following:

In short-term, early-phase training, there is level B evidence to support the effectiveness of instability training in untrained, nonathletic populations,

as it elicits increases in muscle strength and power similar to those with traditional stable resistance exercise.

Program Recommendations

Kathryn should continue to use ground-based, stable-platform exercises as the foundation and the major component of her training programs. However, she should feel comfortable incorporating minimal amounts of instability training into the programs of clients who enjoy a wide variety of exercises and training platforms or who simply request instability training because they think it is fun and challenging. Kathryn should carefully consider each exercise for safety; standing on unstable surfaces in loaded (or unloaded) conditions can be dangerous, result in falls, and cause injury. Although most of the reviewed research evaluated instability training as an exclusive training platform, Kathryn can probably incorporate instability training as either the primary modality for a cycle of training (e.g., 4 weeks) or assign several rotating exercises with instability during every workout in untrained nonathletic clients. At the conclusion of the 4-week period, she should transition to ground-based stable exercises with a rich body of literature on long-term effectiveness.

CASE STUDY 4: Minimalist or Barefoot Running and Running Economy

Wayne is a 32-year-old amateur runner who enjoys competing in 10Ks and half marathons. He is far from a top-level age grouper, but he typically runs a 42-min 10K and half marathons in just over 1:30. His weekly mileage ranges from 25 to 40 miles (40-64 km) depending on the phase of his race season and what distance he is preparing for. As with most amateur runners, with the exception of some track workouts, Wayne logs most of his mileage on asphalt and concrete in typical off-the-shelf running shoes and runs with a rearfoot strike. He is fortunate to have had no significant chronic injuries. Wayne has a background in exercise science and is well aware of the recent surge in popularity of minimalist shoes; he wonders if he might be able to run faster in them.

Background

Our species ran barefoot for millennia and thus, from an evolutionary perspective, humans are apparently anatomically well adapted for running without shoes. With the relatively recent introduction of hard, paved surfaces, which are in contrast to the grass and dirt paths on which our ancestors ran, humans have begun to run in specially designed, cushioned shoes. One of the primary differences between barefoot and shod running is that in general, barefoot runners use a forefoot strike while shod runners typically employ a rearfoot or heel strike; some data suggest that this biomechanical difference is an important aspect of injury prevention (Altman & Davis, 2012; Shih, Lin, & Shiang, 2013).

Some research indicates that barefoot running is more energy economical than running in standard shoes both on a treadmill and overground (Hanson, Berg, Deka, Meendering, & Ryan, 2011); other work has shown that this effect is due to the increased mass of standard shoes (Divert et al., 2008). By this logic, it would seem that the lighter the shoe, the better the running economy. However, other studies have reported no difference in running economy between barefoot and shod running (Burkett, Kohrt, & Buchbinder, 1985; Squadrone & Gallozzi, 2009), suggesting that other factors besides mass also determine the metabolic cost of running. The evidence-based question is as follows:

In amateur runners, do minimalist shoes improve running economy in comparison to traditional cushioned running shoes?

Search Strategy

Typing "running" into PubMed yields a list of MeSH subheadings, one of which is directly relevant to our evidence-based question: "minimalist running." This returns 27 articles. One study was selected as relevant to our evidence-based question (Sobhani et al., 2014). Most of the excluded articles are acute biomechanical analyses or studies that evaluate injury risk. Two other useful articles were found in the reference lists of the initially selected paper (Perl, Daoud, & Leiberman, 2012; Squadrone & Gallozzi, 2009). Finally, two additional articles that did not meet our inclusion criterion of a comparison to traditional cushioned running shoes were nonetheless deemed relevant and included in the analysis; the first was among the 27 articles from the original PubMed search (Tung, Franz, & Kram, 2014), while the second article was referenced in the first (Franz, Wierzbinski, & Kram, 2012).

Discussion of Results

Table 15.4 summarizes the five articles. Sobhani and coworkers (2014) evaluated the effects of different shoes on the metabolic cost of running in young female endurance runners (self-reported 10 km time = 49.6 min). Eighteen subjects completed running trials in three different conditions using a crossover design: (1) rocker bottom shoes (mass = 429 g/shoe), (2) standard running shoes (mass = 271 g/shoe), and (3) minimalist shoes (mass = 161 g/shoe). All subjects were unfamiliar with running in rocker shoes, running in minimalist shoes, and barefoot running. Subjects ran at 7 km/h for 3 min to familiarize themselves with the shoes and then ran at 9 km/h for 6 min, with the last 2 min used for data analysis. Subjects completed all three running trials in one session; trial order was balanced and randomized between subjects. Outcome measures included relative oxygen consumption, respiratory exchange ratio (RER), heart rate (HR), and ratings of perceived exertion (RPE). Oxygen consumption during rocker shoe running was significantly higher than during standard or minimalist shoe running (by 4.5% and 5.6%, respectively). Running economy was not significantly different between standard and minimalist shoes (1.1% lower in minimalist shoes, $P = 0.19$). No differences were found for RER, HR, or RPE. In the discussion section, the authors note that 100 g of shoe mass has been shown to increase oxygen consumption by 1% and thus the weight differences between shoes may have accounted for a large part of the differences in metabolic cost. However, they note that large interindividual differences suggest that other factors (e.g., biomechanics) likely played a role (Sobhani et al., 2014).

Squadrone and Gallozzi (2009) examined the metabolic cost of running in different footwear in a group of runners who were experienced in barefoot running. After completing a familiarization laboratory session, eight male runners (10 km race time = 40.3 min) were given a pair of minimalist shoes and a pair of standard shoes and told to train in them for the next 10 days in order to minimize the effects of a novel shoe on the study outcomes. During a second laboratory session, subjects completed three running trials in a crossover fashion: (1) barefoot run, (2) minimalist shoe run (mass = 148 g/shoe), and (3) standard shoe run (mass = 341 g/shoe). The three experimental

Table 15.4 Summary of Critically Appraised Studies Related to Minimalist Shoes and Running Economy in Amateur Runners

Study	Population	Outcome measures	Findings
Sobhani et al., 2014	18 female endurance runners	Relative oxygen consumption, RER, HR, and RPE	No differences in any outcomes between standard running shoes and minimalist shoes.
Squadrone & Gallozzi, 2009	8 male runners experienced with barefoot running	Biomechanical variables, oxygen consumption, and HR	Minimalist shoes led to lower oxygen consumption, vertical ground impact forces, and contact time.
Perl et al., 2012	15 runners experienced with barefoot or minimalist running (forefoot strikers)	Oxygen consumption	Minimalist shoes led to lower oxygen consumption with both forefoot and rearfoot striking compared to standard running shoes.
Franz et al., 2012	12 male runners experienced with barefoot running (midfoot strikers)	Oxygen consumption	Oxygen consumption did not differ between running barefoot and in lightweight cushioned shoes.
Tung et al., 2014	12 runners experienced with barefoot running (midfoot strikers)	Oxygen consumption	Oxygen cost of barefoot running on cushioned treadmill slats was lower than on a standard rigid deck. Oxygen consumption did not differ between running barefoot and in lightweight cushioned shoes on a standard rigid deck.

RER = respiratory exchange ratio; HR = heart rate; RPE = rating of perceived exertion.

trial conditions were completed in the same session in a randomized order. Subjects ran at 12 km/h for 6 min in each condition using their preferred foot strike technique (i.e., heel or forefoot); the last 2 min of each condition was used for analysis. Outcome measures included biomechanical variables, oxygen consumption, and heart rate. Results showed that oxygen consumption was significantly lower in minimalist shoes (45 ± 2 mL $\cdot$ kg$^{-1} \cdot$ min^{-1}) compared to standard shoes (46.3 ± 2 mL $\cdot$ kg$^{-1} \cdot$ min^{-1}); however, barefoot running was not different from either shod condition (standard or minimalist shoes).

Biomechanical differences were also found for minimalist shoes; minimalist running resulted in lower vertical impact forces and shorter contact time compared to standard shoe running (Squadrone & Gallozzi, 2009).

Perl, Daoud, and Lieberman (2012) assessed biomechanical and metabolic responses to running in minimalist and standard shoes and also employing a forefoot or rearfoot strike in 15 runners (weekly mileage = 33.4 [54 km]) experienced in barefoot or minimalist running (mean = 2.1 years). Subjects reported a preference for forefoot striking, but were comfortable running with a rearfoot strike. After brief familiarization to each condition, subjects completed four trials in a random order: (1) forefoot strike in minimalist shoes, (2) forefoot strike in standard shoes, (3) rearfoot strike in minimalist shoes, and (4) rearfoot strike in standard shoes. Each trial was run at 10.8 km/h and lasted at least 5 min, with a minimum of 1 min of running at a steady state of oxygen consumption. Stride frequency was determined during familiarization and controlled between trials with a metronome. Importantly, ankle weights were added during the minimalist shoe trials to ensure equivalent distal limb segment mass for all trials. The primary outcome was oxygen cost for each condition. Results showed significantly lower oxygen consumption with minimalist shoes compared to standard shoes with both forefoot striking (–2.4%) and rearfoot striking (–3.3%). In a comparison of the two shoes using the striking style commonly associated with them (minimalist: forefoot strike; standard: rearfoot strike), minimalist shoes also showed significantly better running economy (–2.9%) compared to standard shoes (Perl et al., 2012).

Two other studies that did not meet the inclusion criteria (no comparison to a standard shoe) should nevertheless be considered at this point because of the unique insights they provide. Franz, Wierzbinski, and Kram (2012) examined metabolic responses to running barefoot and in lightweight cushioned shoes (~150 g per pair—the weight of minimalist shoes in the previously reviewed studies, but with cushioning like that in standard shoes). Weight was added in the barefoot running condition to permit comparisons with equal distal limb mass. In men with substantial barefoot running experience and a midfoot striking pattern, oxygen consumption did not differ between running barefoot and in minimalist shoes; when mass was equalized, metabolic cost was 3% to 4% lower in the minimalist shoe.

In a more recent study by Tung and colleagues, the investigators demonstrated, in a cohort of experienced barefoot runners employing a midfoot strike, that adding cushioning to the slats of a treadmill decreased the metabolic cost compared to running barefoot on a rigid treadmill deck (Tung et al., 2014). Further, running barefoot and in light cushioned shoes on the rigid treadmill deck elicited metabolic costs that were not different from each other. The authors concluded that cushioning, both on the treadmill and in shoes, reduces the metabolic cost of running.

Conclusion and Strength of Evidence

This case study is complex, as it involves a host of factors that appear to influence metabolic cost (e.g., experience with barefoot or minimalist running, striking pattern, shoe weight, shoe cushioning). In the only study to evaluate runners unaccustomed to barefoot or minimalist running, oxygen consumption was not different between minimalist and standard shoes (Sobhani & Gallozzi, 2014). Squadrone and Gallozzi evaluated experienced barefoot or minimalist runners but did not control for foot strike patterns (Squadrone & Gallozzi, 2009). Although Perl and colleagues compared standard and minimalist shoes while controlling for forefoot and rearfoot striking, the subjects were experienced barefoot or minimalist runners with a forefoot striking preference. Franz and colleagues and Tung and colleagues controlled for foot strike patterns by recruiting midfoot strikers, and also demonstrated the positive metabolic effects of cushioning, both in an artificial experimental fashion (cushioned treadmill slats) and in a hybrid shoe that incorporated the light weight of the previously studied minimalist shoes but with the cushioning of a standard shoe. Overall, we can conclude the following:

> **For inexperienced barefoot or minimalist runners, there is level B evidence to refute the use of minimalist shoes to improve running economy compared to standard shoes.**

Program Recommendations

Because the evidence does not support improved running economy from minimalist shoe running at this time, there is no compelling case to adopt this footwear. Additionally, Wayne has had no injuries in standard running shoes. He should continue using cushioned running shoes but ensure that his shoes are as lightweight as possible, that is, around 150 g per shoe and not the ~250 to 350 g per shoe seen in much off-the-shelf running footwear.

CONCLUSION

The exercise hardware, equipment, and apparel industries, similar to the supplement industry, are rapidly expanding. New products surface on a routine basis with the aim of helping clients, athletes, or patients to gain a competitive advantage. In some cases, there may be evidence supporting the device or the theoretical principles related to the product. In other cases, no evidence aside from anecdotal testimonials may exist. Although some of these products are relatively cheap, many are quite expensive; thus, the decision to purchase and incorporate must be based on solid information. Even if the product is inexpensive, incorporating it costs time. Because training time is almost always limited, one must ensure that it is not better allocated to a different device or product that could produce better results. Therefore, including or excluding a new product is never a trivial decision and should be approached in a systematic, reproducible manner.

PART IV

Integrating Evidence-Based Practice Into Exercise Science

To use the methodology of evidence-based practice, we must first have evidence on which to base our decisions. As noted earlier, there has been an explosion of information in all disciplines; but new trends, devices, and theories appear in practice at an even more rapid pace, often before research is available to support their efficacy. Therefore, there is a need for partnerships between scientists and exercise practitioners; we must work together to develop studies, share information through informal or formal meetings, and discuss new research findings to ensure that the latest information is being used in both practice and scientific study. On a higher level, there is also a need to educate current and future exercise scientists and practitioners on the importance of the evidence-based approach.

In part IV, we discuss practical methods to implement evidence-based practice in our discipline. At the end of this section, readers should be familiar with the various means of information dissemination and appreciate the importance of collaborative relationships among practitioners and scientists. They should also understand the various mechanisms by which the evidence-based approach can be shared within the exercise community. If the exercise community adopts the evidence-based practice paradigm and works in unison to see it flourish, the result will be better outcomes for our athletes, clients, and patients, as well as elevation of the credibility of our profession.

Chapter 16

DISSEMINATING AND SHARING KNOWLEDGE

Learning Objectives

1. Understand the importance of sharing knowledge.
2. Learn to organize a journal club.
3. Discuss various forums for disseminating evidence.
4. Provide some of the strengths and weaknesses associated with each of these mediums of knowledge dissemination.

We have now covered the six steps of the evidence-based practice (EBP) paradigm and worked through a number of case study examples. Any exercise practitioner who is regularly circulating through the six steps will have well-supported and effective training programs. Remember, whether you are an exercise physiologist in a clinic, a personal trainer, or a strength and conditioning coach in a small rural school, it is a team effort to effectively implement an evidence-based approach in your field. Technology and a plethora of communication systems create ways to collaborate and discuss ideas with colleagues worldwide. This chapter discusses an additional step that, although not formally part of the EBP process, is key to the quality and success of the entire field: the sharing of knowledge. Technology and communication systems make it easier than ever to collaborate with and distribute cutting-edge information to as many practitioners as possible. This is a task that falls partly to the large professional societies but can be fully accomplished only through the unified efforts of all exercise professionals working as a team, both to disseminate knowledge that they have acquired or generated on their own and to gain access to similar knowledge from others.

This sort of concerted effort will both improve individual coaches' practice and elevate the quality of exercise programs delivered field-wide. The world of sport is a highly competitive environment in which individual athletes and teams compete against each other not only on the field, court, or rink, but also apart from game day competition. For instance, elite teams develop their own training programs (strength and conditioning), nutrition regimens, sport-specific strategies, scouting assessments, and so on. All of these are created with the intent of giving the athlete or team every possible advantage on the field.

Despite this culture rooted in competition, the world of sport is in many senses a sisterhood or brotherhood with athletes training hard to best each other while simultaneously sharing bonds of friendship with their fellow competitors—witness the hugs and handshakes that often (although not always) follow a National Basketball Association or National Football League contest—and even training together in the off-season. With its inextricable links to competitive sport, the authors envision a similar future for the field of exercise science with the inclusion of EBP. Some exercise professionals working with elite sports will have a strong reluctance to disseminate and share knowledge for fear of giving away a performance advantage. This is understandable and part of the territory, although with the turnover of athletes, sport coaches, strength and conditioning coaches, and athletic trainers from team to team, it is probably inevitable that ideas and practices will naturally diffuse across a sport. In today's world of mass communication, publication, and other forms of information dissemination, there are few secrets but many fine adjustments that can better practice. However, the potential for improvement is threatened by the myths and misinformation perpetuated by poorly informed practitioners or companies that propagate distorted information for profit. Thus, the free exchange of information, evaluated for its legitimacy and strength, is of central importance to forging the future of EBP in exercise science.

This chapter focuses on various forums in which exercise professionals can share knowledge. Table 16.1 provides an overview. Although these forums are appropriate venues in which to discuss peer-reviewed, publicly available research evidence, the chapter focuses primarily on the exchange of evidence that has not yet been published (e.g., an abstract at a research meeting) or even less formalized evidence such as the results of tracking studies (analogous to case series) conducted with a sport team or other training population. Although the authors endorse and strongly encourage the submission of work for peer review and publication, this early vetting process with its inherent criticism and commentary is a vital part of the development of new ideas and paradigms in the exercise science field. On this note, a reminder of the importance of confirming the evidence through systematic testing (discussed in chapter 10) is in order. Although the primary duty and responsibility of exercise professionals is to train (i.e., develop exercise or rehabilitation programs and implement them in their athletes, clients, or patients), ideally, exercise professionals should also gather data on those they train; this will enable greater and more rapid understanding of the effectiveness of various training methods.

LOCAL DISCUSSION AND NETWORKING

The simplest place to begin sharing knowledge is at the local level; this can be on the other side of town or as close as one's own office. These informal exchanges can spark new ideas and initiate practice improvements that belie the simplicity of their beginnings.

Place of Work

One of the greatest potential barriers to EBP is a lack of managerial support (Kitson, Harvey, & McCormack, 1998). The widespread introduction of evidence-based terminology and methodologies into the workplace may influence the company culture or influence management toward the philosophy, providing a strong partner for future initiatives (Aarons & Sawitzky, 2006; Gerrish & Clayton, 2004; Newhouse, 2007). Although many practicing professionals have positive attitudes toward and perspectives on EBP, they simply do not know where to begin in their quest to become an

Table 16.1   Comparison of Forums for Exercise Science Information

Forum	Primary strength	Primary limitation
Journal clubs (workplace)	Quick transfer of information and freedom to discuss and learn from others' opinions	May discuss only one article; thus definitive conclusions should not be made
Area meetings	Avenue of collaboration with established professionals in your area	May be difficult to coordinate, and individuals may be reluctant to share "trade secrets"
State or regional meetings	More formal venue to transfer information and network with larger body of colleagues	The information may not be as rigorously controlled as that in national or international conferences and journal publications
National or international meetings	Cutting-edge research presented that is not published in journals yet	Typically require travel and funding to pay for related expenses
Journal publications	Peer-reviewed information that can be the highest quality of evidence	Depending on the article, it may not always be practically applicable or relevant
Blogs	Informal forum for communication of ideas; discussion can cross, city, state, and even national lines	Anyone can contribute to the discussion, so the information may be high-quality evidence-based material or anecdotal information espoused by a nonexpert with a computer

evidence-based practitioner (Aarons, 2004; Jette et al., 2003). The place to begin the exchange and dissemination of information is your own workplace with your closest colleagues. Whether you work in a physical therapy clinic, a cardiac rehabilitation facility, a commercial gym, or a university athletic facility, your coworkers are an ideal sounding board for generating and discussing new ideas, critically evaluating sensitive data from your athletes (e.g., tracking studies), and discussing relevant published peer-reviewed research. These exchanges can take place in both casual (e.g., lunchtime or around the office) and more established settings.

Journal Clubs

An example of a slightly more formal arrangement with an emphasis on the exchange of published research is a journal club (Linzer, 1987). Journal clubs are ideal for keeping an entire work group abreast of pertinent new research evidence (Fink, Thompson, & Bonnes, 2005). Journal clubs can meet monthly, biweekly, or even weekly. Typically, a research paper is selected by either the lead of the group or the person who will be presenting it. Usually team members rotate presentation duties from meeting to meeting. The paper should be sent out to the group in advance so that everyone can read and become familiar with it; ideally, all members will come with comments or questions that they can contribute to the group discussion. The presenter should summarize the background or introduction, the purpose and hypotheses, the basic methodology employed, prominent results, and the authors' comments and interpretation from the Discussion.

An important note here: A journal club is a great way to expose participants to research in new areas of practice in which they have little previous knowledge and experience. However, in this case, in keeping with the lessons of this book, great care should be taken not to make broad, sweeping conclusions based on the findings of one research study. Instead, let the study under discussion be the first building block in a series that ultimately results in a comprehensive, sophisticated understanding of the state of knowledge on the given topic. Exercise practitioners should not accept a journal club article on a relatively unfamiliar topic at face value—a little knowledge is a dangerous thing. For this reason, journal clubs are probably most ideally suited to introducing and discussing new evidence in an area that the participants are already familiar with.

As the purpose of a journal club is to share knowledge within an organization and thus to improve its evidence base and practice, the presenter should note the level of evidence that the article represents. Another important point of discussion is how the new work fits in with any previous work: Is the study one of the first of its kind or is it one of many? If one of many, how does it compare to previous research? Does it support or overturn previous findings? If it is at odds with related research, what are the possible explanations for the divergent findings; was the study conducted using a different population? Did it employ a novel technology or outcome measure? The answers to these questions help to place a new study in the proper context among similar work and to further refine knowledge and understanding in a particular area. The following five steps outline the process typical of a journal club meeting:

1. **Appointment of an initial journal club lead**
   - Lead member submits original article to group for review.
   - Entire group reads article before journal club meeting.

2. **Short description of the article presented by appointed lead**
   - Discussion should include the hypothesis, methods, and results.
   - The article should be critically appraised and methodological flaws highlighted.

3. **Answering questions related to the article and how it affects practice**
   - Is the study novel, or is there other supporting research?
   - How does it compare to previous research? What are possible explanations for why it contradicts other research?
   - Did it employ a novel technology or outcome measure?

4. **Open discussion of the article**
   - Discussion is open for subjective comments but still based on evidence.
   - Discussion can be moderated by the lead presenter.

5. **Conclusion of meeting and practical or clinical application**
   - Discuss how this research can be incorporated into current practice.
   - Appoint a lead for the next journal club meeting.

After presenting an objective summary of the article and working through a set of fairly standardized questions, the presenter should guide a more subjective (but still evidence-based) discussion of the paper with a particular eye toward its implications for practice for the group. This is the beauty of a journal club: It facilitates a unified, all-hands approach to the judicious incorporation of new research evidence into a particular gym, rehabilitation center, or athletic facility, for example.

Regular or ad hoc meetings are also encouraged for the discussion and development of in-house projects; these may include research or tracking studies, implications or applications of already collected data, and the preparation of abstracts and journal manuscripts. Tracking studies require time, ethics committee approval, and coordination to plan and execute; the resultant data are wasted unless they are examined, reduced, summarized, and then used to inform training decisions for the athlete from which they were collected and, when appropriate, others as well. Of course, for this to be possible, the results must be disseminated beyond the institution where they were collected; this is discussed more in the following sections.

City- or Area-Wide Groups

Occasional meetings with a larger circle of exercise practitioners that is city- or area-wide will expose all participants to a broader array of viewpoints and practices; such meetings are still informal and self-initiated (as opposed to ones organized by a professional society) but nevertheless facilitate knowledge sharing among the group. In a larger city, an example would be a meeting of the strength and conditioning coaches and athletic trainers from the college or professional teams to exchange information on novel training techniques they are using with their athletes. One benefit of this cross-sectional meeting approach is that practitioners are exposed to training ideas not just from their immediate counterparts, but from practitioners in other specialties. For instance, a football strength and conditioning coach could adapt ideas from a basketball strength and conditioning coach that are ideal for training football players; or either of them could gain new ideas and perspective from a physical therapist working with injured populations. Such thinking outside of one's own box is critical for optimizing outcomes for clients, patients, and athletes. An additional, non-EBP benefit of such meetings is that they foster professional networking and relationships that are always valuable in a small field like exercise science.

As with similar discussions among exercise professionals about data they have collected on their clients and athletes, the information exchanged at area-wide meetings may not be peer-reviewed and may not even qualify as consensus expert opinion. How do such discussions fit into the framework of EBP? First, remember that there are many types of evidence and even if a piece of evidence is very low level, it is still evidence (presuming, of course, that it is evidence and not just conjecture). It may seem relatively easy, if one has the fundamental academic background, to read a randomized controlled trial and apply its findings or at least to gain an understanding of the evidence it provides for (or against) certain training practices. However, the refined information that comes from a randomized controlled trial is likely the product of a number of elements: an idea, experimentation with individual athletes, tracking studies, discussion among a wider circle of professionals, more experimentation and replication with clients and athletes, additional tracking studies, perhaps presentation of data at a scientific meeting (discussed later), and then the conduct of a formal study that ultimately yields level A or B evidence for or against the practice. In other words, a substantial amount of work goes into developing and implementing the studies published in peer-reviewed journals. Thus, although a meeting among local exercise professionals may not in itself generate high-level evidence, the exchange of cutting-edge information fostered in these meetings is indispensable to the generation of new research ideas, often carried out by others in the field; these yield high-level research evidence to be added to the canon of knowledge that is the foundation upon which our field practices.

CONFERENCES

Formal conferences represent the next level of information exchange. Conferences, which are usually organized by state, national, or international societies, allow scientists and practitioners to present their latest (typically unpublished) work. More importantly, whether one is presenting new research or simply attending, conferences facilitate face-to-face discussions and collaborations with colleagues who may live on the other side of the world.

Statewide and Regional Meetings

Beyond meetings and journal clubs with coworkers and local colleagues, the next sphere in which knowledge sharing can occur is the state or regional level. Meetings at this level are almost invariably facilitated by a professional organization or society. Both the American College of Sports Medicine (ACSM) and the National Strength and Conditioning Association (NSCA) have well-established state chapter- or regional-level meetings or both. Similar societies are found around the globe, including Europe (European College of Sport Science, ECSS), Southeast Asia (Asian Council of Exercise and Sport Science, ACESS), Japan (Japan Society of Physical Education, Health and Sport Sciences, JSPEHSS), South America (Brazilian College of Sport Science, CBCE), and Australia (Exercise and Sports Science Australia, ESSA). Attending regional meetings gives practitioners a chance to present their data (the bar for abstract acceptance in terms of the sophistication or completeness of the work is somewhat lower at the regional compared to the national or international level); to see research that others are doing; and perhaps most importantly, to network and speak personally with researchers and other practitioners who have similar ideas or are asking the same questions. In addition to these benefits, because they are more formal than those of locally organized groups, regional meetings offer a good "warm-up" experience for presenting your work at the national or international level.

National and International Meetings

National and international meetings are attended by hundreds or even thousands of scientists and practitioners, with a commensurate number of presentations ranging from posters to plenary lectures. For practitioners wishing to obtain cutting-edge information and stimulate creative thinking, these conferences represent a feast of material with clinical and applied options and a nearly exhaustive set of topics (e.g., injury prevention, biomechanics, environmental and occupational physiology, exercise training, fitness assessment, nutrition, sport psychology). In this rich environment of ideas and eminent experts, it is nevertheless important to remember that information presented at a conference is not fully vetted; thus the evidence is not final, and some studies may be poorly designed or executed. Projects with significant findings and impact should be published later in peer-reviewed journals, after more rigorous evaluation and scrutiny of the study hypothesis, methods, and results has taken place.

Beyond providing the obvious benefit of the chance to see and absorb others' cutting-edge research, national meetings offer the ultimate stage on which to present your work. Most large professional societies have web-based abstract submission sites with very specific guidelines and instructions; it is prudent for prospective submitters to examine these requirements before beginning to write an abstract or even before beginning data collection. Abstracts must describe novel research studies with prior ethics committee approval, submitted months in advance of summertime national meetings, so be sure to familiarize yourself with the relevant deadlines.

Upon submission, abstracts are evaluated by a committee for scientific soundness, novelty, and relevance to the conference. A notable exception to this process is having a fellow of the organizing professional society "sponsor" (i.e., be a coauthor of) your abstract; for most meetings, sponsorship by a fellow results in automatic acceptance of the abstract on the rationale that a fellow sponsors only quality research. Of course, from the perspective of the practitioner(s) conducting the research and preparing the abstract, the input of an experienced scientist is invaluable to the quality of the entire project. These are excellent reasons to have close professional relationships with scientists in your field.

Acceptance of an abstract affords the practitioner an opportunity to present the data in the form of a poster or make an oral presentation. Some conferences allow self-selection of the presentation format; others place accepted abstracts into a category. Either provides an opportunity for networking and developing skills in research dissemination (Moule, Judd, & Girot, 1998). Although developing a poster can be intimidating, following some simple guidelines and recommendations should result in a polished product (Biancuzzo, 1994a, 1994b, 1994c; Vogelsang, 1994). Conference presentations can also be rewarding and enable personal and professional growth (Sherbinski & Stroup, 1992). Most importantly, they are a mechanism for sharing important preliminary research information with colleagues around the world (Halligan, 2008).

FORMAL AND INFORMAL PUBLICATIONS

Beyond the previously discussed forums for face-to-face information exchange, knowledge can also be disseminated through a variety of print and audio media. These information sources span a wide range of evidence quality, from peer-reviewed print journals to blogs and podcasts of individuals who do not even qualify as experts in the field. Nevertheless, modern media, in particular the Internet, affords an unprecedented ability to disseminate high-quality evidence that elevates the practice of exercise science.

Journals

As we have discussed throughout this book, peer-reviewed journal publications of scientific research represent the highest levels of evidence due to their reduced levels of bias (Amonette, English, & Ottenbacher, 2010). The primary audience of this book is practitioners—physical therapists, cardiac rehabilitation personnel, athletic trainers, strength and conditioning coaches—whose focus is not generating and publishing research but implementing others' research evidence. Nevertheless, practitioners can, and in many cases should, publish their work (e.g., case or case series studies). Because conducting research and publishing manuscripts is not a typical task for practitioners, they may benefit from collaborating with a research scientist, perhaps from a nearby university (Knight & Ingersoll, 1996a, 1996b; Tipton, 1991). Beyond the benefits of carrying out a single research project, collaborations between researchers and practitioners are vital for the furtherance of our field and its evidence base: Scientists can conduct more relevant research based on ideas and information provided by practitioners, and conversely, practitioners can gain a better understanding of the requirements of conducting a well-controlled research study (Kraemer et al., 2004; Silvestre et al., 2006; Kraemer et al., 2013; Smith et al., 2014).

Many journals publish exercise science research; the appropriate choice is determined by the subspecialty to which the work is applicable (e.g., strength and conditioning, athletic training, cardiac rehabilitation), the nature of the research itself, or both. *Journal Citation Reports* by Thompson-Reuters classifies 81 peer-reviewed journals as "Sport Sciences" journals; these journals cover a wide range of topics from strength and conditioning (*Journal of Strength and Conditioning Research, Strength and Conditioning Journal*) to clinical orthopedics (*Knee Surgery, Sports Traumatology, Arthroscopy*) to nutrition (*Applied Physiology, Nutrition, and Metabolism*) to biomechanics (*Gait & Posture*) to applied sport physiology (*International Journal of Sport Physiology, Journal of Sports Science, Journal of Human Kinetics, European Journal of Applied Physiology*). Other journals identify less with a specific area of research and more with a particular exercise science subspecialty (e.g., *International Journal of Athletic Therapy and Training*).

Several journals were developed specifically for practical exercise science research; their findings and relevance may be particularly useful for exercise practitioners. A strong example of a journal conceived and developed for applied exercise science research is *Journal of Strength and Conditioning Research*. Out of the mission of the NSCA, "bridging the gap between the laboratory science and the practitioner," arose the idea for a research journal addressing applied sport and exercise science topics. In the early 1980s, a task force led by Dr. William J. Kraemer, a former college football player and strength and conditioning coach finishing his doctoral work in 1984, was composed of distinguished scientists including Dr. Jack Wilmore, Dr. David Costill, and members of the NSCA's research committee and future editorial board, such as Dr. Steven Fleck and Dr. Gary Dudley. Each was a strong advocate and champion of the idea for the new journal. As a result of this working group, starting in 1987, the first applied sport science journal was published four times a year by the NSCA: *Journal of Applied Sport Science*. In 1991, Human Kinetics, attempting to better define the journal scope and mission and to facilitate PubMed indexing, changed the name to *Journal of Strength and Conditioning Research*. Since that time it has developed into one of the premier applied sport and exercise science journals, publishing over 440 papers in each of the last 3 years and having been indexed in PubMed since 2001. *Journal of Strength and Conditioning Research* archives provide a wealth of applied research evidence for exercise practitioners and scientists alike. This publication is a key example of a journal born out of the mission to disseminate practical peer-reviewed evidence to the exercise practitioner.

Although most scientific journals, including *Journal of Strength and Conditioning Research*, publish only original research or reviews of previously published research, others, such as *Strength and Conditioning Journal*, publish training programs that are being used by experienced strength and conditioning coaches in the field. These articles are referenced and peer-reviewed but do not present original data, just a scientific rationale (evidence base) for the program and perhaps some qualitative description of its effectiveness. As this is an academic journal, its editorial staff is composed largely of research scientists, but many of these are current and former athletes, sport coaches, or strength and conditioning coaches. *Strength and Conditioning Journal* is essentially a journal by exercise practitioners, for exercise practitioners. Although the training programs and editorials are based on a mix of expert opinion and other, higher-level evidence (e.g., randomized controlled trials), this journal and others like it provide a vital link in the EBP paradigm.

Blogs and Podcasts

Blogs and podcasts (audio blogs) are another outlet for obtaining and disseminating knowledge (Boulos, Maramba, Inocencio, & Wheeler, 2006; Kraft, 2007; Wilson, Petticrew, & Booth, 2009). Exercise and fitness blogs are ubiquitous, but unfortunately, most are filled with all manner of myth, conjecture, and misinformation based chiefly on the opinions of the blogger. Individuals who have a good physique, have had success in sport, or have a winsome (or prickly) personality are presumed by the public (and themselves) to be "experts"; to capitalize on this status, they endorse or sell (or endorse *and* sell) exercise devices, training programs, and nutritional supplements, making dramatic claims for their efficacy when in fact no evidence (or contrary evidence!) exists. Into this sea of misinformation and fallacy enters the evidence-based practitioner.

Blogs and fitness websites can target an audience of other exercise professionals or the general public. It is important to remember that the information on these sites is not peer-reviewed, although the site may reference quality peer-reviewed information. These sites should never be used as a primary source for decisions but may at times provide helpful information on the state of the industry. If you write a blog or website, it is important to incorporate the evidence-based approach in your material; you can present it either explicitly (better suited for an audience of exercise professionals) or implicitly as part of the background. Regardless of whether scientific research is directly presented (i.e., with open discussion of the findings of specific studies) or more subtly embedded in lay-friendly statements with footnoted references, it is critical that we not fall into the trap of disseminating opinions instead of evidence-based knowledge. As with the various types of knowledge dissemination discussed in the previous sections of this chapter, blogs can explain the implications of published scientific articles, discuss new ideas or concepts and how they might fit into the current evidence base, and detail new training approaches, their evidence-based rationale, and any preliminary results on their effectiveness. Unfortunately, in this modern era of social media, some blogs or posts do not enhance understanding of science but instead intentionally or unintentionally misinterpret scientific information. Thus, the ability to critically evaluate information is more vital than ever.

Although a blog begins as one-way communication from the blogger to an audience, comments from readers on a blog post transform it into a bidirectional conversation that is ideal for the sharing of research and new ideas. As a blogger (or commenter), the key is to influence the underlying (or overriding) conversation toward an evidence-based approach to practice. Exchanging baseless, unfounded claims must be avoided at all costs.

CONCLUSION

The dissemination of evidence-based material, although not a formal step in the EBP process, is an integral cog in the translation of evidence from science to the field or clinic. The avenues for information transfer range from informal blogs and journal clubs to international meetings and journal publications; all are useful but each has both strengths and limitations. Perhaps the predominant proposition of this chapter is that scientists and practitioners alike should participate in this conversation. Whether participation is as simple as a discussion about an interesting article with a colleague at work or a publication in a peer-reviewed journal, it contributes to the knowledge base in the discipline. The exercise community as a whole will benefit from transparent and continuous sharing of ideas as we collectively build the evidence base for our discipline.

Chapter 17

THE FUTURE OF EVIDENCE-BASED PRACTICE IN EXERCISE SCIENCE

Learning Objectives

1. Briefly review the rationale for inclusion of evidence-based practice in exercise science.
2. Discuss some practical methods of using evidence-based practice on a daily basis.
3. Propose future initiatives, including ways of incorporating evidence-based practice into academic programs and major certifications.
4. Discuss partnerships between corporations and academic institutions.
5. Discuss the need for partnerships between practitioners and scientists.

On a daily basis, exercise practitioners face a plethora of decisions related to exercise testing, prescription, devices, supplements, and any number of other choices that could affect the health, fitness, and performance of a client, athlete, or patient. These decisions might be seemingly small; for example, should you perform three, six, or nine sets to better optimize strength gains? In other circumstances, the decisions are substantial: Is interval training a safe and effective form of exercise for a patient with chronic heart failure? In every situation, the goal of an exercise practitioner should be to make the decision with the greatest probability of a positive outcome for a client, athlete, or patient. In order to ensure the best decision, practitioners use evidence to make informed choices. In this book, we have described a six-step process, evidence-based practice (EBP). When implemented in its entirety, this process will help exercise practitioners arrive at informed answers to the questions that arise every day in practice. In this chapter we review the rationale for EBP in exercise science and discuss some potential ways to implement the process in practice and in the discipline of exercise science at large.

ABILITY, JUDGMENT, AND EVIDENCE

Newly degreed physicians take a pledge known as the Hippocratic Oath. A commonly quoted component of the affirmation states, ". . . I will prescribe regimens for the good of my patients according to my ability and my judgment and never do harm to anyone." Two essential components in the oath are the affirmation to prescribe regimens for the good of patients and to never do them harm. The oath also affirms that an individual will perform both of these duties in accordance with her abilities and judgment.

Ability suggests competence to do something, and can refer to either a natural aptitude or an acquired skill. In any practice, there are people who have a natural aptitude for performance. In exercise science, some practitioners have natural talents in coaching, instructing, or communicating with athletes, clients, or patients in a way that helps them quickly and successfully perform an exercise or drill. However, it is also apparent that the ability to coach or prescribe exercise is an acquired proficiency. In other words, people become more capable with practice, and their abilities as exercise practitioners improve with experience. Ability is also affected by access to tools, information, and education. This is seen in medicine when a new diagnostic technique or drug emerges that improves physicians' ability to treat a disease.

Consider the sample graph in figure 17.1. This is a pattern commonly seen in the diagnosis of new cases of a disease (incidence) in a population (Gordis, 2009). It appears that there is a steady rise in the number of new cases of the disease from 1990 to 2015 and from 2020 to 2030. However, the number of new cases suddenly decreases between 2015 and 2020. When we observe a pattern such as this, there are three possible reasons.

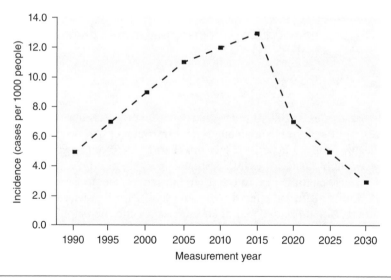

Figure 17.1 Common trend observed in epidemiological medicine when a powerful new preventive treatment option for a disease emerges, such as a vaccine.

(1) The sudden decrease is due to changes in a population's behavior or environment; (2) physicians become more skilled at treatment and prevention of the disease due to experience; or (3) a new technique emerges to prevent the disease (e.g., a vaccination). With a graph like this, the reason for the sudden decline is typically the third reason: A new capability has emerged to prevent the disease within the population. A novel treatment increases the capability of physicians irrespective of any change in their skill level. Thus, ability is not solely talent-based or a skill acquired with practice. In many respects, ability is subject to the technology and treatment options available to practitioners.

Similar to what occurs in medicine, the ability of an exercise practitioner may be significantly affected by technology or information. With no change in skill, ability is increased by the emergence of new programming theories, tools, technologies, and supplements. Modern-day exercise practitioners do not inherently possess greater abilities than a practitioner in the 1920s; however, they have greater capability to help individuals improve their health, fitness, and performance due to technological advances and evidence for new programming techniques, theories, and nutritional supplements, among others.

Judgment is another important word in the Hippocratic Oath. A judgment is an opinion formed after careful consideration and thought. Exercise practitioners develop opinions based on their knowledge (or lack of knowledge) of a particular subject. Knowledge, as discussed previously in this book, is developed from information provided or acquired from academic preparation, scientific and other written information, and experience. The information or evidence acquired through these domains may sway one's opinion or judgment on the efficacy of a treatment or technique in one direction or the other. This book has largely focused on the quality of information or evidence. The continuum of evidence quality in exercise science is vast. It can range from an academic lecture carefully prepared by a discipline expert incorporating a careful synthesis of the literature to a conversation with a person at the gym who gives his opinion on the latest trend. Evidence resulting in the formation of an opinion varies from information derived from many randomized controlled trials with homogeneity of findings, published in well-respected peer-reviewed journals, to information read on a website or in a fitness magazine from the local drug store. There is no shortage of evidence from which people may form their opinions; the key is the quality of the information. If the decision or opinion is based on poor-quality, biased, or unverified evidence, then the judgment may be suboptimal for the client, patient, or athlete. If the opinion is formed based on high-quality information, then it is more likely to result in better outcomes. The pivotal element in a judgment is the quality of the evidence.

REVIEW OF THE RATIONALE FOR EVIDENCE-BASED PRACTICE

Disparate information is common in many fields but perhaps more common in exercise science than in others. Currently, many people are concerned about their health, fitness, and physiques. As people grow older and face more health problems, they often turn to exercise, nutrition, and supplementation as preventive countermeasures. Parents trying to facilitate athletic success for their children are willing to pay for personal training or strength and conditioning (or both) for their children. One characteristic problem in this rush to incorporate exercise is that individuals are often looking for quick, easy solutions. They may be prone to buying into clever marketing based on limited or flawed evidence. Readers of this book have all seen late-night paid programs advertising the most recent piece of exercise equipment that is supposed to accelerate fitness results. Readers have also seen popular magazines with advertisements for newly developed supplements or new delivery mechanisms or formulas for vitamins that are "essential" for health. As discussed previously, the exercise and nutritional industries are poorly regulated, and the statements made by device manufacturers and supplement companies are not verified. The general public may not be aware of this lack of regulation and thus believe the information in the commercial or the magazine.

Although it is easy to point a finger at the public for falling for the latest trend or making decisions based on poor information, it is more appropriate to direct blame back to the exercise profession. Exercise practitioners and scientists are not immune to decisions based on limited or outdated evidence. Experience can be a double-edged sword. Experience is important to developing skills that are essential to the exercise practitioner such as coaching and communicating. With experience comes maturity and an understanding of many logistical and practical constraints associated with exercise prescription. On the other hand, as a professional develops years of experience, she may become accustomed to or vested in certain techniques, programming theories, devices, and so on. The rationale behind their use may be based on academic training, certification material, or mentoring from many years earlier. As new information emerges to support novel techniques, devices, or supplements, there may be a reluctance to change because of success seen with use of the older techniques or products. This resistance to change may stunt the growth of practitioners and prevent their clients, athletes, or patients from access to newer, more useful techniques.

Evidence-based practice is a method that, when implemented, can help to separate the "wheat from the chaff" (English, Amonette, Graham, & Spiering, 2012). It is not always as simple as separating techniques, devices, theories, and supplements that work from those that don't; practical experience can very easily delineate such outcomes. Instead, EBP is in many cases about separating what works from what works best. Evidence-based practice can help exercise practitioners with years of experience to evolve, integrating information learned in practice with the latest scientific findings. The ultimate outcome of this process is a better product for our patients, clients, and athletes.

As a secondary outcome, the incorporation of EBP into mainstream exercise science can only increase the validity of the profession. After the initial evidence-based medicine movement championed by Sackett, Eddy, and their colleagues, there was a robust response from other health professions to follow the lead of medicine (Oxman, Sackett, & Guyatt, 1993; Sackett, 1989). Books and articles are now widely available defining the approach in physical therapy (Bridges, Bierema, & Valentine, 2007), occupational therapy (Bennett & Bennett, 2000), psychology (Spring, 2007), nursing (Burns & Grove, 2010), and social work (Melnyk & Fineout-Overholt, 2011). In truth, there have always been professionals within these disciplines who practiced based on the latest scientific evidence; thus, to say that these disciplines have suddenly become "evidence based" is somewhat erroneous. The change has been the focus on EBP as shown by an abundance of writing, seminars, and academic courses that specifically disseminate information and discuss the EBP process. Although, somewhat ironically, it is nearly impossible to design a study to determine if EBP improves a profession, it is logical to assume that incorporating new, effective methodologies and eliminating older, ineffective techniques can only result in positive outcomes. What is clear is that the increase in dissemination of EBP principles has improved the standing and credibility of these professions among other health professions.

THE FUTURE OF EXERCISE SCIENCE

It is the opinion of these authors that exercise science is at a critical point in its history with respect to its position as an allied health science discipline. With the aging of the population worldwide has come an increased focus on the health care system. In 1900, 1950, and 2000, people over the age of 65 comprised approximately 4%, 8%, and 13% of the total population. It is estimated that by the year 2050, approximately 21% of the total U.S. population will be over the age of 65 (table 17.1, www.census.gov), a trend that is also predicted worldwide. Because people who are older have increased disease and disability, there is a greater need for medical care and prescription medications (Department of Health and Human Services, 2005). Because many of these individuals

Table 17.1 Current (2015) and Projected U.S. Population Growth by Age Group

Age	NUMBERS IN MILLIONS									
	Current	2020	2025	2030	2035	2040	2045	2050	2055	2060
Total	321.4	334.5	347.3	359.4	370.3	380.2	389.4	398.3	407.4	416.8
<14 years	130.5	131.5	133.3	135.8	137.6	138.7	140.1	141.9	144.1	146.3
14-17 years	16.8	16.7	16.7	16.8	17.3	17.6	17.7	17.9	18.1	18.3
18-64 years	199.9	203.9	206.4	209.0	213.7	219.7	225.8	230.4	233.9	236.3
18-24 years	31.2	30.6	30.7	30.8	30.9	31.8	32.4	32.7	32.9	33.3
25-44 years	84.7	89.5	93.4	95.8	97.0	96.9	98.0	99.7	101.2	103.0
45-64 years	84.0	83.9	82.2	82.4	85.8	91.0	95.3	98.1	99.7	100.0
>65 years	47.8	56.4	65.9	74.1	79.2	82.3	84.7	88.0	92.5	98.2
>85 years	6.3	6.7	7.5	9.1	11.9	14.6	17.3	19.0	19.5	19.7
100 years	0.1	0.1	0.1	0.1	0.2	0.2	0.3	0.4	0.5	0.6

Data from U.S. Census Bureau.

use government-provided health insurance, the economic burden is distributed across the entire population. The use of medical care is increasing in conjunction with a rise in the cost of services, further exaggerating the economic impact.

Many in public health have realized the problems that result when a "health care system" is in fact a "sick care system." In other words, individuals do not participate in the health care system until or unless they are sick or disabled. At this point, the cost of treatment is much higher and in the long run, potentially unsustainable due to the ongoing burden of providing such services. There is a renewed interest in preventive medicine and preventive therapeutic interventions. The renewed interest in prevention is motivated by the economic reality that it is less expensive to keep people well than to treat them once they are sick. Many of the common diseases that require long-term treatment are preventable, many through diet and exercise.

The projected increase in the elderly population is due to a number of factors, many related to the quality of medical diagnosis, treatment, and the availability of emergency medical services. A large number of people now survive traumatic health events that in the past would have been fatal. There is also an earlier onset of many hypokinetic diseases (American Diabetes Association, 2000; Sakuragi et al., 2009). The combined effect is a much greater number of people who require long-term medical treatment and monitoring. Exercise science can play a role in reducing this burden of chronic poor health. For example, consider type 2 diabetes, a condition that had until recently been referred to as "adult-onset diabetes." Risk for type 2 diabetes is dramatically increased by physical inactivity and obesity (Colditz, 1999). Compared to physically active, normal-weight controls, individuals who are physically inactive have a 1.3 relative risk for diabetes; people who are obese have an 11.0 relative risk (Colditz, 1999). Thus, diabetes is preventable, in many cases through diet and exercise (Asif, 2014; Halperin & Feig, 2014). Most people with diabetes require lifelong medical treatment and monitoring; many need daily medication. This can have an enormous individual economic impact and may ultimately affect the entire public health system. The earlier the onset of diabetes, the greater the economic costs both for the individual and for society. It is financially advantageous for individuals and insurance companies to delay the onset or prevent the occurrence of diabetes, because exercise treatment and nutritional counseling are much less expensive than lifelong diabetes treatment.

Currently there is a groundswell debate concerning reimbursement for preventive services. The future of reimbursement likely hinges on the evidence supporting the true economic impact of such treatments. For exercise scientists, there is a need to continue to produce and disseminate research that elucidates the impact of exercise and nutrition on health and the prevention of disease and disability. Such support may pave the way for future reimbursement opportunities for exercise and nutrition. The crucial question is, Who should receive the funding for such preventive services? Exercise professionals graduating from academically rigorous degree programs with top-tier certifications would be well-qualified for such an opportunity. However, many other licensed health care professionals would benefit from being identified as qualified to provide these preventive services (e.g., physical therapists, chiropractors, dieticians, nurses). If we as exercise practitioners are to position ourselves for such opportunities, we must begin to differentiate ourselves as a legitimate constituent of the allied health sciences. The integration of EBP as a core value of the profession will improve our credibility among other allied health science professions and better position us for greater opportunities in the health care of tomorrow.

Regardless of the future of health care and reimbursement, there is a need to continue to elevate the reputation of the exercise science profession. Outside of the health care industry, exercise scientists and practitioners have many important vocational roles. Individuals with exercise science degrees work as strength and conditioning coaches, personal trainers, exercise physiologists, and gym managers and owners, among others. In some way, each of these positions functions in a leadership role in relation to the general public; that is, the public trusts that people working in these positions have a level of professional competency. Thus, people place their fitness and health in the hands of exercise practitioners, trusting that they are making the best decisions based on the most up-to-date information—in short, that decisions are evidence-based.

SPREADING THE EVIDENCE-BASED PRACTICE PHILOSOPHY

Several initiatives could facilitate and accelerate the inclusion of EBP as a central concept in exercise science. For this to occur, information regarding the approach must be disseminated to young professionals seeking careers in the discipline, former graduates and current practitioners, and individuals who practice within the field with little or no formal university training in exercise science. Different delivery mechanisms are likely to be required to get the information into the hands of each of these constituents.

Inclusion in Academic Programs

The most obvious delivery mechanism for education regarding the concepts of EBP is the inclusion of such curriculum within academic programs (Amonette, English, & Ottenbacher, 2010). Inclusion of EBP would address the limitation of academic education: Information is obtained and learned at a single point in time despite the reality that information is ever expanding (figure 17.2). The incorporation of EBP could be accomplished through intertwining the philosophical concepts and methodology into

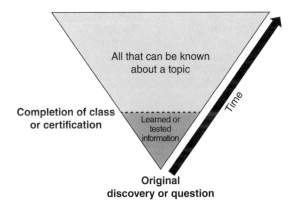

Figure 17.2 Theoretical model depicting information learned or tested in a collegiate or certification course.

the fabric of all courses within academic programs (table 17.2). Case study projects, problem-based learning examples, and directed research assignments could be used in common classes within exercise science programs such as exercise physiology, biomechanics, sports medicine, nutrition, and strength and conditioning. In each of these courses, examples related to the core curriculum could be developed to give students the opportunity to implement the evidence-based process: ask a question, search for evidence, evaluate the evidence, implement the evidence, and confirm the effectiveness

Table 17.2   Strengths and Limitations to Differing Approaches of Integrating EBP Into an Exercise Science Program's Curriculum

	STANDALONE COURSES ON EBP		EBP INTEGRATED INTO CURRICULUM AND PRACTICAL EXPERIENCES	
Domain	**Strength**	**Limitations**	**Strength**	**Limitations**
Information Need	Directed topics provided by instructor.	May lack real-world perspective.	Examples derived from course-specific topics or actual problems that arise in practice.	Takes time away from learning basic material in some courses. Requires substantial work by instructor or mentor.
Teaching and Learning	Strong control over teaching scenarios; can present progressively difficult hypothetical examples.	True integration into discipline-specific scenarios may not be understood by the student.	Problems that arise from actual situations encountered through experience can lead to literature searches and discussions between students and mentors.	No control over the difficulty of the situations presented. Student may encounter advanced problems with less than optimal preparation.
Use of Learning	Instructors can discuss potential strengths and limitations to the approach suggested by student based on literature.	The information may be forgotten because the scenario was not experienced.	Information is applied immediately to practice and student can observe outcome.	Student may not have sufficient practical expertise to successfully apply the evidence. May require additional time from mentor.
Reinforcement of Material	Example can be reinforced by instructor providing additional related examples.	Student will not encounter practical or logistical barriers; thus, they may not truly understand the difficulty of implementation.	Evidence is acquired and applied—actual outcomes are observed. Student is able to work through practical barriers that arise in implementation.	Student may be discouraged by barriers to implementation or potential negative responses of the client, athlete, or patient.

of the evidence. Such inclusion would demonstrate the importance of the process to students and afford many guided practice opportunities over their academic careers.

A second possibility is to teach a course on evidence-based exercise science within academic programs. An informal Internet search quickly demonstrates that this is a common practice in medicine, physical therapy, and nursing programs. These courses typically include a detailed examination of the rationale for EBP and a step-by-step analysis of the associated procedures. A foundational course in which students learn the approach would then allow the principles to be more easily integrated into the curriculum of other courses. Strengths and limitations of stand-alone courses or the intergeneration of EBP into existing courses within a curriculum are provided in table 17.2. As a side note, the accreditation agencies for the academic programs of most other allied health science professions suggest that such courses be offered within a program. The inclusion of EBP in academic programs is also an essential element of "Vision 2020," the professional vision statement of physical therapy (Massey, 2003). Perhaps it is time for exercise science to develop a similar statement to drive a unified vision of the future for the profession.

An important question pertinent to the addition of EBP to the curriculum of exercise science programs is this: At what level should it be included—undergraduate, masters, or doctoral? In short, we recommend all of the above! However, graduate education typically involves significant consumption and generation of research evidence. Accordingly, we would hope that masters and doctoral degreed professionals would intuitively grasp the importance of using research evidence to inform practice; also, graduate students are more likely to work in research or in supervisory roles and less likely to be employed as practitioners. Thus, the inclusion of the basic principles of evidence-based philosophy may be most important at the undergraduate level. Students earning bachelor's degrees in exercise science often serve on the front line of service to the public. They often make the day-to-day decisions on exercise programming, nutritional counseling, and hardware purchase or evaluation. These individuals may benefit most from having the skills to integrate research evidence into practice. In most academic programs, the practical skills of research are not introduced until the graduate level, so having a course or a curriculum that requires the consumption and interpretation of research-based knowledge could be essential to the adoption and propagation of the evidence-based philosophy.

It is important to note that teaching EBP to undergraduate and even graduate students will not immediately create experts in the discipline. It will equip graduates with fundamental skills to practice, learn, and further develop skills in reading and interpreting research. If they are then employed in early career positions on a team with a mentor who is more skilled in practice and research interpretation, they are more likely to successfully mature into strong evidence-based practitioners.

Inclusion in Top-Tier Certifications

The addition of evidence-based courses and curriculum at the university level will ensure that the next generation of practitioners are familiar with the methodology of integrating research-based information with practical and clinical skills. However, there are also individuals working in the field who graduated many years and decades previous. There are other exercise practitioners who never completed an academic degree. Different mechanisms are needed to educate these practitioners. One possible avenue is the inclusion of the EBP process in national and international certification materials.

A large percentage of graduates from exercise science programs seek certifications upon graduation; nearly all exercise practitioners without degrees are certified. Thus, this option may be most beneficial in reaching a large majority of practitioners.

Although the term "evidence-based" appears in some study and marketing materials, to our knowledge the principles of EBP are not currently included in the testable components or key performance indicators (KPI) of any certification exams. In addition, the process is not taught in preparation for the exam. This material could be easily integrated into the current certification material, as curricula are updated on a regular basis. If certification agencies assumed a leadership role in integrating this material, this might serve as motivation for academic programs to follow suit. Some courses in exercise science programs serve, in part, as preparation courses for top-tier certifications. In many of these courses, the exam preparation book may be used as the textbook for the course. To become comprehensive, these courses would need to include material on EBP to effectively prepare students for the exams. Table 17.2 illustrates the theoretical result of completing a class or certification course without the process of EBP included in the curriculum.

Workshops and Seminars for Practitioners

Including evidence-based process material in academic programs and certification materials will reach all students who have yet to graduate from degree programs or to complete certification exams. What can be done about the practitioners and scientists who have already graduated and completed certifications? One possible way to reach these individuals is through workshops and seminars. Because maintenance of most certifications requires the completion of continuing education units (CEUs), offering workshops or courses in the evidence-based process as approved CEU courses would provide opportunities for these individuals to be exposed to the information.

Writing and Publication

To attain the CEUs necessary to maintain a certification, an exercise practitioner has many options for courses, workshops, and so on. Thus, even if EBP courses are offered, not all practitioners will take them. Another way to get information into the hands of practitioners is through publication in peer-reviewed journals, books, and book chapters. The more outlets through which this information is presented, the greater the opportunity for exposure. As more people begin to think, write, and present on the topic, the "sphere of influence" of the philosophy will be increased and propagated to a broader audience within the discipline.

WORKING WITH SCIENTISTS

Exercise practitioners interact with patients, clients, and athletes on a daily basis. Thus, it is critical that practitioners be familiar with the EBP approach. Conversely, because practitioners rely on research-based evidence to inform their decisions, there is a need for scientists to be familiar with the problems encountered in practice. Several initiatives related to science could help improve EBP.

Partnerships Between Practitioners and Scientists

A common theme throughout this book is the need for teams of practitioners, scientists, athletic trainers, physical therapists, and medical doctors with different areas of expertise and knowledge to inform decisions on the treatment of clients, patients, and athletes. The first and perhaps most important of these relationships is the partnership between the scientist and practitioner. Such cooperation gives the scientist a practical view of the current problems and questions in the field. As mentioned in chapter 2, one of the criticisms of EBP is that many scientific studies are not relevant to practice. The countercriticism is that practitioners are not always transparent regarding the programs that they implement in patients, clients, and athletes. Often there is apparent distrust and secrecy. The result is two divorced entities: (a) the strength and conditioning staff with access to elite athletes and a plethora of practical questions to answer and (b) the sport science program staff with state-of-the-art exercise physiology, biomechanics, and motor control laboratories. These are housed in different buildings and have little or no interaction. Both the practitioner and scientist miss an outstanding opportunity to pool resources and work together to generate evidence that can improve practice.

To best facilitate the transfer of legitimate evidence to exercise practitioners, the veil must be removed from both practice and science. Partnerships between the two constituents will help to improve the practicality of research and ultimately the product delivered to the patient, client, or athlete. Often, this relationship would be welcomed by both parties; there is simply a need to reach out and establish a connection. If a practitioner is located at an academic institution, it is reasonable to form a relationship with an academic department, a program, or a professor. Faculty located at institutions with limited practical resources (e.g., no sport teams, gyms, or clinics) may choose to initiate conversations with local organizations serving in exercise practitioner roles. In either case, is important for a dialogue to be opened. The partnership may lead to novel research ideas for the scientist, or perhaps the perspective provided by the scientist may influence or change practice. It may also lead to collaborative research initiatives or independent testing by the science group to confirm the efficacy of the evidence (fifth step in the EBP process).

Partnerships Between Scientists and Product Developers

As discussed previously, the lack of scientific evidence supporting many of the products routinely used in exercise practice is problematic. When searching for evidence, exercise practitioners must often use context or analogue generalizations because there is no research directly evaluating a product, device, or supplement. Thus, an important opportunity for exercise scientists to pursue is partnerships with industry product developers. Because the market changes rapidly, it is not realistic to believe that every product that appears on the shelves or in industry catalogues has strong scientific support before it becomes available. However, legitimate product manufacturers and supplement developers have interests in studying and evaluating their products to ensure that they are safe and effective. Indeed many fund independent laboratory testing to optimize their products to ensure that customers will see solid results and recommend the product to others. Product developers may not be well-networked with the exercise science community and thus may not know whom to contact for unbiased scientific research on their products. It is important for the scientific community to reach out to industry and offer services to help validate the rapidly growing exercise product industry. In many cases, the manufacturers are willing to offer financial support to secure independent testing of their products.

BECOMING AN EVIDENCE-BASED PRACTITIONER

As with any new initiative, the most difficult part of becoming an evidence-based exercise practitioner is the initial steps. It is difficult to approach practice in a new way, to learn how to integrate a new process into an already busy schedule. Also it may be difficult to decide which questions are most important and should be addressed first. McCluskey and colleagues have suggested that there are three levels of evidence-based practitioners: novices, apprentices, and competent nonexperts (Law & MacDermid, 2008; McCluskey, 2004).

Novices in the EBP method tend to engage in the process very infrequently (Law & MacDermid, 2008). Because they are just beginning, their knowledge and skills are somewhat limited. Thus, answering simple questions may require much time, creating a somewhat pessimistic attitude about the process. For novices the key is to battle through the growing pains, develop strong mentoring relationships, and realize that as with any skill, it takes time to master the skills of an evidence-based practitioner.

Competent nonexperts engage frequently in the evidence-based process (Law & MacDermid, 2008). When questions arise that influence their daily practice, the first place they go for answers is the scientific literature. Critical appraisal is easier for these individuals, and the use of peer-reviewed evidence is woven into the fabric of their practice. They have extensive skills and are very optimistic about using the process since it has become easier and they have seen positive results in their patients, clients, and athletes.

Apprentices, obviously, lie somewhere in the middle (Law & MacDermid, 2008). The good news is that with a moderate commitment, the amount of time necessary to move from novice to apprentice is small and the method becomes increasingly easier with practice. It is also reinforcing to see the positive effects of evaluating practice using the evidence. In the next sections we suggest a few considerations as you begin your journey as an evidence-based practitioner.

Develop Skills

Perhaps the most important aspect of becoming a competent evidence-based practitioner is developing your abilities. Specific skills are associated with every element of the process. Developing questions, searching and appraising evidence, incorporating evidence into practice, and confirming the evidence all become easier with practice. As noted earlier in the chapter, there is a need within the exercise science community for workshops and seminars regarding EBP techniques; seminars would be particularly helpful for the novice evidence-based practitioner. It is beneficial to build relationships with other individuals who are just beginning while simultaneously being mentored by an apprentice or a competent nonexpert. A small amount of guidance can go a long way toward improving and smoothing the learning curve. But workshops and mentorship are no substitute for practice. The best way to improve your skills as an evidence-based practitioner is to practice. Ask lots of questions, search for evidence on a daily basis, critically appraise the evidence, and search diligently for ways to improve your practice. There is no substitute for experience: Begin building your experience base in the process now.

Stay Focused

Another important tip is to stay focused and remember the rationale for the EBP approach (Law & MacDermid, 2008). As you begin your journey as an evidence-based practitioner, you will face barriers. These might be opposition from coworkers who would prefer to "do it the way it's always been done," time barriers as professional commitments increase, and problems with access to databases or journals, among others. All of these obstacles can be discouraging and cause young practitioners to lose focus on finding, evaluating, and incorporating evidence. When barriers arise, it is important to be grounded in the purpose of the process and to remain confident that the ultimate product is a better outcome for your patients, clients, and athletes. If you are convinced of this and want to be a top-level practitioner, you must remain focused.

Commit Time

Time is the most commonly cited barrier to being an evidenced-based practitioner. It is a reality in any exercise vocation that most of the time in a day is already committed to other responsibilities; it may be difficult to find time to allocate to asking questions and searching for literature. However, people who are convinced of the beneficial outcomes of EBP will make the time. At first it may be advantageous to simply set aside an hour per week. Some groups have used weekly journal clubs as a way to start the process and collectively involve coworkers. Perhaps 15 min at lunch for asking questions and searching for and evaluating evidence is all the time you can commit. Start with the time you have, but make sure that it is scheduled time not interrupted by other responsibilities. The 15 min per day will become more productive as your skills increase and you become creative in finding ways to commit more time to the process.

Identify the Most Important Questions

When time is limited, determine which questions are most important to your practice. It could be that initially it is most important to evaluate your current procedures and protocols. This can be an enlightening experience and can motivate further searching to ensure that you are implementing the best protocols for your patients, athletes, and clients. Another possibility is to begin by evaluating a new device, supplement, or protocol that is trending within the industry or of interest to your practice. Such questions can arise from conversations with coworkers, managers, or clients. Finding evidence and providing answers for management may be a way to start the process.

Seek Professional Support

Along with time, professional support for EBP methodology can be one of the most challenging barriers to starting. A manager most likely will not approach an employee unsolicited and offer an hour off per day to search for and read research literature. When approaching management to seek support, it is vital to develop a strong rationale—to have evidence. The evidence you use to seek managerial support may be an analysis of how the process will save money, improve the efficacy of equipment purchases, or improve exercise outcomes. All of these may motivate management to provide dedicated time in your daily or weekly schedule. Although you probably feel strongly that the process is beneficial, do not be surprised or discouraged if your request is met with reluctance. Commit any time you can, and as you produce evidence of the efficacy of EBP, managerial support may increase.

Seek support from your coworkers. It is easier to stay encouraged if your colleagues are also engaged in the process and if you can discuss ideas and findings with them. An in-service presentation to colleagues that discusses EBP could win the support of a small group. With time and commitment, the small group may turn into a larger supporting group that eventually includes management.

Speak the Language

"How strong is the evidence supporting the use of this protocol?" "There is level B evidence to support the use of this protocol." Such questions or statements may have been foreign to you before reading this book, but they are part of the language of the evidence-based practitioner. As discussed in this book and elsewhere, the term "evidence-based" is often used inconsistently with its actual meaning (English et al., 2012). When people proclaim that something is "evidence-based," ask them, "What level of evidence?" or "How strong is the evidence?" Speaking the language may provide opportunities to discuss the true meaning of the term and to explain how the process can be integrated into practice.

Stay Positive and Committed

It is important to stay positive about EBP and to believe in the final outcome. When an individual runs, bikes, swims, or lifts weights, results are not observed immediately. In fact, when a novice engages in exercise, what is immediately realized is discomfort, fatigue, and other negative consequences. From an initial bout of exercise, you do not immediately have more energy, improve your physical function, strength or power, or increase the size of your muscles. You will never actually "see" an increase in your cardiac stroke volume or amino acids incorporated into your muscle. However, through longitudinal observation and research evidence, knowledgeable exercisers know that if they commit to exercise training, the temporary discomfort from an acute exercise bout will result in positive physiological and functional changes. If the exerciser quits at the first sign of discomfort, he will never realize the positive results from exercise.

The same is true of EBP. The immediate result of a decision to commit to becoming an evidence-based exercise practitioner is more work. The practitioner will commit more time to finding information, will put more thought into decisions for programming, will carefully consider whether to use a device or a supplement, will need to evaluate and implement long-term testing strategies, will struggle to assemble a team of discipline experts, and will likely face barriers at every step. As the practitioner endures the initial discomfort of change, there may be no immediate improvements. However, the authors of this book and practitioners who have implemented EBP in exercise and other disciplines realize that the evidence-based philosophy leads to better results. We encourage you to endure the temporary discomfort of change to reap the long-term benefits of being a better exercise professional.

CONCLUSION

The defendant nervously waits in the courtroom for the jury to return. After a long, thorough trial with many expert witnesses from both sides, the jury deliberates to determine the fate of the defendant. Ultimately, the decision will be based on the evidence presented throughout the trial. Undoubtedly, the decision to convict or exonerate will be based on the strength of the case presented by the legal teams prosecuting and representing the defendant.

In this book and in previous writings, the authors have attempted to present a strong case for the inclusion of EBP in the discipline of exercise science. In many ways, the EBP approach is already on trial in medicine and other allied health sciences. Since the late 1980s, experts for and against have presented their cases. Those who fight for EBP argue that there must be a systematic methodology to evaluate and incorporate the newest, latest treatments into the plans of their clients and patients. This is especially true today, with the rapid expansion and propagation of scientific information. A refusal to implement EBP will result in practice based on outdated and perhaps invalid information. The experts who argue against EBP suggest that science is biased, limited, and impractical with regard to their professional responsibilities. After a foundation of formal education, they contend that with experience comes knowledge and that ultimately this experience is what leads to best practices for clients and patients. There are intelligent people on both sides of the argument—and each exercise practitioner is ultimately the jury that must make the decision.

Most important to the argument, for or against, is that exercise practitioners must constantly search for the best and safest options for their athletes, clients, and patients. Although the EBP approach requires a greater commitment on the part of the practitioner and more time to develop exercise prescriptions, assess treatments for clients with special needs, and evaluate exercise hardware and nutritional supplements, we firmly believe that the time is well spent and that the inclusion of EBP will benefit our clients, patients, and athletes while simultaneously elevating our profession to a new level of prestige.

References

Part I

DeLorme, T.L. (1946). Heavy resistance exercises. *Arch Phys Med Rehabil*, 27, 607–630.

Todd, J.S., Shurley, J.P., & Todd, T.C. (2012). Thomas L. DeLorme and the science of progressive resistance exercise. *J Strength Cond Res*, 26 (11), 2913-2923.

Chapter 1

Aibiki, M., Maekawa, S., Ogura, S., Kinoshita, Y., Kawai, N., & Yokono, S. (1999). Effect of moderate hypothermia on systemic and internal jugular plasma IL-6 levels after traumatic brain injury in humans. *J Neurotrauma*, 16(3), 225-232.

Amonette, W.E., English, K.L., & Ottenbacher, K.J. (2010). Nullius in verba: a call for the incorporation of evidence-based practice into the discipline of exercise science. *Sports Med*, 40(6), 449-457.

Amonette, W.E., English, K.L., Spiering, B.A., & Kraemer, W.J. (2013). Evidence-based practice in strength and conditioning. In T.J. Chandler & L.E. Brown (Eds.), *Conditioning for Strength and Human Performance* (pp. 285-303). Philadelphia: Lippincott Williams & Wilkins.

Bayne-Jones, S., Burdette, W.J., Cochran, W.G., Farber, E., Fieser, L.F., Furth, J., . . . Seevers, M.H. (1964). *Smoking and Health: Report of the Advisory Committee to the Surgeon General of the Public Health Service*. Washington, DC: U.S. Government Printing Office.

Cappuccino, A., Bisson, L.J., Carpenter, B., Marzo, J., Dietrich, W.D. 3rd, & Cappuccino, H. (2010). The use of systemic hypothermia for the treatment of an acute cervical spinal cord injury in a professional football player. *Spine*, 35(2), E57-E62.

Carchidi, S. (2008). *Standing Tall: The Kevin Everett Story*. Chicago: Random House.

Davidoff, F., Haynes, B., Sackett, D., & Smith, R. (1995). Evidence based medicine. *BMJ*, 310(6987), 1085-1086.

Dixon, S.R., Whitbourn, R.J., Dae, M.W., Grube, E., Sherman, W., Schaer, G.L., . . . O'Neill, W.W. (2002). Induction of mild systemic hypothermia with endovascular cooling during primary percutaneous coronary intervention for acute myocardial infarction. *J Am Coll Cardiol*, 40(11), 1928-1934.

Eddy, D.M. (1990). Practice policies: where do they come from? *JAMA*, 263(9), 1265-1275.

Ellis, J., & Mulligan, I. (1995). Inpatient general medicine is evidence based. *Lancet*, 346(8972), 407.

Ervin, N.E. (2002). Evidence-based nursing practice: are we there yet? *J N Y State Nurses Assoc*, 33(2), 11-16.

Kraemer, W.J., Dunn-Lewis, C., Comstock, B.A., Thomas, G.A., Clark, J.E., & Nindl, B.C. (2010). Growth hormone, exercise, and athletic performance: a continued evolution of complexity. *Curr Sports Med Rep*, 9(4), 242-252.

Lavin, M.A., Krieger, M.M., Meyer, G.A., Spasser, M.A., Cvitan, T., Reese, C.G., . . . McNary, P. (2005). Development and evaluation of evidence-based nursing (EBN) filters and related databases. *J Med Libr Assoc*, 93(1), 104-115.

Law, M., & MacDermid, J. (Eds.) (2008). *Evidence-Based Rehabilitation: A Guide to Practice* (2nd ed.). Thorofare, NJ: Slack.

Maher, C.G., Sherrington, C., Elkins, M., Herbert, R.D., & Moseley, A.M. (2004). Challenges for evidence-based physical therapy: accessing and interpreting high-quality evidence on therapy. *Phys Ther*, 84(7), 644-654.

Mummaneni, P.V. (2010). Use of systemic hypothermia for patients with spinal cord injury. *Neurosurgery*, 66(6), E1217.

Ottenbacher, K.J., Tickle-Degnen, L., & Hasselkus, B.R. (2002). Therapists awake! The challenge of evidence-based occupational therapy. *Am J Occup Ther,* 56(3), 247-249.

Oxman, A.D., Sackett, D.L., & Guyatt, G.H. (1993). Users' guides to the medical literature. I. How to get started. *JAMA,* 270(17), 2093-2095.

Richards, D., & Lawrence, A. (1995). Evidence based dentistry. *Br Dent J,* 179(7), 270-273.

Sackett, D.L. (1989a). Inference and decision at the bedside. *J Clin Epidemiol,* 42(4), 309-316.

Sackett, D.L. (1989b). Rules of evidence and clinical recommendations on the use of antithrombotic agents. *Chest,* 95(2 Suppl), 2S-4S.

Sackett, D.L., Richardson, W.S., Rosenberg, W.M., & Haynes, R.B. (1997). *Evidence-Based Medicine: How to Practice and Teach EBM.* New York: Pearson Professional Limited.

Sackett, D.L., & Rosenberg, W.M. (1995). The need for evidence-based medicine. *J Roy Soc Med,* 88(11), 620-624.

Sackett, D.L., Rosenberg, W.M., Gray, J.A., Haynes, R.B., & Richardson, W.S. (1996). Evidence based medicine: what it is and what it isn't. *BMJ,* 312(7023), 71-72.

Sackett, D.L., Straus, S.E., Richardson, S.W., Rosenberg, W., & Haynes, R.B. (2000). *Evidence-Based Medicine: How to Practice and Teach EBM* (2nd ed.). Edinburgh: Hancourt.

Satterfield, J.M., Spring, B., Brownson, R.C., Mullen, E.J., Newhouse, R.P., Walker, B.B., & Whitlock, E.P. (2009). Toward a transdisciplinary model of evidence-based practice. *Milbank Q,* 87(2), 368-390.

Snegireff, L.S., & Lombard, O.M. (1959). Smoking Habits of Massachusetts Physicians: Five-Year Follow-Up Study (1954–1959). *New Eng J Med,* 261(12), 603-604.

von Zweck, C. (1999). The promotion of evidence-based occupational therapy practice in Canada. *Can J Occup Ther,* 66(5), 208-213.

Westergren, H., Farooque, M., Olsson, Y., & Holtz, A. (2000). Motor function changes in the rat following severe spinal cord injury. Does treatment with moderate systemic hypothermia improve functional outcome? *Acta Neurochir (Wien),* 142(5), 567-573.

Weston, K.S., Wisloff, U., & Coombes, J.S. (2014). High-intensity interval training in patients with lifestyle-induced cardiometabolic disease: a systematic review and meta-analysis. *Br J Sports Med,* 48(16), 1227-1234.

Chapter 2

American College of Sports Medicine. (2009). American College of Sports Medicine position stand. Progression models in resistance training for healthy adults. *Med Sci Sports Exerc,* 41(3), 687-708.

Amonette, W.E., English, K.L., & Ottenbacher, K.J. (2010). Nullius in verba: a call for the incorporation of evidence-based practice into the discipline of exercise science. *Sports Med,* 40(6), 449-457.

Bruyere, O., Varela, A.R., Adami, S., Detilleux, J., Rabenda, V., Hiligsmann, M., & Reginster, J.Y. (2009). Loss of hip bone mineral density over time is associated with spine and hip fracture incidence in osteoporotic postmenopausal women. *Eur J Epidemiol,* 24(11), 707-712.

Candow, D.G., Chilibeck, P.D., Abeysekara, S., & Zello, G.A. (2011). Short-term heavy resistance training eliminates age-related deficits in muscle mass and strength in healthy older males. *J Strength Cond Res,* 25(2), 326-333.

Cohen, M.A., Stavri, Z.P., & Hersh, W.R. (2004). A categorization and analysis of the criticisms of evidence-based medicine. *Int J Med Inform,* 73(1), 35-43.

Cook, M. (2004). Evidence-based medicine and experience-based practice—clash or consensus? *Med Law,* 23(4), 735-743.

Crewther, B. T., Kilduff, L. P., Cook, C. J., Middleton, M. K., Bunce, P. J., & Yang, G. Z. (2011). The acute potentiating effects of back squats on athlete performance. *J Strength Cond Res,* 25(12), 3319-3325.

Cussler, E.C., Lohman, T.G., Going, S.B., Houtkooper, L.B., Metcalfe, L.L., Flint-Wagner, H.G., . . . Teixeira, P.J. (2003). Weight lifted in strength training predicts bone change in postmenopausal women. *Med Sci Sports Exerc,* 35(1), 10-17.

Daly, P.A., Krieger, D.R., Dulloo, A.G., Young, J.B., & Landsberg, L. (1993). Ephedrine, caffeine and aspirin: safety and efficacy for treatment of human obesity. *Int J Obes Relat Metab Disord,* 17 Suppl 1, S73-S78.

Davidoff, F., Case, K., & Fried, P.W. (1995). Evidence-based medicine: why all the fuss? *Ann Int Med*, 122(9), 727.

Dickerman, R.D., Pertusi, R., & Smith, G.H. (2000). The upper range of lumbar spine bone mineral density? An examination of the current world record holder in the squat lift. *Int J Sports Med*, 21(7), 469-470.

Dobek, J., Winters-Stone, K.M., Bennett, J.A., & Nail, L. (2014). Musculoskeletal changes after 1 year of exercise in older breast cancer survivors. *J Cancer Surviv*, 8(2), 304-311.

English, K.L., Amonette, W.E., Graham, M., & Spiering, B.A. (2012). What is "evidence-based" strength and conditioning? *Strength Cond J*, 34(3), 19-24.

English, K.L., Loehr, J.A., Lee, S.M., & Smith, S.M. (2014). Early-phase musculo-skeletal adaptations to different levels of eccentric resistance after 8 weeks of lower body training. *Eur J Appl Physiol*, 114(11), 2263-2280.

Fletcher, R.H., & Fletcher, S.W. (1997). Evidence-based approach to the medical literature. *J Gen Int Med*, 12 Suppl 2, S5-S14.

Gray, M., Di Brezzo, R., & Fort, I.L. (2013). The effects of power and strength training on bone mineral density in premenopausal women. *J Sports Med Phys Fitness*, 53(4), 428-436.

Gurley, B.J., Steelman, S.C., & Thomas, S.L. (2014). Multi-ingredient, caffeine-containing dietary supplements: history, safety, and efficacy. *Clin Ther*, 37(2), 275-301.

Guyatt, G.H., Haynes, R.B., Jaeschke, R.Z., Cook, D.J., Green, L., Naylor, C.D., . . . Richardson, W.S. (2000). Users' guides to the medical literature: XXV. Evidence-based medicine: principles for applying the Users' Guides to patient care. Evidence-Based Medicine Working Group. *JAMA*, 284(10), 1290-1296.

Guyatt, G.H., Sackett, D.L., & Cook, D.J. (1993). Users' guides to the medical literature. II. How to use an article about therapy or prevention. A. Are the results of the study valid? Evidence-Based Medicine Working Group. *JAMA*, 270(21), 2598-2601.

Guyatt, G.H., Sackett, D.L., & Cook, D.J. (1994). Users' guides to the medical literature. II. How to use an article about therapy or prevention. B. What were the results and will they help me in caring for my patients? Evidence-Based Medicine Working Group. *JAMA*, 271(1), 59-63.

Guyatt, G.H., Sackett, D.L., Sinclair, J.C., Hayward, R., Cook, D.J., & Cook, R.J. (1995). Users' guides to the medical literature. IX. A method for grading health care recommendations. Evidence-Based Medicine Working Group. *JAMA*, 274(22), 1800-1804.

Hakkinen, K., & Hakkinen, A. (1995). Neuromuscular adaptations during intensive strength training in middle-aged and elderly males and females. *Electromyogr Clin Neurophysiol*, 35(3), 137-147.

Hedlund, L.R., & Gallagher, J.C. (1989). Increased incidence of hip fracture in osteoporotic women treated with sodium fluoride. *J Bone Miner Res*, 4(2), 223-225.

Humphries, B., Newton, R.U., Bronks, R., Marshall, S., McBride, J., Triplett-McBride, T., Humphries, N. (2000). Effect of exercise intensity on bone density, strength, and calcium turnover in older women. *Med Sci Sports Exerc*, 32(6), 1043-1050.

Jensen, R.L., & Ebben, W.P. (2003). Kinetic analysis of complex training rest interval effect on vertical jump performance. *J Strength Cond Res*, 17(2), 345-349.

Judge, J.O., Kleppinger, A., Kenny, A., Smith, J.A., Biskup, B., & Marcella, G. (2005). Home-based resistance training improves femoral bone mineral density in women on hormone therapy. *Osteoporos Int*, 16(9), 1096-1108.

Koukoura, O., & Hajiioannou, I. (2014). Exception, evidence, experience-based medicine: the evolution of medical practice and the Greek paradox. *Med Teach*, 36(8), 730-731.

Kraemer, W.J., Adams, K., Cafarelli, E., Dudley, G.A., Dooly, C., Feigenbaum, M.S., . . . American College of Sports Medicine. (2002). American College of Sports Medicine position stand. Progression models in resistance training for healthy adults. *Med Sci Sports Exerc*, 34(2), 364-380.

Kraemer, W.J., Fleck, S.J., & Deschenes, M.R. (2016). *Exercise Physiology: Integrating Theory and Applications* (2nd ed.). Baltimore: Lippincott Williams & Wilkins.

Kraemer, W.J., & Ratamess, N.A. (2004). Fundamentals of resistance training: progression and exercise prescription. *Med Sci Sports Exerc,* 36(4), 674-688.

Krieger, D., Newman, M.A., Parse, R.R., & Phillips, J.R. (1994). Current issues of science-based practice. *NLN Publ,* 15-2610, 37-59.

Law, M., & MacDermid, J. (Eds.) (2008). *Evidence-Based Rehabilitation: A Guide to Practice* (2nd ed.). Thorofare, NJ: Slack.

Macaluso, A., & De Vito, G. (2004). Muscle strength, power and adaptations to resistance training in older people. *Eur J Appl Physiol,* 91(4), 450-472.

McCann, M.R., & Flanagan, S.P. (2010). The effects of exercise selection and rest interval on postactivation potentiation of vertical jump performance. *J Strength Cond Res,* 24(5), 1285-1291.

Pereira, A., Izquierdo, M., Silva, A.J., Costa, A.M., Bastos, E., Gonzalez-Badillo, J.J., & Marques, M.C. (2012). Effects of high-speed power training on functional capacity and muscle performance in older women. *Exp Gerontol,* 47(3), 250-255.

Phillips, B. (1997). *Sports Supplement Review* (1st ed.). Golden, CO: Mile High.

Ross, P.D., Genant, H.K., Davis, J.W., Miller, P.D., & Wasnich, R.D. (1993). Predicting vertebral fracture incidence from prevalent fractures and bone density among non-black, osteoporotic women. *Osteoporos Int,* 3(3), 120-126.

Sackett, D.L., Rosenberg, W.M., Gray, J.A., Haynes, R.B., & Richardson, W.S. (1996). Evidence based medicine: what it is and what it isn't. *BMJ,* 312(7023), 71-72.

Sackett, D.L., Straus, S.E., Richardson, S.W., Rosenberg, W., & Haynes, R.B. (2000). *Evidence-Based Medicine: How to Practice and Teach EBM* (2nd ed.). Edinburgh: Hancourt.

Schneider, S.M., Amonette, W.E., Blazine, K., Bentley, J., Lee, S.M.C., Loehr, J.A., . . . Smith, S.M. (2003). Training with the International Space Station interim resistive exercise device. *Med Sci Sports Exerc,* 35(11), 1935-1945.

Smith, G., & Pell, J.P. (2003). Parachute use to prevent death and major trauma related to gravitational challenge: systematic review of randomised controlled trials. *BMJ,* 327(7429), 1459-1461.

Straus, S.E., & McAlister, F.A. (2000). Evidence-based medicine: a commentary on common criticisms. *Can Med Assoc J,* 163(7), 837-841.

Straus, S.E., & Sackett, D.L. (1999). Applying evidence to the individual patient. *Ann Oncol,* 10(1), 29-32.

Suominen, H. (2006). Muscle training for bone strength. *Aging Clin Exp Res,* 18(2), 85-93.

Swift, J.M., Gasier, H.G., Swift, S.N., Wiggs, M.P., Hogan, H.A., Fluckey, J.D., & Bloomfield, S.A. (2010). Increased training loads do not magnify cancellous bone gains with rodent jump resistance exercise. *J Appl Physiol,* 109, 1600-1607.

Thiebaud, R.S., Loenneke, J.P., Fahs, C.A., Rossow, L.M., Kim, D., Abe, T., . . . Bemben, M.G. (2013). The effects of elastic band resistance training combined with blood flow restriction on strength, total bone-free lean body mass and muscle thickness in postmenopausal women. *Clin Physiol Funct Imaging,* 33(5), 344-352.

U.S. Food and Drug Administration. (2004). FDA Issues Regulation Prohibiting Sale of Dietary Supplements Containing Ephedrine Alkaloids and Reiterates Its Advice That Consumers Stop Using These Products [news release]. Silver Spring, MD: U.S. Food and Drug Administration.

Wilk, R., Skrzypek, M., Kowalska, M., Kusz, D., Wielgorecki, A., Horyniecki, M., . . . Pluskiewicz, W. (2014). Standardized incidence and trend of osteoporotic hip fracture in Polish women and men: a nine year observation. *Maturitas,* 77(1), 59-63.

Winters-Stone, K.M., Dobek, J.C., Bennett, J.A., Maddalozzo, G.F., Ryan, C.W., & Beer, T.M. (2014). Skeletal response to resistance and impact training in prostate cancer survivors. *Med Sci Sports Exerc,* 46(8), 1482-1488.

Winters-Stone, K.M., Dobek, J., Nail, L., Bennett, J.A., Leo, M.C., Naik, A., & Schwartz, A. (2011). Strength training stops bone loss and builds muscle in postmenopausal breast cancer survivors: a randomized, controlled trial. *Breast Cancer Res Treat,* 127(2), 447-456.

Winters-Stone, K.M., & Snow, C.M. (2006). Site-specific response of bone to exercise in premenopausal women. *Bone*, 39(6), 1203-1209.

Witzke, K.A., & Snow, C.M. (2000). Effects of plyometric jump training on bone mass in adolescent girls. *Med Sci Sports Exerc*, 32(6), 1051-1057.

Chapter 3

Amonette, W.E., English, K.L., & Ottenbacher, K.J. (2010). Nullius in verba: a call for the incorporation of evidence-based practice into the discipline of exercise science. *Sports Med*, 40(6), 449-457.

Amonette, W.E., English, K.L., Spiering, B.A., & Kraemer, W.J. (2012). Evidence-based practice in strength and conditioning. In T.J. Chandler & L.E. Brown (Eds.), *Conditioning for Strength and Human Performance* (2nd ed.). Baltimore: Lippincott Williams & Wilkins.

Anderson, F.H. (1971). *The Philosophy of Francis Bacon*. New York: Octagon Books.

Bacon, F. (1999). *Selected Philosophical Works* (Vol. 290). Indianapolis, IN: Hackett Publishing Company, Inc.

Compton, J., & Robinson, M. (1995). The move towards evidence-based medicine [comment letter]. *Med J Aust*, 163(6), 333.

Corvi, R. (1997). *An Introduction to the Thought of Karl Popper*. New York: Routledge.

Eddy, D.M. (1990). Practice policies: where do they come from? *JAMA*, 263(9), 1265-1275.

Eddy, D.M. (2005). Evidence-based medicine: a unified approach. *Health Aff (Millwood)*, 24(1), 9-17.

Eddy, D.M. (2011). The origins of evidence-based medicine—a personal perspective. *Virtual Mentor*, 13(1), 55-60.

English, K.L., Amonette, W.E., Graham, M., & Spiering, B.A. (2012). What is "evidence-based" strength and conditioning? *Strength Cond J*, 34(3), 19-24.

Fox, C., Kay, E.J., & Anderson, R. (2014). Evidence-based dentistry—overcoming the challenges for the UK's dental practitioners. *Br Dent J*, 217(4), 191-194.

Gibson, A.B. (1967). *The Philosophy of Descartes*. New York: Russell.

Gulley, N. (2013). *Plato's Theory of Knowledge* (Routledge Revivals). New York: Routledge.

Hanson, B.P., Bhandari, M., Audige, L., & Helfet, D. (2004). The need for education in evidence-based orthopedics: an international survey of AO course participants. *Acta Orthop Scand*, 75(3), 328-332.

Irwin, T. (Ed.). (1995). *Classical Philosophy: Aristotle: Metaphysics, Epistemology, Natural Philosophy*. New York: Taylor & Francis.

Madhok, R., & Stothard, J. (2002). Promoting evidence based orthopaedic surgery. An English experience. *Acta Orthop Scand Suppl*, 73(305), 26-29.

McGinn, C. (1975, January). "A priori" and "a posteriori" knowledge. In *Proceedings of the Aristotelian Society*, 76(1975-76), 195-208.

Medawar, P.B. (2013). *Induction and Intuition in Scientific Thought* (Vol. 22). London and New York: Routledge.

Mordecai, S.C., Al-Hadithy, N., Ware, H.E., & Gupte, C.M. (2014). Treatment of meniscal tears: an evidence based approach [review]. *World J Orthop*, 5(3), 233-241.

Partridge, C. (1996). Evidence based medicine—implications for physiotherapy? [review] *Physiother Res Int*, 1(2), 69-73.

Partridge, C., & Edwards, S. (1996). The bases of practice—neurological physiotherapy [review]. *Physiother Res Int*, 1(3), 205-208.

Poolman, R.W., Sierevelt, I.N., Farrokhyar, F., Mazel, J.A., Blankevoort, L., & Bhandari, M. (2007). Perceptions and competence in evidence-based medicine: are surgeons getting better? A questionnaire survey of members of the Dutch Orthopaedic Association. *J Bone Joint Surg Am*, 89(1), 206-215.

Popper, K. (2014). *Conjectures and Refutations: The Growth of Scientific Knowledge*. London and New York: Routledge.

Reichenbach, H. (1968). *The Rise of Scientific Philosophy*. Berkeley and Los Angeles: University of California Press.

Sackett, D. (1995). Evidence-based medicine. *Lancet*, 346(8983), 1171.

Sackett, D.L. (1997). Evidence-based medicine and treatment choices. *Lancet*, 349(9051), 572.

Sackett, D.L., Richardson, W.S., Rosenberg, W.M., & Haynes, R.B. (1997). *Evidence-Based Medicine: How to Practice and Teach EBM*. New York: Pearson Professional Limited.

Sackett, D.L., & Rosenberg, W.M. (1995). On the need for evidence-based medicine. *J Public Health*, 17(3), 330-334.

Sackett, D.L., Rosenberg, W.M., Gray, J.A., Haynes, R.B., & Richardson, W.S. (1996). Evidence based medicine: what it is and what it isn't. *BMJ*, 312(7023), 71-72.

Sackett, D.L., & Straus, S.E. (1998). Finding and applying evidence during clinical rounds: the "evidence cart." *JAMA*, 280(15), 1336-1338.

Schoenfeld, P.S. (2008). Pro: EBM: an invaluable tool for medical practice. *Am J Gastroenterol*, 103(12), 2965-2967.

Shahar, E. (1997). A Popperian perspective of the term "evidence-based medicine." *J Eval Clin Pract*, 3(2), 109-116.

Shorten, A., & Wallace, M. (1996). Evidence based practice. The future is clear. *Aust Nurs J*, 4(6), 22-24.

Simpson, B. (1996). Evidence-based nursing practice: the state of the art. *Can Nurse*, 92(10), 22-25.

Sproul, R.C. (2000). *The Consequences of Ideas: Understanding the Concepts That Shaped Our World*. Wheaton, IL: Crossway Books.

Chapter 4

Amonette, W.E., English, K.L., & Ottenbacher, K.J. (2010). Nullius in verba: a call for the incorporation of evidence-based practice into the discipline of exercise science. *Sports Med*, 40(6), 449-457.

Centers for Disease Control and Prevention. (2004). 150th anniversary of John Snow and the pump handle. *MMWR Morb Mortal Wkly Rep*, 53(34), 783.

Cerhan, J.R., Fredericksen, Z.S., Wang, A.H., Habermann, T.M., Kay, N.E., Macon, W.R., . . . Liebow, M. (2011). Design and validity of a clinic-based case-control study on the molecular epidemiology of lymphoma. *Int J Mol Epidemiol Genet*, 2(2), 95-113.

Coyle, E.F. (2005). Improved muscular efficiency displayed as Tour de France champion matures. *J Appl Physiol*, 98(6), 2191-2196.

Flanders, W.D., & Longini, I.M. Jr. (1990). Estimating benefits of screening from observational cohort studies. *Stat Med*, 9(8), 969-980.

Gordis, L. (2009). *Epidemiology* (4th ed.). Philadelphia: Saunders Elsevier.

Kennedy, M.M. (1979). Generalizing from single case studies. *Eval Rev*, 3(4), 661-678.

Kono, S., Toyomura, K., Yin, G., Nagano, J., & Mizoue, T. (2004). A case-control study of colorectal cancer in relation to lifestyle factors and genetic polymorphisms: design and conduct of the Fukuoka colorectal cancer study. *Asian Pac J Cancer Prev*, 5(4), 393-400.

Kooistra, B., Dijkman, B., Einhorn, T.A., & Bhandari, M. (2009). How to design a good case series. *J Bone Joint Surg Am*, 91 Suppl 3, 21-26.

Levin, K.A. (2006). Study design V. Case–control studies. *Evid Based Dent*, 7(3), 83-84.

Martyn, C. (2002). Case reports, case series and systematic reviews. *QJM*, 95(4), 197-198.

Meldrum, M.L. (2000). A brief history of the randomized controlled trial. From oranges and lemons to the gold standard. *Hematol Oncol Clin North Am*, 14(4), 745-760, vii.

Paneth, N. (2004). Assessing the contributions of John Snow to epidemiology: 150 years after removal of the broad street pump handle. *Epidemiology*, 15(5), 514-516.

Riffenburgh, R.H. (2012). *Statistics in Medicine* (3rd ed.). San Diego: Saunders Elsevier.

Sackett, D.L., Straus, S.E., Richardson, S.W., Rosenberg, W., & Haynes, R.B. (2000). *Evidence-Based Medicine: How to Practice and Teach EBM* (2nd ed.). Edinburgh: Hancourt.

Stanwell-Smith, R. (2002). The making of an epidemiologist: John Snow before the episode of the Broad Street pump. *Commun Dis Public Health*, 5(4), 269-270.

Vandenbroucke, J.P. (2001). In defense of case reports and case series. *Ann Int Med*, 134(4), 330-334.

Chapter 5

Alexandrov, A.V., & Hennerici, M.G. (2007). Writing good abstracts. *Cerebrovasc Dis*, 23(4), 256-259.

Amonette, W.E., English, K.L., Spiering, B.A., & Kraemer, W.J. (2013). Evidence-based practice in strength and conditioning. In T.J.

Chandler & L.E. Brown (Eds.), *Conditioning for Strength and Human Performance* (pp. 285-303). Philadelphia: Lippincott Williams & Wilkins.

Dobkin, B.H. (1989). Focused stroke rehabilitation programs do not improve outcome. *Arch Neurol*, 46(6), 701-703.

Jewell, D.V. (2011). *Guide to Evidence-Based Therapist Practice*. Sudbury, MA: Jones & Bartlett Learning.

Knight, K.L., & Ingersoll, C.D. (1996a). Optimizing scholarly communication: 30 tips for writing clearly. *J Athl Train*, 31(3), 209.

Knight, K.L., & Ingersoll, C.D. (1996b). Structure of a scholarly manuscript: 66 tips for what goes where. *J Athl Train*, 31(3), 201-206.

Kraemer, W.J., Fleck, S.J. & Deschenes, M.R. (2016). *Exercise Physiology: Integrating Theory and Applications* (2nd ed.). Baltimore: Lippincott Williams & Wilkins.

Moher, D., Dulberg, C.S., & Wells, G.A. (1994). Statistical power, sample size, and their reporting in randomized controlled trials. *JAMA*, 272(2), 122-124.

National Institutes of Health. (2013, March 20). Grants & Funding: Glossary and Acronym List. Retrieved June 1, 2014. http://grants.nih.gov/grants/glossary.htm

Ohwovoriole, A.E. (2011). Writing biomedical manuscripts part I: fundamentals and general rules. *West Afr J Med*, 30(3), 151-157.

Ploutz-Snyder, R.J., Fiedler, J., & Feiveson, A.H. (2014). Justifying small-n research in scientifically amazing settings: challenging the notion that only "big-n" studies are worthwhile. *J Appl Physiol*, 116(9), 1251-1252.

Reding, M.J., & McDowell, F.H. (1989). Focused stroke rehabilitation programs improve outcome. *Arch Neurol*, 46(6), 700-701.

Rottensteiner, C., Konttinen, N., & Laakso, L. (2015). Sustained participation in youth sports related to coach-athlete relationship and coach-created motivational climate. *Int Sport Coach J*, 2(1), 29-38.

Sifft, J.M. (1983). Research—reading and understanding: utilizing descriptive statistics in sport performance. *Natl Strength Cond Assoc J*, 5(5), 26-28.

Sifft, J.M. (1984). Research—reading and understanding: guidelines for selecting a sample. *Natl Strength Cond Assoc J*, 6(1), 26-27.

Sifft, J.M. (1986). Research—reading and understanding: statistics for sport performance–basic inferential analysis. *Natl Strength Cond Assoc J*, 8(6), 46-48.

Sifft, J.M. (1990a). Research—reading and understanding #2: utilizing descriptive statistics in sport performance. *Natl Strength Cond Assoc J*, 12(3), 38-41.

Sifft, J.M. (1990b). Research—reading and understanding #4: statistics for sport performance—basic inferential analysis. *Natl Strength Cond Assoc J*, 12(6), 70.

Sifft, J.M., & Kraemer, W.J. (1982). Research—reading and understanding: introduction, review of literature and methods. *Natl Strength Coaches Assoc J*, 4(4), 24-25.

Starck, A., & Fleck, S. (1982). Research—reading and understanding: the discussion section. *Natl Strength Coaches Assoc J*, 4(6), 40-41.

Suchanek, P., Poledne, R., & Hubacek, J.A. (2011). Dietary intake reports fidelity—fact or fiction? *Neuro Endocrinol Lett*, 32 Suppl 2, 29-31.

Tipton, C.M. (1991). Publishing in peer-reviewed journals. Fundamentals for new investigators. *Physiologist*, 34(5), 275, 278-279.

Trabulsi, J., Schall, J.I., Ittenbach, R.F., Olsen, I.E., Yudkoff, M., Daikhin, Y., ... Stallings, V.A. (2006). Energy balance and the accuracy of reported energy intake in preadolescent children with cystic fibrosis. *Am J Clin Nutr*, 84(3), 523-530.

Valentini, N., Getchell, N., Logan, S.W., Liang, L.Y., Golden, D., Rudisill, M.E., & Robinson, L.E. (2015). Exploring associations between motor skill assessments in children with, without, and at-risk for developmental coordination disorder. *J Motor Learn Dev*, 3(1), 39-52.

Chapter 6

Amonette, W.E., English, K.L., & Ottenbacher, K.J. (2010). Nullius in verba: a call for the incorporation of evidence-based practice into the discipline of exercise science. *Sports Med*, 40(6), 449-457.

Bar-Or, O., Lundegren, H.M., & Buskirk, E.R. (1969). Heat tolerance of exercising obese and lean women. *J Appl Physiol*, 26(4), 403-409.

Bickel, C.S., Cross, J.M., & Bamman, M.M. (2011). Exercise dosing to retain resistance training adaptations in young and older adults. *Med Sci Sports Exerc*, 43(7), 1177-1187.

Bottaro, M., Machado, S.N., Nogueira, W., Scales, R., & Veloso, J. (2007). Effect of high versus low-velocity resistance training on muscular fitness and functional performance in older men. *Eur J Appl Physiol*, 99(3), 257-264.

Ceddia, M.A., Price, E.A., Kohlmeier, C.K., Evans, J.K., Lu, Q., McAuley, E., & Woods, J.A. (1999). Differential leukocytosis and lymphocyte mitogenic response to acute maximal exercise in the young and old. *Med Sci Sports Exerc*, 31(6), 829-836.

Charette, S.L., McEvoy, L., Pyka, G., Snow-Harter, C., Guido, D., Wiswell, R.A., & Marcus, R. (1991). Muscle hypertrophy response to resistance training in older women. *J Appl Physiol*, 70(5), 1912-1916.

Corbeil, P., Simoneau, M., Rancourt, D., Tremblay, A., & Teasdale, N. (2001). Increased risk for falling associated with obesity: mathematical modeling of postural control. *IEEE Neural Syst Rehabil Eng*, 9(2), 126-136.

Cowley, H.R., Ford, K.R., Myer, G.D., Kernozek, T.W., & Hewett, T.E. (2006). Differences in neuromuscular strategies between landing and cutting tasks in female basketball and soccer athletes. *J Athl Train*, 41(1), 67-73.

DeLorey, D.S., Wyrick, B.L., & Babb, T.G. (2005). Mild-to-moderate obesity: implications for respiratory mechanics at rest and during exercise in young men. *Int J Obesity*, 29(9), 1039-1047.

Felson, D.T., & Chaisson, C.E. (1997). Understanding the relationship between body weight and osteoarthritis. *Clin Rheumatol*, 11(4), 671-681.

Ferketich, A.K., Kirby, T.E., & Alway, S.E. (1998). Cardiovascular and muscular adaptations to combined endurance and strength training in elderly women. *Acta Physiol Scand*, 164(3), 259-267.

Fielding, R.A., LeBrasseur, N.K., Cuoco, A., Bean, J., Mizer, K., & Fiatarone Singh, M.A. (2002). High-velocity resistance training increases skeletal muscle peak power in older women. *J Am Geriatr Soc*, 50(4), 655-662.

Frontera, W.R., Meredith, C.N., O'Reilly, K.P., Knuttgen, H.G., & Evans, W.J. (1988). Strength conditioning in older men: skeletal muscle hypertrophy and improved function. *J Appl Physiol*, 64(3), 1038-1044.

Haddad, F., & Adams, G.R. (2006). Aging-sensitive cellular and molecular mechanisms associated with skeletal muscle hypertrophy. *J Appl Physiol (1985)*, 100(4), 1188-1203.

Hakkinen, K. (1985). Factors influencing trainability of short term and prolonged training. *NSCA J*, 7(2), 32-37.

Hakkinen, K., Pakarinen, A., Newton, R.U., & Kraemer, W.J. (1998). Acute hormone responses to heavy resistance lower and upper extremity exercise in young versus old men. *Eur J Appl Physiol Occup Physiol*, 77(4), 312-319.

Henwood, T.R., & Taaffe, D.R. (2005). Improved physical performance in older adults undertaking a short-term programme of high-velocity resistance training. *Gerontology*, 51(2), 108-115.

Hewett, T.E., Ford, K.R., & Myer, G.D. (2006). Anterior cruciate ligament injuries in female athletes: part 2, a meta-analysis of neuromuscular interventions aimed at injury prevention. *Am J Sports Med*, 34(3), 490-498.

Hewett, T.E., Myer, G.D., & Ford, K.R. (2006). Anterior cruciate ligament injuries in female athletes: part 1, mechanisms and risk factors. *Am J Sports Med*, 34(2), 299-311.

Hunter, G.R., McCarthy, J.P., & Bamman, M.M. (2004). Effects of resistance training on older adults. *Sports Med*, 34(5), 329-348.

Kosek, D.J., Kim, J.S., Petrella, J.K., Cross, J.M., & Bamman, M.M. (2006). Efficacy of 3 days/wk resistance training on myofiber hypertrophy and myogenic mechanisms in young vs. older adults. *J Appl Physiol (1985)*, 101(2), 531-544.

Kraemer, W.J., Hakkinen, K., Newton, R.U., McCormick, M., Nindl, B.C., Volek, J.S., . . . Evans, W.J. (1998). Acute hormonal responses to heavy resistance exercise in younger and older men. *Eur J Appl Physiol Occup Physiol*, 77(3), 206-211.

Law, M., & MacDermid, J. (Eds.). (2008). *Evidenced-Based Rehabilitation: A Guide to Practice.* Thorofare, NJ: Slack.

Miszko, T.A., Cress, M.E., Slade, J.M., Covey, C.J., Agrawal, S.K., & Doerr, C.E. (2003). Effect of strength and power training on physical function in community-dwelling older adults. *J Gerontol A Biol Sci Med Sci,* 58(2), 171-175.

Myer, G.D., Ford, K.R., Barber Foss, K.D., Liu, C., Nick, T.G., & Hewett, T.E. (2009). The relationship of hamstrings and quadriceps strength to anterior cruciate ligament injury in female athletes. *Clin J Sports Med,* 19(1), 3-8.

Renstrom, P., Ljungqvist, A., Arendt, E., Beynnon, B., Fukubayashi, T., Garrett, W., .. . Engebretsen, L. (2008). Non-contact ACL injuries in female athletes: an International Olympic Committee current concepts statement. *Br J Sports Med,* 42(6), 394-412.

Riffenburgh, R.H. (2012). *Statistics in Medicine* (3rd ed.). San Diego: Saunders Elsevier.

Roelants, M., Delecluse, C., & Verschueren, S.M. (2004). Whole-body-vibration training increases knee-extension strength and speed of movement in older women. *J Am Geriatr Soc,* 52(6), 901-908.

Sackett, D.L., Straus, S.E., Richardson, S.W., Rosenberg, W., & Haynes, R.B. (2000). *Evidence-Based Medicine: How to Practice and Teach EBM* (2nd ed.). Edinburgh: Hancourt.

Shinkai, S., Konishi, M., & Shephard, R.J. (1998). Aging and immune response to exercise. *Can J Physiol Pharmacol,* 76(5), 562-572.

Singh, M.A., Ding, W., Manfredi, T.J., Solares, G.S., O'Neill, E.F., Clements, K.M., ... Evans, W.J. (1999). Insulin-like growth factor I in skeletal muscle after weight-lifting exercise in frail elders. *Am J Physiol,* 277(1 Pt 1), E135-E143.

Stone, M.H., Sands, W.A., & Stone, M.E. (2004). The downfall of sports science in the United States. *Strength Cond J,* 26(2), 72-75.

Vos, R., Houtepen, R., & Horstman, K. (2002). Evidence-based medicine and power shifts in health care systems. *Health Care Anal,* 10(3), 319-328.

Chapter 7

Amonette, W.E., English, K.L., & Ottenbacher, K.J. (2010). Nullius in verba: a call for the incorporation of evidence-based practice into the discipline of exercise science. *Sports Med,* 40(6), 449-457.

Anders, M.E., & Evans, D.P. (2010). Comparison of PubMed and Google Scholar literature searches. *Respir Care,* 55(5), 578-583.

Coomarasamy, A., & Khan, K.S. (2004). What is the evidence that postgraduate teaching in evidence based medicine changes anything? A systematic review. *BMJ,* 329(7473), 1017.

Doig, G.S., & Simpson, F. (2003). Efficient literature searching: a core skill for the practice of evidence-based medicine. *Intensive Care Med,* 29(12), 2119-2127.

Ebbert, J.O., Dupras, D.M., & Erwin, P.J. (2003). Searching the medical literature using PubMed: a tutorial. *Mayo Clin Proc,* 78(1), 87-91.

Falagas, M.E., Pitsouni, E.I., Malietzis, G.A., & Pappas, G. (2008). Comparison of PubMed, Scopus, Web of Science, and Google Scholar: strengths and weaknesses. *FASEB J,* 22(2), 338-342.

Haynes, R.B., McKibbon, K, Wilczynski, N.L., Walter, S.D., & Werre, S.R. (2005). Optimal search strategies for retrieving scientifically strong studies of treatment from Medline: analytical survey. *BMJ,* 330(7501), 1179.

Haynes, R.B., Wilczynski, N., McKibbon, K., Walker, C.J., & Sinclair, J.C. (1994). Developing optimal search strategies for detecting clinically sound studies in MEDLINE. *J Am Med Inform Assoc,* 1(6), 447-458.

International Committee of Medical Journal Editors. (1997). Uniform requirements for manuscripts submitted to biomedical journals. *Pathology,* 29(4), 441-447.

Kraemer, W.J., Dunn-Lewis, C., Comstock, B.A., Thomas, G.A., Clark, J.E., & Nindl, B.C. (2010). Growth hormone, exercise, and athletic performance: a continued evolution of complexity. *Curr Sports Med Rep,* 9(4), 242-252.

Kraemer, W.J., Fry, A.C., Rubin, M.R., Triplett-McBride, T., Gordon, S.E., Koziris, L.P., . . . Fleck, S.J. (2001). Physiological and performance responses to tournament wrestling. *Med Sci Sports Exerc*, 33(8), 1367-1378.

Kulkarni, A.V., Aziz, B., Shams, I., & Busse, J.W. (2009). Comparisons of citations in Web of Science, Scopus, and Google Scholar for articles published in general medical journals. *JAMA*, 302(10), 1092-1096.

Law, M., & MacDermid, J. (Eds.). (2008). *Evidenced-Based Rehabilitation: A Guide to Practice*. Thorofare, NJ: Slack.

Lu, Z. (2011). PubMed and beyond: a survey of web tools for searching biomedical literature. *Database (Oxford)*, 2011, baq036.

McAlister, F.A., Graham, I., Karr, G.W., & Laupacis, A. (1999). Evidence-based medicine and the practicing clinician. *J Gen Intern Med*, 14(4), 236-242.

Neuhaus, C., Neuhaus, E., Asher, A., & Wrede, C. (2006). The depth and breadth of Google Scholar: an empirical study. *portal: Libraries and the Academy*, 6(2), 127-141.

Nindl, B.C., Kraemer, W.J., Marx, J.O., Tuckow, A.P., & Hymer, W.C. (2003). Growth hormone molecular heterogeneity and exercise. *Exerc Sport Sci Rev*, 31(4), 161-166.

Oxman, A.D., Sackett, D.L., & Guyatt, G.H. (1993). Users' guides to the medical literature. I. How to get started. *JAMA*, 270(17), 2093-2095.

Rosenberg, W., & Donald, A. (1995). Evidence based medicine: an approach to clinical problem-solving. *BMJ*, 310, 1122-1126.

Sackett, D.L. (1997). Evidence-based medicine. *Semin Perinatol*, 21(1), 3-5.

Thomas, G.A., Kraemer, W.J., Comstock, B.A., Dunn-Lewis, C., Maresh, C.M., & Volek, J.S. (2013). Obesity, growth hormone and exercise. *Sports Med*, 43(9), 839-849.

Chapter 8

Bayne-Jones, S., Burdette, W.J., Cochran, W.G., Farber, E., Fieser, L.F., Furth, J., . . . Seevers, M.H. (1964). *Smoking and Health: Report of the Advisory Committee to the Surgeon General of the Public Health Service*. Washington, DC: U.S. Government Printing Office.

Daly, C., McCarthy Persson, U., Twycross-Lewis, R., Woledge, R.C., & Morrissey, D. (2015). The biomechanics of running in athletes with previous hamstring injury: a case-control study. *Scand J Med Sci Sports*, 2015. doi: 10.1111/sms.12464. [Epub ahead of print]

Dickerman, R.D., Pertusi, R., & Smith, G.H. (2000). The upper range of lumbar spine bone mineral density? An examination of the current world record holder in the squat lift. *Int J Sports Med*, 21(7), 469-470.

Glass, D.C., Gray, C.N., Jolley, D.J., Gibbons, C., Sim, M.R., Fritschi, L., . . . Manuell, R. (2003). Leukemia risk associated with low-level benzene exposure. *Epidemiology*, 14(5), 569-577.

Gordis, L. (2009). *Epidemiology* (4th ed.). Philadelphia: Saunders Elsevier.

Hill, A.B. (1965). The environment and disease: association or causation? *Proc R Soc Med*, 58, 295-300.

Hutto, B., Sharpe, P.A., Granner, M.L., Addy, C.L., & Hooker, S. (2008). The effect of question order on reporting physical activity and walking behavior. *J Phys Act Health*, 5 Suppl 1, S16-S29.

James, P.A., Oparil, S., Carter, B.L., Cushman, W.C., Dennison-Himmelfarb, C., Handler, J., . . . Ogedegbe, O. (2014). 2014 evidence-based guideline for the management of high blood pressure in adults: report from the panel members appointed to the Eighth Joint National Committee (JNC 8). *JAMA*, 311(5), 507-520.

Ksir, C., Shank, M., Kraemer, W., & Noble, B. (1986). Effects of chewing tobacco on heart rate and blood pressure during exercise. *J Sports Med Phys Fitness*, 26(4), 384-389.

Law, M., & MacDermid, J. (Eds.). (2008). *Evidenced-Based Rehabilitation: A Guide to Practice*. Thorofare, NJ: Slack.

Lee, S.M., Schneider, S.M., Boda, W.L., Watenpaugh, D., Macias, B., Meyer, R., & Hargens, A.R. (2007). Supine lower body negative pressure exercise maintains upright exercise capacity in male twins during 30 days of bed rest. *Med Sci Sports Exerc*, 39(8), 1315-1326.

Levey, A.S., Coresh, J., Balk, E., Kausz, A.T., Levin, A., Steffes, M.W., . . . Eknoyan, G. (2003). National Kidney Foundation practice guidelines for chronic kidney disease: evaluation, classification, and stratification. *Ann Intern Med*, 139(2), 137-147.

Monga, M., Macias, B., Groppo, E., Kostelec, M., & Hargens, A. (2006). Renal stone risk in a simulated microgravity environment: impact of treadmill exercise with lower body negative pressure. *J Urol*, 176(1), 127-131.

Mossberg, K.A., Orlander, E.E., & Norcross, J.L. (2008). Cardiorespiratory capacity after weight-supported treadmill training in patients with traumatic brain injury. *Phys Ther*, 88(1), 77-87.

Ohkubo, T., Kikuya, M., Metoki, H., Asayama, K., Obara, T., Hashimoto, J., . . . Imai, Y. (2005). Prognosis of "masked" hypertension and "white-coat" hypertension detected by 24-h ambulatory blood pressure monitoring10-year follow-up from the Ohasama study. *J Am Col Cardiol*, 46(3), 508-515.

Owens, P., Atkins, N., & O'Brien, E. (1999). Diagnosis of white coat hypertension by ambulatory blood pressure monitoring. *Hypertension*, 34(2), 267-272.

Pittaluga, M., Casini, B., & Parisi, P. (2004). Physical activity and genetic influences in risk factors and aging: a study on twins. *Int J Sports Med*, 25(5), 345-350.

Ratamess, N.A., Alvar, B.A., Evetoch, T.K., Housh, T.J., Kibler, B., Kraemer, W.J., & Triplett, T.N. (2009). Progression models in resistance training for healthy adults. *Med Sci Sports Exerc*, 41(3), 687-708.

Riffenburgh, R.H. (2012). *Statistics in Medicine* (3rd. ed.). San Diego: Saunders Elsevier.

Rossow, L.M., Fukuda, D.H., Fahs, C.A., Loenneke, J.P., & Stout, J.R. (2013). Natural bodybuilding competition preparation and recovery: a 12-month case study. *Int J Sports Physiol Perform*, 8(5), 582-592.

Sackett, D.L. (1979). Bias in analytic research. *J Chronic Dis*, 32(1-2), 51-63.

Smith, S.M., Davis-Street, J.E., Fesperman, J.V., Calkins, D.S., Bawa, M., Macias, B.R., . . . Hargens, A.R. (2003). Evaluation of treadmill exercise in a lower body negative pressure chamber as a countermeasure for weightlessness-induced bone loss: a bed rest study with identical twins. *J Bone Miner Res*, 18(12), 2223-2230.

Tsai, H.M., Chee, W., & Im, E.O. (2006). Internet methods in the study of women's physical activity. *Stud Health Technol Inform*, 122, 396-400.

Turner, R.C., Millns, H., Neil, H.A.W., Stratton, I.M., Manley, S.E., Matthews, D.R., & Holman, R.R. (1998). Risk factors for coronary artery disease in non-insulin dependent diabetes mellitus: United Kingdom Prospective Diabetes Study (UKPDS: 23). *BMJ*, 316(7134), 823-828.

Ventura, A.K., Loken, E., Mitchell, D.C., Smiciklas-Wright, H., & Birch, L.L. (2006). Understanding reporting bias in the dietary recall data of 11-year-old girls. *Obesity (Silver Spring)*, 14(6), 1073-1084.

Weber, M.A., Schiffrin, E.L., White, W.B., Mann, S., Lindholm, L.H., Kenerson, J.G., . . . Ram, C.V. (2014). Clinical practice guidelines for the management of hypertension in the community. *J Clin Hyperten*, 16(1), 14-26.

Young, J.M., & Solomon, M.J. (2009). How to critically appraise an article. *Nat Clin Pract Gastroenterol Hepatol*, 6(2), 82-91.

Chapter 9

Castro, A.V., Kolka, C.M., Kim, S.P., & Bergman, R.N. (2014). Obesity, insulin resistance and comorbidities? Mechanisms of association. *Arq Bras Endocrinol Metabol*, 58(6), 600-609.

Charles, C., Gafni, A., & Whelan, T. (1997). Shared decision-making in the medical encounter: what does it mean? (or it takes at least two to tango). *Soc Sci Med*, 44(5), 681-692.

Degner, L.F., & Sloan, J.A. (1992). Decision making during serious illness: what role do patients really want to play? *J Clin Epidemiol*, 45(9), 941-950.

Despres, J.P., Lemieux, I., Bergeron, J., Pibarot, P., Mathieu, P., Larose, E., . . . Poirier, P. (2008). Abdominal obesity and the metabolic syndrome: contribution to global cardiometabolic risk. *Arterioscler Thromb Vasc Biol*, 28(6), 1039-1049.

Emanuel, E.J., & Emanuel, L.L. (1992). Four models of the physician-patient relationship. *JAMA*, 267(16), 2221-2226.

Grimshaw, J., Eccles, M.T., MacLennan, R., Graeme, R., Fraser, C., & Vale, L. (2006). Toward evidence-based quality improvement. *J Gen Intern Med*, 21(S2), S14-S20.

Janz, N.K., & Becker, M.H. (1984). The health belief model: a decade later. *Health Educ Q*, 11(1), 1-47.

Kazis, L.E., Anderson, J.J., & Meenan, R.F. (1989). Effect sizes for interpreting changes in health status. *Med Care*, 27(3), S178-S189.

Law, M., & MacDermid, J. (Eds.). (2008). *Evidenced-Based Rehabilitation: A Guide to Practice*. Thorofare, NJ: Slack.

Morone, G., Bragoni, M., Iosa, M., De Angelis, D., Venturiero, V., Coiro, P., . . . Paolucci, S. (2011). Who may benefit from robotic-assisted gait training? A randomized clinical trial in patients with subacute stroke. *Neurorehabil Neural Repair*, 25(7), 636-644.

Riffenburgh, R.H. (2012). *Statistics in Medicine* (3rd ed.). San Diego: Saunders Elsevier.

Rosenstock, I.M. (1974). Historical origins of the health belief model. *Health Educ Behav*, 2(4), 328-335.

Sackett, D.L., Rosenberg, W., Gray, J.A., Haynes, R.B., & Richardson, S.W. (1996). Evidence based medicine: what it is and what it isn't. *Br Med J*, 312(7023), 71-72.

Sackett, D.L., Straus, S.E., Richardson, S.W., Rosenberg, W., & Haynes, R.B. (2000). *Evidence-Based Medicine: How to Practice and Teach EBM* (2nd ed.). Edinburgh: Hancourt.

Westlake, K.P., & Patten, C. (2009). Pilot study of Lokomat versus manual-assisted treadmill training for locomotor recovery post-stroke. *J Neuroeng Rehabil*, 6, 18.

Wyrwich, K.W., Nienaber, N.A., Tierney, W.M., & Wolinsky, F.D. (1999). Linking clinical relevance and statistical significance in evaluating intra-individual changes in health-related quality of life. *Med Care*, 37(5), 469-478.

Chapter 10

Amonette, W.E., Brown, L.E., De Witt, J.K., Dupler, T.L., Tran, T.T., Tufano, J.J., & Spiering, B.A. (2012). Peak vertical jump power estimations in youths and young adults. *J Strength Cond Res*, 26(7), 1749-1755.

Atkinson, G., & Nevill, A.M. (1998). Statistical methods for assessing measurement error (reliability) in variables relevant to sports medicine. *Sports Med*, 26(4), 217-238.

Baechle, T.R., Earle, R.W., & Wathen, D. (2008). Resistance training. In T.R. Baechle & R.W. Earle (Eds.), *Essentials of Strength Training and Conditioning* (3rd ed., pp. 382-412). Champaign, IL: Human Kinetics.

Brown, L.E., Khamoui, A.V., & Jo, E. (2013). Test administration and interpretation. In T.J. Chandler & L.E. Brown (Eds.), *Conditioning for Strength and Human Performance* (2nd ed., pp. 165-193). Philadelphia: Lippincott Williams & Wilkins.

Coleman, A.E., & Amonette, W.E. (2015). Sprint accelerations to first base among major league baseball players with different years of career experience. *J Strength Cond Res*, 29(7), 1759-1765.

De Witt, J.K., Perusek, G.P., Lewandowski, B.E., Gilkey, K.M., Savina, M.C., Samorezov, S., & Edwards, W.B. (2010). Locomotion in simulated and real microgravity: horizontal suspension vs. parabolic flight. *Aviat Space Environ Med*, 81(12), 1092-1099.

Dreyer, H.C., Drummond, M.J., Pennings, B., Fujita, S., Glynn, E.L., Chinkes, D.L., . . . Rasmussen, B.B. (2008). Leucine-enriched essential amino acid and carbohydrate ingestion following resistance exercise enhances mTOR signaling and protein synthesis in human muscle. *Am J Physiol Endocrinol Metab*, 294(2), E392-E400.

Drummond, M.J., Dreyer, H.C., Pennings, B., Fry, C.S., Dhanani, S., Dillon, E.L., . . . Rasmussen, B.B. (2008). Skeletal muscle protein anabolic response to resistance exercise and essential amino acids is delayed with aging. *J Appl Physiol*, 104(5), 1452-1461.

Fitts, R.H., Trappe, S.W., Costill, D.L., Gallagher, P.M., Creer, A.C., Colloton, P.A., . . . Riley, D.A. (2010). Prolonged space flight-induced alterations in the structure and function of human skeletal muscle fibres. *J Physiol*, 588(Pt 18), 3567-3592.

Fleck, S.J., & Kraemer, W.J. (2014). *Designing Resistance Training Programs* (4th ed.). Champaign, IL: Human Kinetics.

Guyatt, G.H., Heyting, A., Jaeschke, R., Keller, J., Adachi, J.D., & Roberts, R.S. (1990). N of 1 randomized trials for investigating new drugs. *Control Clin Trials*, 11(2), 88-100.

Guyatt, G.H., Keller, J.L., Jaeschke, R., Rosenbloom, D., Adachi, J.D., & Newhouse, M.T. (1990). The n-of-1 randomized controlled trial: clinical usefulness. Our three-year experience. *Ann Intern Med*, 112(4), 293-299.

Guyatt, G., Sackett, D., Adachi, J., Roberts, R., Chong, J., Rosenbloom, D., & Keller, J. (1988). A clinician's guide for conducting randomized trials in individual patients. *Can Med Assoc J*, 139(6), 497.

Guyatt, G., Sackett, D., Taylor, D., Chong, J., Roberts, R., & Pugsley, S. (1986). Determining optimal therapy—randomized trials in individual patients. *New Engl J Med*, 314(14), 889-892.

Jost, P.D. (2008). Simulating human space physiology with bed rest. *Hippokratia*, 12 Suppl 1, 37-40.

Keller, J.L., Guyatt, G.H., Roberts, R.S., Adachi, J.D., & Rosenbloom, D. (1988). An N of 1 service: applying the scientific method in clinical practice. *Scand J Gastroenterol Suppl*, 147, 22-29.

Magee, D.J., Quillen, W.S., Amonette, W.E., & Spiering, B.A. (2012). Preparticipation physical examination. In D.J. Magee, R.C. Manske, J.E. Zachazewski, & Quillen, W.S. (Eds.), *Musculoskeletal Rehabilitation Series, Volume IV: Selected Topics in Sports Injuries and Rehabilitation.* Baltimore: Elsevier.

Moore, A.D. Jr., Downs, M.E., Lee, S.M., Feiveson, A.H., Knudsen, P., & Ploutz-Snyder, L. (2014). Peak exercise oxygen uptake during and following long-duration spaceflight. *J Appl Physiol (1985)*, 117(3), 231-238.

Perhonen, M.A., Franco, F., Lane, L.D., Buckey, J.C., Blomqvist, C.G., Zerwekh, J.E., . . . Levine, B.D. (2001). Cardiac atrophy after bed rest and spaceflight. *J Appl Physiol (1985)*, 91(2), 645-653.

Phillips, S.M. (2014). A brief review of critical processes in exercise-induced muscular hypertrophy. *Sports Med*, 44 Suppl 1, S71-S77.

Rickham, P.P. (1964). Human experimentation. Code of Ethics of the World Medical Association. Declaration of Helsinki. *Br Med J*, 2(5402), 177.

Riffenburgh, R.H. (2012). *Statistics in Medicine* (3rd ed.). San Diego: Saunders Elsevier.

Schneider, S.M., Amonette, W.E., Blazine, K., Bentley, J., Lee, S.M.C., Loehr, J.A., . . . Smith, S.M. (2003). Training with the International Space Station interim resistive exercise device. *Med Sci Sports Exerc*, 35(11), 1935-1945.

Smith, S.M., Abrams, S.A., Davis-Street, J.E., Heer, M., O'Brien, K.O., Wastney, M.E., & Zwart, S.R. (2014). Fifty years of human space travel: implications for bone and calcium research. *Annu Rev Nutr*, 34, 377-400.

Stone, M.H., Sands, W.A., & Stone, M.E. (2004). The downfall of sports science in the United States. *Strength Cond J*, 26(2), 72-75.

Tsang, E., & Williams, J.N. (2012). Generalization and induction: misconceptions, clarifications, and a classification of induction. *MIS Q*, 36(3), 729-748.

Verschueren, S., Bogaerts, A., Delecluse, C., Claessens, A.L., Haentjens, P., Vanderschueren, D., & Boonen, S. (2011). The effects of whole-body vibration training and vitamin D supplementation on muscle strength, muscle mass, and bone density in institutionalized elderly women: a 6-month randomized, controlled trial. *J Bone Miner Res*, 26(1), 42-49.

Zucker, D.R., Schmid, C.H., McIntosh, M.W., D'Agostino, R.B., Selker, H.P., & Lau, J. (1997). Combining single patient (N-of-1) trials to estimate population treatment effects and to evaluate individual patient responses to treatment. *J Clin Epidemiol*, 50(4), 401-410.

Chapter 11

Amonette, W.E., Bentley, J.R., Lee, S., Loehr, J.A., & Schneider, S. (2004). Ground reaction force and mechanical differences between the interim resistive exercise device (iRED) and Smith machine while performing a squat. Technical Report (2004–212063). Washington, DC: National Aeronautics and Space Administration.

Archer, M.C. (2004). Creatine: a safety concern. *Toxicol Lett*, 152(3), 275.

Bjurstedt, H., & Eiken, O. (1995). Graded restriction of blood flow in exercising leg muscles: a human model. *Adv Exp Med Biol*, 381, 147-156.

Burgomaster, K.A., Moore, D.R., Schofield, L.M., Phillips, S.M., Sale, D.G., & Gibala, M.J. (2003). Resistance training with vascular occlusion: metabolic adaptations in human muscle. *Med Sci Sports Exerc*, 35(7), 1203-1208.

Cook, S.B., Brown, K.A., Deruisseau, K., Kanaley, J.A., & Ploutz-Snyder, L.L. (2010). Skeletal muscle adaptations following blood flow-restricted training during 30 days of muscular unloading. *J Appl Physiol*, 109(2), 341-349.

Cook, S.B., Clark, B.C., & Ploutz-Snyder, L.L. (2007). Effects of exercise load and blood-flow restriction on skeletal muscle function. *Med Sci Sports Exerc*, 39(10), 1708-1713.

Credeur, D.P., Hollis, B.C., & Welsch, M.A. (2010). Effects of handgrip training with venous restriction on brachial artery vasodilation. *Med Sci Sports Exerc*, 42(7), 1296-1302.

Eiken, O., & Bjurstedt, H. (1987). Dynamic exercise in man as influenced by experimental restriction of blood flow in the working muscles. *Acta Physiol Scand*, 131(3), 339-345.

Ellis S., Vodian M.A., Grindeland R.E. (1978) Studies on the bioassayable growth hormone-like activity of plasma. Recent Prog Horm Res. 34:213-38.

Evans, C., Vance, S., & Brown, M. (2010). Short-term resistance training with blood flow restriction enhances microvascular filtration capacity of human calf muscles. *J Sports Sci*, 28(9), 999-1007.

Fujita, S., Abe, T., Drummond, M.J., Cadenas, J.G., Dreyer, H.C., Sato, Y., . . . Rasmussen, B.B. (2007). Blood flow restriction during low-intensity resistance exercise increases S6K1 phosphorylation and muscle protein synthesis. *J Appl Physiol*, 103(3), 903-910.

Hile, A.M., Anderson, J.M., Fiala, K.A., Stevenson, J.H., Casa, D.J., & Maresh, C.M. (2006). Creatine supplementation and anterior compartment pressure during exercise in the heat in dehydrated men. *J Athl Train*, 41(1), 30-35.

Hymer, W.C., Kraemer, W.J., Nindl, B.C., Marx, J.O., Benson, D.E., Welsch, J.R., Mazzetti, S.A., Volek, J.S., & Deaver, D.R. (2001). Characteristics of circulating growth hormone in women after acute heavy resistance exercise. *Am J Physiol Endocrinol Metab*, 281(4), E878-E887.

Jager, R., Purpura, M., Shao, A., Inoue, T., & Kreider, R.B. (2011). Analysis of the efficacy, safety, and regulatory status of novel forms of creatine. *Amino Acids*, 40(5), 1369-1383.

Karabulut, M., Abe, T., Sato, Y., & Bemben, M.G. (2010). The effects of low-intensity resistance training with vascular restriction on leg muscle strength in older men. *Eur J Appl Physiol*, 108(1), 147-155.

Kraemer, W.J., Dunn-Lewis, C., Comstock, B.A., Thomas, G.A., Clark, J.E., & Nindl, B.C. (2010). Growth hormone, exercise, and athletic performance: a continued evolution of complexity. *Curr Sports Med Rep*, 9(4), 242-252.

Liphardt, A.-M., Schipilow, J.D., Macdonald, H.M., Kan, M., Zieger, A., Boyd, S.K. (2015). Bone micro-architecture of elite alpine skiers is not reflected by bone mineral density. *Osteoporos Int*, 2015. doi: 10.1007/s00198-015-3133-y. [Epub ahead of print]

Loehr, J.A., Lee, S.M., English, K.L., Sibonga, J., Smith, S.M., Spiering, B.A., & Hagan, R.D. (2011). Musculoskeletal adaptations to training with the advanced resistive exercise device. *Med Sci Sports Exerc*, 43(1), 146-156.

McCall, G.E., Grindeland, R.E., Roy, R.R., & Edgerton, V.R. (2000). Muscle afferent activity modulates bioassayable growth hormone in human plasma. *J Appl Physiol*, 89(3),1137-1141.

Moore, D.R., Burgomaster, K.A., Schofield, L.M., Gibala, M.J., Sale, D.G., & Phillips, S.M. (2004). Neuromuscular adaptations in human muscle following low intensity resistance training with vascular occlusion. *Eur J Appl Physiol*, 92(4-5), 399-406.

Nindl, B.C., Kraemer, W.J., Marx, J.O., Tuckow, A.P., & Hymer, W.C. (2003). Growth hormone molecular heterogeneity and exercise. *Exerc Sport Sci Rev*, 31(4), 61-66.

Organov, V., Schneider, V., Bakulin, A., Voronin, V., Morgun, L., Shackelford, A., LeBlanc, L., Murashko, V., & Novikov, V. (1997). Human bone tissue changes after long-term space flight: Phenomenology and possible mechanics [abstract]. 12th Man in Space Symposium: The Future of Humans in Space, June 8-13. Retrieved September 01, 2015 from http://ntrs.nasa.gov/archive/nasa/casi.ntrs.nasa.gov/19980024339.pdf

Patterson, S.D., & Ferguson, R.A. (2010). Increase in calf post-occlusive blood flow and strength following short-term resistance exercise training with blood flow restriction in young women. *Eur J Appl Physiol*, 108(5), 1025-1033.

Polyviou, T.P., Pitsiladis, Y.P., Lee, W.C., Pantazis, T., Hambly, C., Speakman, J.R., & Malkova, D. (2012). Thermoregulatory and cardiovascular responses to creatine, glycerol and alpha lipoic acid in trained cyclists. *J Int Soc Sports Nutr*, 9(1), 29.

Poortmans, J.R., & Francaux, M. (2000). Adverse effects of creatine supplementation: fact or fiction? *Sports Med*, 30(3), 155-170.

Rawson, E.S., & Volek, J.S. (2003). Effects of creatine supplementation and resistance training on muscle strength and weightlifting performance. *J Strength Cond Res*, 17(4), 822-831.

Schneider, S.M., Amonette, W.E., Blazine, K., Bentley, J., Lee, S.M.C., Loehr, J.A., . . . Smith, S.M. (2003). Training with the International Space Station interim resistive exercise device. *Med Sci Sports Exerc*, 35(11), 1935-1945.

Takarada, Y., Nakamura, Y., Aruga, S., Onda, T., Miyazaki, S., & Ishii, N. (2000). Rapid increase in plasma growth hormone after low-intensity resistance exercise with vascular occlusion. *J Appl Physiol*, 88(1), 61-65.

Volek, J.S. (2003). Strength nutrition. *Curr Sports Med Rep*, 2(4), 189-193.

Watson, G., Casa, D.J., Fiala, K.A., Hile, A., Roti, M.W., Healey, J.C., . . . Maresh, C.M. (2006). Creatine use and exercise heat tolerance in dehydrated men. *J Athl Train*, 41(1), 18-29.

Chapter 12

Apel, J., Lacey, R.M., & Kell, R.T. (2011). A comparison of traditional and weekly undulating periodized strength training programs with total volume and intensity equated. *J Strength Cond Res*, 25(3), 694-703.

Bandy, W.D., & Irion, J.M. (1997). The effect of time and frequency of static stretching on flexibility of the hamstring muscles. *Phys Ther*, 77(10), 1090.

Bastiaans, J., Diemen, A.V., Veneberg, T., & Jeukendrup, A. (2001). The effects of replacing a portion of endurance training by explosive strength training on performance in trained cyclists. *Eur J Appl Physiol*, 86(1), 79-84.

Bartolomei, S., Hoffman, J.R., Merni, F., & Stout, J.R. (2014). A comparison of traditional and block periodized strength training programs in trained athletes. *J Strength Cond Res*, 28(4), 990-997.

Beckett, J.R., Schneiker, K.T., Wallman, K.E., Dawson, B.T., & Guelfi, K.J. (2009). Effects of static stretching on repeated sprint and change of direction performance. *Med Sci Sports Exerc*, 41(2), 444-450.

Beutler, A., de la Motte, S., Marshall, S., Padua, D., & Boden, B. (2009). Muscle Strength and Qualitative Jump-Landing Differences in Male and Female Military Cadets: The Jump-Acl Study. *J Sport Sci Med*, 8, 663-671.

Bien, D.P. (2011). Rationale and implementation of anterior cruciate ligament injury prevention warm-up programs in female athletes. *J Strength Cond Res*, 25(1), 271-285.

Bishop, D., Jenkins, D.G., Mackinnon, T., McEniery, M., & Carey, M.F. (1999). The effects of strength training on endurance performance and muscle characteristics. *Med Sci Sports Exerc*, 31(6), 886-891.

Buford, T.W., Rossi, S.J., Smith, D.B., & Warren, A.J. (2007). A comparison of periodization models during nine weeks with equated volume and intensity for strength. *J Strength Cond Res*, 21(4), 1245-1250.

Byrd, R., Chandler, T.J., Conley, M.S., Fry, A.C., Haff, G.G., Koch, A., . . . Wathen, D. (1999). Strength training: single versus multiple sets. *Sports Med*, 27(6), 409-416.

Carpinelli, R.N., & Otto, R.M. (1998). Strength training. Single versus multiple sets. *Sports Med*, 26(2), 73-84.

Chappell, J.D., Herman, D.C., Knight, B.S., Kirkendall, D.T., Garrett, W.E., & Yu, B. (2005). Effect of fatigue on knee kinetics and kinematics in stop-jump tasks. *Am J Sports Med*, 33(7), 1022-1029.

DeLorme, T.L., & Watkins, A.L. (1948). Technics of progressive resistance exercise. *Arch Phys Med Rehabil*, 29(5), 263-273.

Favero, J.P., Midgley, A.W., & Bentley, D.J. (2009). Effects of an acute bout of static stretching on 40 m sprint performance:

influence of baseline flexibility. *Res Sports Med*, 17(1), 50-60.

Ford, K.R., Myer, G.D., Schmitt, L.C., Uhl, T.L., & Hewett, T.E. (2011). Preferential quadriceps activation in female athletes with incremental increases in landing intensity. *J Appl Biomech*, 27(3), 215-222.

Foster, C., Hoyos, J., Earnest, C., & Lucia, A. (2005). Regulation of energy expenditure during prolonged athletic competition. *Med Sci Sports Exerc*, 37(4), 670-675.

Franchini, E., Branco, B.M., Agostinho, M.F., Calmet, M., Candau, R., & Paulo, S.P. (2015). Influence of linear and undulating strength periodization on physical fitness, physiological and performance responses to simulated judo matches, *J Strength Cond Res*, 29(2), 358-367.

Gilchrist, J., Mandelbaum, B.R., Melancon, H., Ryan, G.W., Silvers, H.J., Griffin, L.Y., ... Dvorak, J. (2008). A randomized controlled trial to prevent noncontact anterior cruciate ligament injury in female collegiate soccer players. *Am J Sports Med*, 36(8), 1476-1483.

Gokeler, A., Hof, A.L., Arnold, M.P., Dijkstra, P.U., Postema, K., & Otten, E. (2010). Abnormal landing strategies after ACL reconstruction. *Scand J Med Sci Sports*, 20(1), e12-e19.

Hewett, T.E., Ford, K.R., & Myer, G.D. (2006). Anterior cruciate ligament injuries in female athletes: part 2, a meta-analysis of neuromuscular interventions aimed at injury prevention. *Am J Sports Med*, 34(3), 490-498.

Hewett, T.E., Lynch, T.R., Myer, G.D., Ford, K.R., Gwin, R.C., & Heidt, R.S. Jr. (2010). Multiple risk factors related to familial predisposition to anterior cruciate ligament injury: fraternal twin sisters with anterior cruciate ligament ruptures. *Br J Sports Med*, 44(12), 848-855.

Hewett, T.E., Myer, G.D., & Ford, K.R. (2005). Reducing knee and anterior cruciate ligament injuries among female athletes: a systematic review of neuromuscular training interventions. *J Knee Surg*, 18(1), 82-88.

Hewett, T.E., Myer, G.D., & Ford, K.R. (2006). Anterior cruciate ligament injuries in female athletes: part 1, mechanisms and risk factors. *Am J Sports Med*, 34(2), 299-311.

Hewett, T.E., Zazulak, B.T., Myer, G.D., & Ford, K.R. (2005). A review of electromyographic activation levels, timing differences, and increased anterior cruciate ligament injury incidence in female athletes. *Br J Sports Med*, 39(6), 347-350.

Hickson, R.C., Dvorak, B.A., Gorostiaga, E.M., Kurowski, T.T., & Foster, C. (1988). Potential for strength and endurance training to amplify endurance performance. *J Appl Physiol (1985)*, 65(5), 2285-2290.

Hubscher, M., Zech, A., Pfeifer, K., Hänsel, F., Vogt, L., & Banzer, W. (2010). Neuromuscular training for sports injury prevention: a systematic review. *Med Sci Sports Exerc*, 42(3), 413-421.

Jackson, N.P., Hickey, M.S., & Reiser, R.F. (2007). High resistance/low repetitions vs. low resistance/ high repetition training: effects on performance of trained cyclists. *J Strength Cond Res*, 21(1), 289-295.

Kelly, A.K. (2008). Anterior cruciate ligament injury prevention. *Curr Sports Med Rep*, 7(5), 255-262.

Kinser, A.M., Ramsey, M.W., O'Bryant, H.S., Ayres, C.A., Sands, W.A., & Stone, M.H. (2008). Vibration and stretching effects on flexibility and explosive strength in young gymnasts. *Med Sci Sports Exerc*, 40(1), 133-140.

Kistler, B.M., Walsh, M.S., Horn, T.S., & Cox, R.H. (2010). The acute effects of static stretching on the sprint performance of collegiate men in the 60- and 100-m dash after a dynamic warm-up. *J Strength Cond Res*, 24(9), 2280-2284.

Krieger, J.W. (2009). Single versus multiple sets of resistance exercise: a meta-regression. *J Strength Cond Res*, 23(6), 1890-1901.

Krieger, J.W. (2010). Single vs. multiple sets of resistance exercise for muscle hypertrophy: a meta-analysis. *J Strength Cond Res*, 24(4), 1150-1159.

Malinzak, R.A., Colby, S.M., Kirkendall, D.T., Yu, B., & Garrett, W.E. (2001). A comparison of knee joint motion patterns between men and women in selected athletic tasks. *Clin Biomech*, 16(5), 438-445.

Mentzer, M., & Little, J. (2002). *High-Intensity Training the Mike Mentzer Way*. Columbus, OH: McGraw-Hill.

Paradisis, G.P., Pappas, P.T., Theodorou, A.S., Zacharogiannis, E.G., Skordilis, E.K., & Smirniotou, A.S. (2014). Effects of static and dynamic stretching on sprint and jump performance in boys and girls. *J Strength Cond Res*, 28(1), 154-160.

Paton, C.D., & Hopkins, W.G. (2005). Combining explosive and high-resistance training improves performance in competitive cyclists. *J Strength Cond Res*, 19(4), 826-830.

Peterson, M.D., Dodd, D.J., Alvar, B.A., Rhea, M.R., & Favre, M. (2008). Undulation training for development of hierarchical fitness and improved firefighter job performance. *J Strength Cond Res*, 22(5), 1683-1695.

Peterson, M.D., Rhea, M.R., & Alvar, B.A. (2004). Maximizing strength development in athletes: a meta-analysis to determine the dose-response relationship. *J Strength Cond Res*, 18(2), 377-382.

Pfile, K.R., Hart, J.M., Herman, D.C., Hertel, J., Kerrigan, D.C., & Ingersoll, C.D. (2013). Different exercise training interventions and drop-landing biomechanics in high school female athletes. *J Athl Train*, 48(4), 450-462.

Ratamess, N.A., Alvar, B.A., Evetoch, T.K., Housh, T.J., Kibler, W.B., Kraemer, W.J., & Triplett, T.N. (2009). Progression models in resistance training for healthy adults. *Med Sci Sports Exerc*, 41(3), 687-708.

Rhea, M.R., Alvar, B.A., Burkett, L.N., & Ball, S.D. (2003). A meta-analysis to determine the dose response for strength development. *Med Sci Sports Exerc*, 35(3), 456-464.

Rhea, M.R., Ball, S.D., Phillips, W.T., & Burkett, L.N. (2002). A comparison of linear and daily undulating periodized programs with equated volume and intensity for strength. *J Strength Cond Res*, 16(2), 250-255.

Rogan, S., Wust, D., Schwitter, T., & Schmidtbleicher, D. (2013). Static stretching of the hamstring muscle for injury prevention in football codes: a systematic review. *Asian J Sports Med*, 4(1), 1-9.

Ronnestad, B.R., Hansen, E.A., & Raastad, T. (2010). In-season strength maintenance training increases well-trained cyclists' performance. *Eur J Appl Physiol*, 110(6), 1269-1282.

Sayers, A.L., Farley, R.S., Fuller, D.K., Jubenville, C.B., & Caputo, J.L. (2008). The effect of static stretching on phases of sprint performance in elite soccer players. *J Strength Cond Res*, 22(5), 1416-1421.

Selye, H., & Fortier, C. (1950). Adaptive reaction to stress. *Psychosom Med*, 12(3), 149-157.

Shrier, I. (1999). Stretching before exercise does not reduce the risk of local muscle injury: a critical review of the clinical and basic science literature. *Clin J Sport Med*, 9(4), 221-227.

Sim, A.Y., Dawson, B.T., Guelfi, K.J., Wallman, K.E., & Young, W.B. (2009). Effects of static stretching in warm-up on repeated sprint performance. *J Strength Cond Res*, 23(7), 2155-2162.

Small, K., Mc Naughton, L., & Matthews, M. (2008). A systematic review into the efficacy of static stretching as part of a warm-up for the prevention of exercise-related injury. *Res Sports Med*, 16(3), 213-231.

Stone, M.H., O'Bryant, H., & Garhammer, J. (1981). Hypothetical model for strength training. *J Sports Med Phys Fitness*, 21(4), 342-351.

Stone, M.H., O'Bryant, H., Garhammer, J., McMillan, J., & Rozenek, R. (1982). Theoretical model of strength training. *NSCA J*, 4(4), 36-39.

Sugimoto, D., Myer, G.D., McKeon, J.M., & Hewett, T.E. (2012). Evaluation of the effectiveness of neuromuscular training to reduce anterior cruciate ligament injury in female athletes: a critical review of relative risk reduction and numbers-needed-to-treat analyses. *Br J Sports Med*, 46(14), 979-988.

Ter Stege, M.H., Dallinga, J.M., Benjaminse, A., & Lemmink, K.A. (2014). Effect of interventions on potential, modifiable risk factors for knee injury in team ball sports: a systematic review. *Sports Med*, 44(10), 1403-1426.

Thacker, S.B., Gilchrist, J., Stroup, D.F., & Kimsey Jr., C.D. (2004). The impact of stretching on sports injury risk: a systematic review of the literature. *Med Sci Sports Exerc*, 36(3), 371-378.

Torres, E.M., Kraemer, W.J., Vingren, J.L., Volek, J.S., Hatfield, D.L., Spiering, B.A., Ho, J.Y., Fragala, M.S., Thomas, G.A., Anderson, J.M., Häkkinen, K., & Maresh, C.M. (2008). Effects of stretching on upper-body muscular performance. *J Strength Cond Res*, 22(4), 1279-1285.

Weldon, S.M., & Hill, R.H. (2003). The efficacy of stretching for prevention of exercise-related injury: a systematic review of the literature. *Man Ther*, 8(3), 141-150.

Winchester, J.B., Nelson, A.G., Landin, D., Young, M.A., & Schexnayder, I.C. (2008). Static stretching impairs sprint performance in collegiate track and field athletes. *J Strength Cond Res*, 22(1), 13-19.

Wojtys, E.M., Huston, L.J., Taylor, P.D., & Bastian, S.D. (1996). Neuromuscular adaptations in isokinetic, isotonic, and agility training programs. *Am J Sports Med*, 24(2), 187-192.

Yamamoto, L.M., Klau, J.F., Casa, D.J., Kraemer, W.J., Armstrong, L.E., & Maresh, C.M. (2010). The effects of resistance training on road cycling performance among highly trained cyclists: a systematic review. *J Strength Cond Res*, 24(2), 560-566.

Yu, B., Lin, C., & Garrett, W.E. (2006). Lower extremity biomechanics during the landing of a stop-jump task. *Clin Biomech*, 21(3), 297-305.

Chapter 13

Amonette, W.E., & Mossberg, K.A. (2013). Ventilatory anaerobic thresholds of individuals recovering from traumatic brain injury compared with noninjured controls. *J Head Trauma Rehabil*, 28(5), E13-E20.

Bassey, E.J., Fiatarone, M.A., O'Neill, E.F., Kelly, M., Evans, W.J., & Lipsitz, L.A. (1992). Leg extensor power and functional performance in very old men and women. *Clin Sci (Lond)*, 82(3), 321-327.

Bhagia, V., Gilkison, C., Fitts, R.H., Zgaljardic, D.J., High, W.M. Jr., Masel, B.E., ... Mossberg, K.A. (2010). Effect of recombinant growth hormone replacement in a growth hormone deficient subject recovering from mild traumatic brain injury: a case report. *Brain Inj*, 24(3), 560-567.

Bhambhani, Y., Rowland, G., & Farag, M. (2005). Effects of circuit training on body composition and peak cardiorespiratory responses in patients with moderate to severe traumatic brain injury. *Arch Phys Med Rehabil*, 86(2), 268-276.

Bottaro, M., Machado, S.N., Nogueira, W., Scales, R., & Veloso, J. (2007). Effect of high versus low-velocity resistance training on muscular fitness and functional performance in older men. *Eur J Appl Physiol*, 99(3), 257-264.

Caldroney, R.D., & Radike, J. (2010). Experience with mild traumatic brain injuries and postconcussion syndrome at Kandahar, Afghanistan. *US Army Med Dep J*, 22-30.

Cantarero-Villanueva, I., Fernandez-Lao, C., Cuesta-Vargas, A.I., Del Moral-Avila, R., Fernandez-de-Las-Penas, C., & Arroyo-Morales, M. (2013). The effectiveness of a deep water aquatic exercise program in cancer-related fatigue in breast cancer survivors: a randomized controlled trial. *Arch Phys Med Rehabil*, 94(2), 221-230.

Cormie, P., Pumpa, K., Galvao, D.A., Turner, E., Spry, N., Saunders, C., ... Newton, R.U. (2013). Is it safe and efficacious for women with lymphedema secondary to breast cancer to lift heavy weights during exercise: a randomised controlled trial. *J Cancer Surviv*, 7(3), 413-424.

Correa-de-Araujo, R., & Hadley, E. (2014). Skeletal muscle function deficit: a new terminology to embrace the evolving concepts of sarcopenia and age-related muscle dysfunction. *J Gerontol A Biol Sci Med Sci*, 69(5), 591-594.

De Backer, I.C., Van Breda, E., Vreugdenhil, A., Nijziel, M.R., Kester, A.D., & Schep, G. (2007). High-intensity strength training improves quality of life in cancer survivors. *Acta Oncol*, 46(8), 1143-1151.

Department of Economic and Social Affairs Population Division. (2001). *World Population Ageing*. New York: United Nations.

Department of Health and Human Services, Administration on Aging. (2005). *A Profile of Older Americans: 2005*. Washington, DC: U.S. Department of Health and Human Services.

Engels, H.J., Drouin, J., Zhu, W., & Kazmierski, J.F. (1998). Effects of low-impact, moderate-intensity exercise training with and without wrist weights on functional capacities and mood states in older adults. *Gerontology*, 44(4), 239-244.

Eyigor, S., & Kanyilmaz, S. (2014). Exercise in patients coping with breast cancer: an overview. *World J Clin Oncol*, 5(3), 406-411.

Foldvari, M., Clark, M., Laviolette, L.C., Bernstein, M.A., Kaliton, D., Castaneda, C., . . . Singh, M.A. (2000). Association of muscle power with functional status in community-dwelling elderly women. *J Gerontol A Biol Sci Med Sci*, 55(4), M192-M199.

Gauchard, G.C., Tessier, A., Jeandel, C., & Perrin, P.P. (2003). Improved muscle strength and power in elderly exercising regularly. *Int J Sports Med*, 24(1), 71-74.

Harada, A. (2014). Epidemiology of bone and joint disease—the present and future [in Japanese]. *Clin Calcium*, 24(5), 669-678.

Hassett, L.M., Moseley, A.M., Tate, R., & Harmer, A.R. (2008). Fitness training for cardiorespiratory conditioning after traumatic brain injury. *Cochrane Database Syst Rev*, 2, CD006123.

Helfer, T.M., Jordan, N.N., Lee, R.B., Pietrusiak, P., Cave, K., & Schairer, K. (2011). Noise-induced hearing injury and comorbidities among postdeployment U.S. Army soldiers: April 2003-June 2009. *Am J Audiol*, 20(1), 33-41.

Herrero, F., San Juan, A.F., Fleck, S.J., Balmer, J., Perez, M., Canete, S., . . . Lucia, A. (2006). Combined aerobic and resistance training in breast cancer survivors: a randomized, controlled pilot trial. *Int J Sports Med*, 27(7), 573-580.

High, W.M. Jr., Briones-Galang, M., Clark, J.A., Gilkison, C., Mossberg, K.A., Zgaljardic, D.J., . . . Urban, R.J. (2010). Effect of growth hormone replacement therapy on cognition after traumatic brain injury. *J Neurotrauma*, 27(9), 1565-1575.

Holviala, J., Kraemer, W.J., Sillanpää, E., Karppinen, H., Avela, J., Kauhanen, A., Häkkinen, A., & Häkkinen, K. (2012). Effects of strength, endurance and combined training on muscle strength, walking speed and dynamic balance in aging men. *Eur J Appl Physiol*, 112(4), 1335-1347.

Jankowski, L.W., & Sullivan, S.J. (1990). Aerobic and neuromuscular training: effect on the capacity, efficiency, and fatigability of patients with traumatic brain injuries. *Arch Phys Med Rehabil*, 71(7), 500-504.

Kozminski, M. (2010). Combat-related post-traumatic headache: diagnosis, mechanisms of injury, and challenges to treatment. *J Am Osteopath Assoc*, 110(9), 514-519.

LaChapelle, D.L., & Finlayson, M.A.J. (1998). An evaluation of subjective and objective measures of fatigue in patients with brain injury and healthy controls. *Brain Inj*, 12(8), 649-659.

Leon-Carrion, J. (2002). Dementia due to head trauma: an obscure name for a clear neurocognitive syndrome. *Neurorehabil*, 17(2), 115.

Liu, C.J., & Latham, N.K. (2009). Progressive resistance strength training for improving physical function in older adults. *Cochrane Database Syst Rev*, 3, CD002759.

Maggio, M., Lauretani, F., De Vita, F., Basaria, S., Lippi, G., Butto, V., . . . Ceda, G.P. (2014). Multiple hormonal dysregulation as determinant of low physical performance and mobility in older persons. *Curr Pharm Des*, 20(19), 3119-3148.

Marcell, T.J. (2003). Sarcopenia: causes, consequences, and preventions. *J Gerontol A Biol Sci Med Sci*, 58(10), M911-M916.

Mero, A.A., Hulmi, J.J., Salmijärvi, H., Katajavuori, M., Haverinen, M., Holviala, J., Ridanpää, T., Häkkinen, K., Kovanen, V., Ahtiainen, J.P., & Selänne, H. (2013). Resistance training induced increase in muscle fiber size in young and older men. *Eur J Appl Physiol*, 113(3), 641-650.

Meuleman, J.R., Brechue, W.F., Kubilis, P.S., & Lowenthal, D.T. (2000). Exercise training in the debilitated aged: strength and functional outcomes. *Arch Phys Med Rehabil*, 81(3), 312-318.

Mossberg, K.A., Ayala, D., Baker, T., Heard, J., & Masel, B. (2007). Aerobic capacity after traumatic brain injury: comparison with a nondisabled cohort. *Arch Phys Med Rehabil*, 88(3), 315-320.

Mossberg, K.A., Masel, B.E., Gilkison, C.R., & Urban, R.J. (2008). Aerobic capacity and growth hormone deficiency after traumatic brain injury. *J Clin Endocrinol Metab*, 93(7), 2581-2587.

Mossberg, K.A., Orlander, E.E., & Norcross, J.L. (2008). Cardiorespiratory capacity after weight-supported treadmill training in patients with traumatic brain injury. *Phys Ther*, 88(1), 77-87.

Nakano, M.M., Otonari, T.S., Takara, K.S., Carmo, C.M., & Tanaka, C. (2014). Physical performance, balance, mobility, and muscle strength decline at different rates in elderly people. *J Phys Ther Sci*, 26(4), 583-586.

National Center for Health Statistics. (2006). *United States, 2006 with Chartbook on Trends in the Health of Americans.* Hyattsville, MD.

Newton, R.U., Hakkinen, K., Hakkinen, A., McCormick, M., Volek, J., & Kraemer, W.J. (2002) Mixed-methods resistance training increases power and strength of young and older men. *Med Sci Sports Exerc*, 34(8), 1367-1375.

Orr, R., de Vos, N.J., Singh, N.A., Ross, D.A., Stavrinos, T.M., & Fiatarone-Singh, M.A. (2006). Power training improves balance in healthy older adults. *J Gerontol A Biol Sci Med Sci*, 61(1), 78-85.

Raymond, M.J., Bramley-Tzerefos, R.E., Jeffs, K.J., Winter, A., & Holland, A.E. (2013). Systematic review of high-intensity progressive resistance strength training of the lower limb compared with other intensities of strength training in older adults. *Arch Phys Med Rehabil*, 94(8), 1458-1472.

Reid, K.F., & Fielding, R.A. (2012). Skeletal muscle power: a critical determinant of physical functioning in older adults. *Exerc Sport Sci Rev*, 40(1), 4-12.

Sallinen, J., Fogelholm, M., Volek, J.S., Kraemer, W.J., Alen, M., & Häkkinen, K. (2007). Effects of strength training and reduced training on functional performance and metabolic health indicators in middle-aged men. *Int J Sports Med*, 28(10), 815-822.

Sayer, N.A., Chiros, C.E., Sigford, B., Scott, S., Clothier, B., Pickett, T., & Lew, H.L. (2008). Characteristics and rehabilitation outcomes among patients with blast and other injuries sustained during the Global War on Terror. *Arch Phys Med Rehabil*, 89(1), 163-170.

Scherer, M. (2007). Gait rehabilitation with body weight-supported treadmill training for a blast injury survivor with traumatic brain injury. *Brain Inj*, 21(1), 93-100.

Scherer, M.R., Burrows, H., Pinto, R., Littlefield, P., French, L.M., Tarbett, A.K., & Schubert, M.C. (2011). Evidence of central and peripheral vestibular pathology in blast-related traumatic brain injury. *Otol Neurotol*, 32(4), 571-580.

Schmidt, T., Weisser, B., Jonat, W., Baumann, F.T., & Mundhenke, C. (2012). Gentle strength training in rehabilitation of breast cancer patients compared to conventional therapy. *Anticancer Res*, 32(8), 3229-3233.

Seynnes, O., Fiatarone Singh, M.A., Hue, O., Pras, P., Legros, P., & Bernard, P.L. (2004). Physiological and functional responses to low-moderate versus high-intensity progressive resistance training in frail elders. *J Gerontol A Biol Sci Med Sci*, 59(5), 503-509.

Snell, F. I., & Halter, M. J. (2010). A Signature Wound of War. *J Psychosoc Nurs*, 48(2,:22-28.

Steib, S., Schoene, D., & Pfeifer, K. (2010). Dose-response relationship of resistance training in older adults: a meta-analysis. *Med Sci Sports Exerc*, 42(5), 902-914.

Theeler, B.J., Flynn, F.G., & Erickson, J.C. (2010). Headaches after concussion in US soldiers returning from Iraq or Afghanistan. *Headache*, 50(8), 1262-1272.

Thompson, W. (Ed.). (2010). *ACSM's Guidelines for Exercise Testing and Prescription* (8th ed.). Baltimore: Lippincott Williams & Wilkins.

Tisdale, M.J. (2002). Cachexia in cancer patients. *Nat Rev Cancer*, 2(11), 862-871.

Walker, S., Santolamazza, F., Kraemer, W., & Häkkinen, K. (2014). Effects of prolonged hypertrophic resistance training on acute endocrine responses in young and older men. *J Aging Phys Act*.

Chapter 14

Bolster, D.R., Vary, T.C., Kimball, S.R., & Jefferson, L.S. (2004). Leucine regulates translation initiation in rat skeletal muscle via enhanced eIF4G phosphorylation. *J Nutr*, 134(7), 1704-1710.

Branch, J.D. (2003). Effect of creatine supplementation on body composition and performance: a meta-analysis. *Int J Sport Nutr Exerc Metab*, 13(2), 198-226.

Brosnan, J.T., & Brosnan, M.E. (2007). Creatine: endogenous metabolite, dietary, and therapeutic supplement. *Annu Rev Nutr*, 27, 241-261.

Brouns, F., Saris, W.H., Stroecken, J., Beckers, E., Thijssen, R., Rehrer, N.J., & ten Hoor, F. (1989a). Eating, drinking, and cycling. A controlled Tour de France simulation study, part I. *Int J Sports Med*, 10 Suppl 1, S32-S40.

Brouns, F., Saris, W.H., Stroecken, J., Beckers, E., Thijssen, R., Rehrer, N.J., & ten Hoor, F. (1989b). Eating, drinking, and cycling. A controlled Tour de France simulation study, part II. Effect of diet manipulation. *Int J Sports Med*, 10 Suppl 1, S41-S48.

Carraro, F., Hartl, W.H., Stuart, C.A., Layman, D.K., Jahoor, F., & Wolfe, R.R. (1990). Whole body and plasma protein synthesis in exercise and recovery in human subjects. *Am J Physiol*, 258(5 Pt 1), E821-E831.

Carraro, F., Stuart, C.A., Hartl, W.H., Rosenblatt, J., & Wolfe, R.R. (1990). Effect of exercise and recovery on muscle protein synthesis in human subjects. *Am J Physiol*, 259(4 Pt 1), E470-E476.

Coyle, E.F., Coggan, A.R., Hopper, M.K., & Walters, T.J. (1988). Determinants of endurance in well-trained cyclists. *J Appl Physiol*, 64(6), 2622-2630.

Cuthbertson, D., Smith, K., Babraj, J., Leese, G., Waddell, T., Atherton, P., . . . Rennie, M.J. (2005). Anabolic signaling deficits underlie amino acid resistance of wasting, aging muscle. *FASEB J*, 19(3), 422-424.

Dalbo, V.J., Roberts, M.D., Stout, J.R., & Kerksick, C.M. (2008). Putting to rest the myth of creatine supplementation leading to muscle cramps and dehydration. *Br J Sports Med*, 42(7), 567-573.

Dawson-Hughes, B. (2003). Calcium and protein in bone health. *Proc Nutr Soc*, 62(2), 505-509.

Deutz, N.E., & Wolfe, R.R. (2013). Is there a maximal anabolic response to protein intake with a meal? *Clin Nutr*, 32(2), 309-313.

Doherty, M., & Smith, P.M. (2005). Effects of caffeine ingestion on rating of perceived exertion during and after exercise: a meta-analysis. *Scand J Med Sci Sports*, 15(2), 69-78.

EFSA Panel on Dietetic Products, Nutrition and Allergies (2012). Scientific opinion on dietary reference values for protein. *EFSA J*, 10(2), 2557.

Finn, J.P., Ebert, T.R., Withers, R.T., Carey, M.F., Mackay, M., Phillips, J.W., & Febbraio, M.A. (2001). Effect of creatine supplementation on metabolism and performance in humans during intermittent sprint cycling. *Eur J Appl Physiol*, 84(3), 238-243.

Friedman, J.E., & Lemon, P.W. (1989). Effect of chronic endurance exercise on retention of dietary protein. *Int J Sports Med*, 10(2), 118-123.

Glynn, E.L., Fry, C.S., Drummond, M.J., Timmerman, K.L., Dhanani, S., Volpi, E., & Rasmussen, B.B. (2010). Excess leucine intake enhances muscle anabolic signaling but not net protein anabolism in young men and women. *J Nutr*, 140(11), 1970-1976.

Graham, T.E., Hibbert, E., & Sathasivam, P. (1998). Metabolic and exercise endurance effects of coffee and caffeine ingestion. *J Appl Physiol*, 85(3), 883-889.

Green, A.L., Hultman, E., Macdonald, I.A., Sewell, D.A., & Greenhaff, P.L. (1996). Carbohydrate ingestion augments skeletal muscle creatine accumulation during creatine supplementation in humans. *Am J Physiol*, 271(5 Pt 1), E821-E826.

Harris, R.C., Soderlund, K., & Hultman, E. (1992). Elevation of creatine in resting and exercised muscle of normal subjects by creatine supplementation. *Clin Sci (Lond)*, 83(3), 367-374.

Hodgson, A.B., Randell, R.K., & Jeukendrup, A.E. (2013). The metabolic and performance effects of caffeine compared to coffee during endurance exercise. *PLoS One*, 8(4), e59561.

Hoffman, J.R., Cooper, J., Wendell, M., Im, J., & Kang, J. (2004). Effects of beta-hydroxy beta-methylbutyrate on power performance and indices of muscle damage and stress during high-intensity training. *J Strength Cond Res*, 18(4), 747-752.

Institute of Medicine. (2005). *Dietary Reference Intakes for Energy, Carbohydrate, Fiber, Fatty Acids, Cholesterol, Protein and Amino Acids.* Washington, DC: National Academy Press.

Juhn, M.S., & Tarnopolsky, M. (1998). Oral creatine supplementation and athletic performance: a critical review. *Clin J Sport Med*, 8(4), 286-297.

Kim, H.J., Kim, C.K., Carpentier, A., & Poortmans, J.R. (2011). Studies on the safety of creatine supplementation. *Amino Acids*, 40(5), 1409-1418.

Kraemer, W.J., Hatfield, D.L., Volek, J.S., Fragala, M.S., Vingren, J.L., Anderson, J.M., Spiering, B.A., Thomas, G.A., Ho, J.Y., Quann, E.E., Izquierdo, M., Häkkinen, K., Maresh, C.M. (2009). Effects of amino acids supplement on physiological adaptations to resistance training. *Med Sci Sports Exerc*, 41(5):1111-1121.

Kreider, R.B., & Campbell, B. (2009). Protein for exercise and recovery. *Phys Sportsmed*, 37(2), 13-21.

Kreider, R.B., Ferreira, M., Wilson, M., & Almada, A.L. (1999). Effects of calcium beta-hydroxy-beta-methylbutyrate (HMB) supplementation during resistance-training on markers of catabolism, body composition and strength. *Int J Sports Med*, 20(8), 503-509.

Mamerow, M.M., Mettler, J.A., English, K.L., Casperson, S.L., Arentson-Lantz, E., Sheffield-Moore, M.,... Paddon-Jones, D. (2014). Dietary protein distribution positively influences 24-h muscle protein synthesis in healthy adults. *J Nutr*, 144(6), 876-880.

Martin, J.C., Farrar, R.P., Wagner, B.M., & Spirduso, W.W. (2000). Maximal power across the lifespan. *J Gerontol A Biol Sci Med Sci*, 55(6), M311-M316.

McLellan, T.M., & Bell, D.G. (2004). The impact of prior coffee consumption on the subsequent ergogenic effect of anhydrous caffeine. *Int J Sport Nutr Exerc Metab*, 14(6), 698-708.

Menaspa, P., Abbiss, C.R., & Martin, D.T. (2013). Performance analysis of a world-class sprinter during cycling grand tours. *Int J Sports Physiol Perform*, 8(3), 336-340.

Mohr, M., Nielsen, J.J., & Bangsbo, J. (2011). Caffeine intake improves intense intermittent exercise performance and reduces muscle interstitial potassium accumulation. *J Appl Physiol*, 111(5), 1372-1379.

O'Connor, D.M., & Crowe, M.J. (2007). Effects of six weeks of beta-hydroxy-beta-methylbutyrate (HMB) and HMB/creatine supplementation on strength, power, and anthropometry of highly trained athletes. *J Strength Cond Res*, 21(2), 419-423.

Paddon-Jones, D., & Rasmussen, B.B. (2009). Dietary protein recommendations and the prevention of sarcopenia. *Curr Opin Clin Nutr Metab Care*, 12(1), 86-90.

Palisin, T., & Stacy, J.J. (2005). Beta-hydroxy-beta-methylbutyrate and its use in athletics. *Curr Sports Med Rep*, 4(4), 220-223.

Panton, L.B., Rathmacher, J.A., Baier, S., & Nissen, S. (2000). Nutritional supplementation of the leucine metabolite beta-hydroxy-beta-methylbutyrate (hmb) during resistance training. *Nutrition*, 16(9), 734-739.

Phillips, S.M. (2012). Dietary protein requirements and adaptive advantages in athletes. *Br J Nutr*, 108 Suppl 2, S158-S167.

Pincivero, D.M., & Bompa, T.O. (1997). A physiological review of American football. *Sports Med*, 23(4), 247-260.

Preen, D., Dawson, B., Goodman, C., Beilby, J., & Ching, S. (2003). Creatine supplementation: a comparison of loading and maintenance protocols on creatine uptake by human skeletal muscle. *Int J Sport Nutr Exerc Metab*, 13(1), 97-111.

Rand, W.M., Pellett, P.L., & Young, V.R. (2003). Meta-analysis of nitrogen balance studies for estimating protein requirements in healthy adults. *Am J Clin Nutr*, 77(1), 109-127.

Ransone, J., Neighbors, K., Lefavi, R., & Chromiak, J. (2003). The effect of beta-hydroxy beta-methylbutyrate on muscular strength and body composition in collegiate football players. *J Strength Cond Res*, 17(1), 34-39.

Rodriguez, N.R., Di Marco, N.M., & Langley, S. (2009). American College of Sports Medicine position stand. Nutrition and athletic performance. *Med Sci Sports Exerc*, 41(3), 709-731.

Slater, G., Jenkins, D., Logan, P., Lee, H., Vukovich, M., Rathmacher, J.A., & Hahn, A.G. (2001). Beta-hydroxy-beta-methylbutyrate (HMB) supplementation does not affect changes in strength or body composition during resistance training in trained men. *Int J Sport Nutr Exerc Metab*, 11(3), 384-396.

Sokmen, B., Armstrong, L.E., Kraemer, W.J., Casa, D.J., Dias, J.C., Judelson, D.A., & Maresh, C.M. (2008). Caffeine use in sports: considerations for the athlete. *J Strength Cond Res*, 22(3), 978-986.

Stamler, J., Elliott, P., Kesteloot, H., Nichols, R., Claeys, G., Dyer, A.R., & Stamler, R. (1996). Inverse relation of dietary protein markers with blood pressure. Findings for 10,020 men and women in the INTERSALT Study. INTERSALT Cooperative Research Group. INTERnational study of SALT and blood pressure. *Circulation*, 94(7), 1629-1634.

Symons, T.B., Sheffield-Moore, M., Wolfe, R.R., & Paddon-Jones, D. (2009). A moderate serving of high-quality protein maximally stimulates skeletal muscle protein synthesis in young and elderly subjects. *J Am Diet Assoc*, 109(9), 1582-1586.

Tallis, J., James, R.S., Cox, V.M., & Duncan, M.J. (2012). The effect of physiological concentrations of caffeine on the power output of maximally and submaximally stimulated mouse EDL (fast) and soleus (slow) muscle. *J Appl Physiol*, 112(1), 64-71.

Tarnopolsky, M.A. (2004). Protein requirements for endurance athletes. *Nutrition*, 20(7-8), 662-668.

Tarnopolsky, M.A. (2008). Effect of caffeine on the neuromuscular system—potential as an ergogenic aid. *Appl Physiol Nutr Metab*, 33(6), 1284-1289.

Tarnopolsky, M.A., & Cupido, C. (2000). Caffeine potentiates low frequency skeletal muscle force in habitual and nonhabitual caffeine consumers. *J Appl Physiol*, 89(5), 1719-1724.

Tarnopolsky, M.A., MacDougall, J.D., & Atkinson, S.A. (1988). Influence of protein intake and training status on nitrogen balance and lean body mass. *J Appl Physiol*, 64(1), 187-193.

Terjung, R.L., Clarkson, P., Eichner, E.R., Greenhaff, P.L., Hespel, P.J., Israel, R.G., ... Williams, M.H. (2000). American College of Sports Medicine roundtable. The physiological and health effects of oral creatine supplementation. *Med Sci Sports Exerc*, 32(3), 706-717.

Tunnicliffe, J.M., Erdman, K.A., Reimer, R.A., Lun, V., & Shearer, J. (2008). Consumption of dietary caffeine and coffee in physically active populations: physiological interactions. *Appl Physiol Nutr Metab*, 33(6), 1301-1310.

Vandebuerie, F., Vanden Eynde, B., Vandenberghe, K., & Hespel, P. (1998). Effect of creatine loading on endurance capacity and sprint power in cyclists. *Int J Sports Med*, 19(7), 490-495.

Van Schuylenbergh, R., Van Leemputte, M., & Hespel, P. (2003). Effects of oral creatine-pyruvate supplementation in cycling performance. *Int J Sports Med*, 24(2), 144-150.

Volek, J.S., Duncan, N.D., Mazzetti, S.A., Staron, R.S., Putukian, M., Gomez, A.L., ... Kraemer, W.J. (1999). Performance and muscle fiber adaptations to creatine supplementation and heavy resistance training. *Med Sci Sports Exerc*, 31(8), 1147-1156.

Volek, J.S., Mazzetti, S.A., Farquhar, W.B., Barnes, B.R., Gomez, A.L., & Kraemer, W.J. (2001). Physiological responses to short-term exercise in the heat after creatine loading. *Med Sci Sports Exerc*, 33(7), 1101-1108.

Wilson, J.M., Fitschen, P.J., Campbell, B., Wilson, G.J., Zanchi, N., Taylor, L., ... Antonio, J. (2013). International Society of Sports Nutrition position stand: beta-hydroxy-beta-methylbutyrate (HMB). *J Int Soc Sports Nutr*, 10(1), 6.

Wilson, J.M., Lowery, R.P., Joy, J.M., Andersen, J.C., Wilson, S.M., Stout, J.R., ... Rathmacher, J. (2014). The effects of 12 weeks of beta-hydroxy-beta-methylbutyrate free acid supplementation on muscle mass, strength, and power in resistance-trained individuals: a randomized, double-blind, placebo-controlled study. *Eur J Appl Physiol*, 114(6), 1217-1227.

Wolfe, R.R. (2006). The underappreciated role of muscle in health and disease. *Am J Clin Nutr*, 84(3), 475-482.

Wolfe, R.R., & Chinkes, D.L. (2005). *Isotope Tracers in Metabolic Research: Principles and Practice of Kinetic Analysis* (2nd ed.). Hoboken, NJ: Wiley.

Wolfe, R.R., & Miller, S.L. (2008). The recommended dietary allowance of protein: a misunderstood concept. *JAMA*, 299(24), 2891-2893.

Zanchi, N.E., Gerlinger-Romero, F., Guimaraes-Ferreira, L., de Siqueira Filho, M.A., Felitti, V., Lira, F.S., ... Lancha, A.H. Jr. (2011). HMB supplementation: clinical and athletic performance-related effects and mechanisms of action. *Amino Acids*, 40(4), 1015-1025.

Chapter 15

Abercromby, A.F., Amonette, W.E., Layne, C.S., McFarlin, B.K., Hinman, M.R., & Paloski, W.H. (2007). Variation in neuromuscular responses during acute whole-body vibration exercise. *Med Sci Sports Exerc*, 39(9), 1642-1650.

Altman, A.R., & Davis, I.S. (2012). Barefoot running: biomechanics and implications for running injuries. *Curr Sports Med Rep*, 11(5), 244-250.

Ataee, J., Koozehchian, M.S., Kreider, R.B., & Zuo, L. (2014). Effectiveness of accommodation and constant resistance training on maximal strength and power in trained athletes. *PeerJ*, 2, e441.

Behm, D.G., Drinkwater, E.J., Willardson, J.M., & Cowley, P.M. (2010). Canadian Society for Exercise Physiology position stand: the use of instability to train the core in athletic and nonathletic conditioning. *Appl Physiol Nutr Metab*, 35(1), 109-112.

Berning, J.M., Coker, C.A., & Adams, K.J. (2004). Using chains for strength and conditioning. *Strength Cond J*, 26(5), 80-84.

Blottner, D., Salanova, M., Puttmann, B., Schiffl, G., Felsenberg, D., Buehring, B., & Rittweger, J. (2006). Human skeletal muscle structure and function preserved by vibration muscle exercise following 55 days of bed rest. *Eur J Appl Physiol*, 97(3), 261-271.

Burkett, L.N., Kohrt, W.M., & Buchbinder, R. (1985). Effects of shoes and foot orthotics on $\dot{V}O_2$ and selected frontal plane knee kinematics. *Med Sci Sports Exerc*, 17(1), 158-163.

Coleman, A.E., & Amonette, W.E. (2012). Pure acceleration is the primary determinant of speed to first-base in major-league baseball game situations. *J Strength Cond Res*, 26(6), 1455-1460.

Cowley, P.M., Swensen, T., & Sforzo, G.A. (2007). Efficacy of instability resistance training. *Int J Sports Med*, 28(10), 829-835.

Delecluse, C., Roelants, M., Diels, R., Koninckx, E., & Verschueren, S. (2005). Effects of whole body vibration training on muscle strength and sprint performance in sprint-trained athletes. *Int J Sports Med*, 26(8), 662-668.

Divert, C., Mornieux, G., Freychat, P., Baly, L., Mayer, F., & Belli, A. (2008). Barefoot-shod running differences: shoe or mass effect? *Int J Sports Med*, 29(6), 512-518.

Edge, J., Mundel, T., Weir, K., & Cochrane, D.J. (2009). The effects of acute whole body vibration as a recovery modality following high-intensity interval training in well-trained, middle-aged runners. *Eur J Appl Physiol*, 105(3), 421-428.

Escamilla, R.F., Ionno, M., deMahy, M.S., Fleisig, G.S., Wilk, K.E., Yamashiro, K., . . . Andrews, J.R. (2012). Comparison of three baseball-specific 6-week training programs on throwing velocity in high school baseball players. *J Strength Cond Res*, 26(7), 1767-1781.

Fagnani, F., Giombini, A., Di Cesare, A., Pigozzi, F., & Di Salvo, V. (2006). The effects of a whole-body vibration program on muscle performance and flexibility in female athletes. *Am J Phys Med Rehabil*, 85(12), 956-962.

Franz, J.R., Wierzbinski, C.M., & Kram, R. (2012). Metabolic cost of running barefoot versus shod: is lighter better? *Med Sci Sports Exerc*, 44(8), 1519-1525.

Ghigiarelli, J.J., Nagle, E.F., Gross, F.L., Robertson, R.J., Irrgang, J.J., & Myslinski, T. (2009). The effects of a 7-week heavy elastic band and weight chain program on upper-body strength and upper-body power in a sample of division 1-AA football players. *J Strength Cond Res*, 23(3), 756-764.

Hanson, N.J., Berg, K., Deka, P., Meendering, J.R., & Ryan, C. (2011). Oxygen cost of running barefoot vs. running shod. *Int J Sports Med*, 32(6), 401-406.

Issurin, V.B., Liebermann, D.G., & Tenenbaum, G. (1994). Effect of vibratory stimulation training on maximal force and flexibility. *J Sports Sci*, 12(6), 561-566.

Issurin, V.B., & Tenenbaum, G. (1999). Acute and residual effects of vibratory stimulation on explosive strength in elite and amateur athletes. *J Sports Sci*, 17(3), 177-182.

Keogh, J.W., Aickin, S.E., & Oldham, A.R. (2010). Can common measures of core stability distinguish performance in a shoulder pressing task under stable and unstable conditions? *J Strength Cond Res*, 24(2), 422-429.

Kibele, A., & Behm, D.G. (2009). Seven weeks of instability and traditional resistance training effects on strength, balance and functional performance. *J Strength Cond Res*, 23(9), 2443-2450.

Liebermann, D.G., & Issurin, V.B. (1997). Effort perception during isotonic muscle contractions with superimposed mechanical vibration stimulation. *J Hum Mov Stud*, 32, 171-186.

Luo, J., McNamara, B., & Moran, K. (2005). The use of vibration training to enhance muscle strength and power. *Sports Med*, 35(1), 23-41.

Marshall, P.W., & Murphy, B.A. (2006). Increased deltoid and abdominal muscle activity during Swiss ball bench press. *J Strength Cond Res*, 20(4), 745-750.

Mate-Munoz, J.L., Monroy, A.J., Jodra Jimenez, P., & Garnacho-Castano, M.V. (2014). Effects of instability versus traditional resistance training on strength, power and velocity in untrained men. *J Sports Sci Med*, 13(3), 460-468.

McCurdy, K., Langford, G., Ernest, J., Jenkerson, D., & Doscher, M. (2009). Comparison of chain- and plate-loaded bench press training on strength, joint pain, and muscle soreness in Division II baseball players. *J Strength Cond Res*, 23(1), 187-195.

Miyaguchi, K., & Demura, S. (2012). Relationship between upper-body strength and bat swing speed in high-school baseball players. *J Strength Cond Res*, 26(7), 1786-1791.

Moezy, A., Olyaei, G., Hadian, M., Razi, M., & Faghihzadeh, S. (2008). A comparative study of whole body vibration training and conventional training on knee proprioception and postural stability after anterior cruciate ligament reconstruction. *Br J Sports Med*, 42(5), 373-378.

Mulder, E.R., Gerrits, K.H., Rittweger, J., Felsenberg, D., Stegeman, D.F., & de Haan, A. (2008). Characteristics of fast voluntary and electrically evoked isometric knee extensions during 56 days of bed rest with and without exercise countermeasure. *Eur J Appl Physiol*, 103(4), 431-440.

Mulder, E.R., Kuebler, W.M., Gerrits, K.H., Rittweger, J., Felsenberg, D., Stegeman, D.F., & de Haan, A. (2007). Knee extensor fatigability after bedrest for 8 weeks with and without countermeasure. *Muscle Nerve*, 36(6), 798-806.

Perl, D.P., Daoud, A.I., & Lieberman, D.E. (2012). Effects of footwear and strike type on running economy. *Med Sci Sports Exerc*, 44(7), 1335-1343.

Preatoni, E., Colombo, A., Verga, M., Galvani, C., Faina, M., Rodano, R., & Cardinale, M. (2012). The effects of whole-body vibration in isolation or combined with strength training in female athletes. *J Strength Cond Res*, 26(9), 2495-2506.

Prokopy, M.P., Ingersoll, C.D., Nordenschild, E., Katch, F.I., Gaesser, G.A., & Weltman, A. (2008). Closed-kinetic chain upper-body training improves throwing performance of NCAA Division I softball players. *J Strength Cond Res*, 22(6), 1790-1798.

Rittweger, J., Belavy, D., Hunek, P., Gast, U., Boerst, H., Feilcke, B., . . . Felsenberg, D. (2006). Highly demanding resistive vibration exercise program is tolerated during 56 days of strict bed-rest. *Int J Sports Med*, 27(7), 553-559.

Saeterbakken, A.H., & Fimland, M.S. (2013a). Electromyographic activity and 6RM strength in bench press on stable and unstable surfaces. *J Strength Cond Res*, 27(4), 1101-1107.

Saeterbakken, A.H., & Fimland, M.S. (2013b). Muscle force output and electromyographic activity in squats with various unstable surfaces. *J Strength Cond Res*, 27(1), 130-136.

Sekendiz, B., Cug, M., & Korkusuz, F. (2010). Effects of Swiss-ball core strength training on strength, endurance, flexibility, and balance in sedentary women. *J Strength Cond Res*, 24(11), 3032-3040.

Shih, Y., Lin, K.L., & Shiang, T.Y. (2013). Is the foot striking pattern more important than barefoot or shod conditions in running? *Gait Posture*, 38(3), 490-494.

Silva-Batista, C., Kanegusuku, H., Roschel, H., Souza, E.O., Cunha, T.F., Laurentino, G.C., . . . Ugrinowitsch, C. (2014). Resistance training with instability in multiple system atrophy: a case report. *J Sports Sci Med*, 13(3), 597-603.

Sobhani, S., Bredeweg, S., Dekker, R., Kluitenberg, B., van den Heuvel, E., Hijmans, J., & Postema, K. (2014). Rocker shoe, minimalist shoe, and standard running shoe: a comparison of running economy. *J Sci Med Sport*, 17(3), 312-316.

Sparkes, R., & Behm, D.G. (2010). Training adaptations associated with an 8-week instability resistance training program with recreationally active individuals. *J Strength Cond Res*, 24(7), 1931-1941.

Squadrone, R., & Gallozzi, C. (2009). Biomechanical and physiological comparison of barefoot and two shod conditions in experienced barefoot runners. *J Sports Med Phys Fitness*, 49(1), 6-13.

Tung, K.D., Franz, J.R., & Kram, R. (2014). A test of the metabolic cost of cushioning hypothesis during unshod and shod running. *Med Sci Sports Exerc*, 46(2), 324-329.

Uribe, B.P., Coburn, J.W., Brown, L.E., Judelson, D.A., Khamoui, A.V., & Nguyen, D. (2010). Muscle activation when performing the chest press and shoulder press on a stable bench vs. a Swiss ball. *J Strength Cond Res*, 24(4), 1028-1033.

Chapter 16

Aarons, G.A. (2004). Mental health provider attitudes toward adoption of evidence-based practice: the Evidence-Based Practice Attitude Scale (EBPAS). *Ment Health Serv Res*, 6(2), 61-74.

Aarons, G.A., & Sawitzky, A.C. (2006). Organizational culture and climate and mental health provider attitudes toward evidence-based practice. *Psychol Serv*, 3(1), 61.

Amonette, W.E., English, K.L., & Ottenbacher, K.J. (2010). Nullius in verba: a call for the incorporation of evidence-based practice into the discipline of exercise science. *Sports Med*, 40(6), 449-457.

Biancuzzo, M. (1994a). Developing a poster about a clinical innovation. Part I: ideas and abstract. *Clin Nurse Spec*, 8(3), 153-155, 172.

Biancuzzo, M. (1994b). Developing a poster about a clinical innovation. Part II: creating the poster. *Clin Nurse Spec*, 8(4), 203-207.

Biancuzzo, M. (1994c). Developing a poster about a clinical innovation. Part III: presentation and evaluation. *Clin Nurse Spec*, 8(5), 262-264.

Boulos, M., Maramba, N.K., Inocencio, N., & Wheeler, S. (2006). Wikis, blogs and podcasts: a new generation of Web-based tools for virtual collaborative clinical practice and education. *BMC Med Educ*, 6(1), 41.

Fink, R., Thompson, C.J., & Bonnes, D. (2005). Overcoming barriers and promoting the use of research in practice. *J Nurs Admin*, 35(3), 121-129.

Gerrish, K., & Clayton, J. (2004). Promoting evidence-based practice: an organizational approach. *J Nurs Manag*, 12(2), 114-123.

Halligan, P. (2008). Poster presentations: valuing all forms of evidence. *Nurse Educ Pract*, 8(1), 41-45.

Jette, D.U., Bacon, K., Batty, C., Carlson, M., Ferland, A., Hemingway, R.D., . . . Volk, D. (2003). Evidence-based practice: beliefs, attitudes, knowledge, and behaviors of physical therapists. *Phys Ther*, 83(9), 786-805.

Kitson, A., Harvey, G., & McCormack, B. (1998). Enabling the implementation of evidence based practice: a conceptual framework. *Qual Health Care*, 7(3), 149-158.

Knight, K.L., & Ingersoll, C.D. (1996a). Optimizing scholarly communication: 30 tips for writing clearly. *J Athl Train*, 31(3), 209.

Knight, K.L., & Ingersoll, C.D. (1996b). Structure of a scholarly manuscript: 66 tips for what goes where. *J Athl Train*, 31(3), 201-206.

Kraemer, W.J., French, D.N., Paxton, N.J., Häkkinen, K., Volek, J.S., Sebastianelli, W.J., Putukian, M., Newton, R.U., Rubin, M.R., Gómez, A.L., Vescovi, J.D., Ratamess, N.A., Fleck, S.J., Lynch, J.M., & Knuttgen, H.G. (2004). Changes in exercise performance and hormonal concentrations over a big ten soccer season in starters and nonstarters. *J Strength Cond Res*, 18(1), 121-128.

Kraemer, W.J., Looney, D.P., Martin, G.J., Ratamess, N.A., Vingren, J.L., French, D.N., Hatfield, D.L., Fragala, M.S., Spiering, B.A., Howard, R.L., Cortis, C., Szivak, T.K., Comstock, B.A., Dunn-Lewis, C., Hooper, D.R., Flanagan, S.D., Volek, J.S., Anderson, J.M., Maresh, C.M., & Fleck, S.J. (2013). Changes in creatine kinase and cortisol in National Collegiate Athletic Association Division I American football players during a season. *J Strength Cond Res*, 27(2), 434-441.

Kraft, M. (2007). Integrating and promoting medical podcasts into the library collection. *Med Ref Serv Q*, 26(1), 27-35.

Linzer, M. (1987). The journal club and medical education: over one hundred years of unrecorded history. *Postgrad Med J*, 63(740), 475-478.

Moule, P., Judd, M., & Girot, E. (1998). The poster presentation: what value to the teaching and assessment of research in pre- and post-registration nursing courses? *Nurse Educ Today*, 18(3), 237-242.

Newhouse, R.P. (2007). Creating infrastructure supportive of evidence-based nursing practice: leadership strategies. *Worldviews Evid Based Nurs*, 4(1), 21-29.

Sherbinski, L.A., & Stroup, D.R. (1992). Developing a poster for disseminating research findings. *AANA J*, 60, 567-572.

Silvestre, R., Kraemer, W.J., West, C., Judelson, D.A., Spiering, B.A., Vingren, J.L., Hatfield, D.L., Anderson, J.M., & Maresh, C.M. (2006). Body composition and physical performance during a National Collegiate Athletic Association Division I men's soccer season. *J Strength Cond Res*, 20(4), 962-970.

Smith, R.A., Martin, G.J., Szivak, T.K., Comstock, B.A., Dunn-Lewis, C., Hooper, D.R., Flanagan, S.D., Looney, D.P., Volek, J.S., Maresh, C.M., & Kraemer, W.J. (2014). The effects of resistance training prioritization in NCAA Division I Football summer training. *J Strength Cond Res*, 28(1), 14-22.

Tipton, C.M. (1991). Publishing in peer-reviewed journals. Fundamentals for new investigators. *Physiologist*, 34(5):275, 278–279.

Vogelsang, J. (1994). Guidelines for developing a research poster for presentation. *J Post Anesth Nurs*, 9(2), 126-128.

Wilson, P., Petticrew, M., & Booth, A. (2009). After the gold rush? A systematic and critical review of general medical podcasts. *J R Soc Med*, 102(2), 69-74.

Chapter 17

American Diabetes Association. (2000). Type 2 diabetes in children and adolescents. *Pediatrics*, 105(3), 671-680.

Amonette, W.E., English, K.L., & Ottenbacher, K.J. (2010). Nullius in verba: a call for the incorporation of evidence-based practice into the discipline of exercise science. *Sports Med*, 40(6), 449-457.

Asif, M. (2014). The prevention and control the type-2 diabetes by changing lifestyle and dietary pattern. *J Educ Health Promot*, 3, 1.

Bennett, S., & Bennett, J.W. (2000). The process of evidence-based practice in occupational therapy: informing clinical decisions. *Aust Occup Ther J*, 47(4), 171-180.

Bridges, P.H., Bierema, L.L., & Valentine, T. (2007). The propensity to adopt evidence-based practice among physical therapists. *BMC Health Serv Res*, 7(1), 103.

Burns, N., & Grove, S.K. (2010). *Understanding Nursing Research: Building an Evidence-Based Practice*. 5th ed. Maryland Heights, MO: Elsevier Health Sciences.

Colditz, G.A. (1999). Economic costs of obesity and inactivity. *Med Sci Sports Exerc*, 31(11 Suppl), S663-S667.

Department of Health and Human Services, Administration on Aging. (2005). *A Profile of Older Americans: 2005*. Washington, DC: U.S. Department of Health and Human Services.

English, K.L., Amonette, W.E., Graham, M., & Spiering, B.A. (2012). What is "evidence-based" strength and conditioning? *Strength Cond J*, 34(3), 19-24.

Gordis, L. (2009). *Epidemiology* (4th ed.). Philadelphia: Saunders Elsevier.

Halperin, I.J., & Feig, D.S. (2014). The role of lifestyle interventions in the prevention of gestational diabetes. *Curr Diab Rep*, 14(1), 452.

Law, M., & MacDermid, J. (2008). *Evidence-Based Rehabilitation: A Guide to Practice* (2nd ed.). Thorofare, NJ: Slack.

Massey, B.F. (2003). Making vision 2020 a reality. *Phys Ther*, 83(11), 1023-1026.

McCluskey, A. (2004). *Increasing the Use of Research Evidence by Occupational Therapists*. Penrith, South Sidney: University of Western Sydney.

Melnyk, B.M., & Fineout-Overholt, E. (2011). *Evidence-Based Practice in Nursing & Healthcare: A Guide to Best Practice*. Philadelphia: Lippincott Williams & Wilkins.

Oxman, A.D., Sackett, D.L., & Guyatt, G.H. (1993). Users' guides to the medical literature. I. How to get started. *JAMA*, 270(17), 2093-2095.

Sackett, D.L. (1989). Inference and decision at the bedside. *J Clin Epidemiol*, 42(4), 309-316.

Sakuragi, S., Abhayaratna, K., Gravenmaker, K.J., O'Reilly, C., Srikusalanukul, W., Budge, M.M., …Abhayaratna, W.P. (2009). Influence of adiposity and physical activity on arterial stiffness in healthy children: the lifestyle of our kids study. *Hypertension*, 53(4), 611-616.

Spring, B. (2007). Evidence-based practice in clinical psychology: what it is, why it matters; what you need to know. *J Clin Psychol*, 63(7), 611-631.

Index

Note: The italicized *f* and *t* following page numbers refer to figures and tables, respectively.

About the Authors

William E. Amonette, PhD, is an assistant professor and director of the exercise and health sciences program in the Department of Clinical Health and Applied Sciences at the University of Houston – Clear Lake. Prior to becoming an academician, Amonette served as an assistant strength and conditioning coach for the Chinese national basketball team at the Beijing Olympic Training Center. He was also previously the assistant strength and conditioning coach and rehabilitation coordinator for the NBA's Houston Rockets, an astronaut strength, conditioning, and rehabilitation specialist, an exercise physiologist, and an integrated testing specialist for the Countermeasures Evaluation and Validation Project for Wyle Laboratories at NASA–Johnson Space Center.

Amonette earned his PhD at the University of Texas Medical Branch in rehabilitation sciences, with a research emphasis in clinical exercise physiology. He is a certified strength and conditioning specialist (CSCS) through the National Strength and Conditioning Association, an Associate Editor for the *Journal of Strength and Conditioning Research*, and ad hoc peer-reviewer for many scientific journals related to exercise and sport science.

Amonette's research interests include physiological and mechanical predictors of sports performance and injury. He also has clinical research interest in neuroendocrine and metabolic responses to exercise in patients with traumatic brain injuries and the effect of novel exercise interventions on rehabilitation outcomes in people with disabilities. He has published numerous scientific and academic peer-reviewed journal articles, reports, and book chapters and has presented his work nationally and internationally.

Kirk L. English, PhD, is a senior scientist with JES Tech LLC, a NASA contractor, and works in the Exercise Physiology and Countermeasures Laboratory at NASA – Johnson Space Center. He is also a research scientist in the Department of Nutrition and Metabolism at the University of Texas Medical Branch (UTMB) and an adjunct professor at the University of Houston – Clear Lake, where he teaches a graduate course.

English, who is a member of the National Strength and Conditioning Association, American College of Sports Medicine, and American Physiological Society, received his PhD in rehabilitation sciences from

UTMB. During his graduate studies, he was awarded a competitive three-year NASA/ Texas Space Grant Consortium Graduate Fellowship. English has published numerous peer-reviewed articles, technical reports, conference abstracts, and book chapters on exercise, nutrition, aging, spaceflight, and evidence-based practice in the field of exercise science.

In his work with NASA, English's research focuses on the prevention of spaceflight-induced decreases in skeletal muscle mass, strength, and performance. His work includes the development and validation of novel exercise protocols and hardware that are used both on the ground and during spaceflight. He also conducts all pre- and post-flight strength testing of American, European, Canadian, and Japanese International Space Station crewmembers and serves as the liaison and subject matter expert on this topic to NASA's international partners.

William J. Kraemer, PhD, is a full professor in the Department of Human Sciences at The Ohio State University. He has also held full professorships at the University of Connecticut, Ball State University, and The Pennsylvania State University, including each medical school. Dr. Kraemer is a fellow of the American College of Sports Medicine, the National Strength and Conditioning Association (NSCA), and the American College of Nutrition. Among many of his professional achievements, he is a recipient of the NSCA's Lifetime Achievement Award. He is the editor in chief of the NSCA's *Journal of Strength and Conditioning Research*, an editor of the *European Journal of Applied Physiology* and an associate editor of the *Journal of the American College of Nutrition*. He holds many other editorial board positions in the field. Kraemer has published more than 450 peer reviewed papers in scientific literature and has published 12 books. The NSCA also named the Outstanding Sport Scientist Award in his honor in 2006. He received the 2014 Expertscape Award, which named him the nation's top expert in resistance training research over the past ten years. With almost 40,000 citations on Harzing's Publish or Perish lists, his scholarly impact is impressive.

Courtesy of William Kraemer